Mathematics for Industry

Volume 19

Aims & Scope

The meaning of "Mathematics for Industry" (sometimes abbreviated as MI or MfI) is different from that of "Mathematics in Industry" (or of "Industrial Mathematics"). The latter is restrictive: it tends to be identified with the actual mathematics that specifically arises in the daily management and operation of manufacturing. The former, however, denotes a new research field in mathematics that may serve as a foundation for creating future technologies. This concept was born from the integration and reorganization of pure and applied mathematics in the present day into a fluid and versatile form capable of stimulating awareness of the importance of mathematics in industry, as well as responding to the needs of industrial technologies. The history of this integration and reorganization indicates that this basic idea will someday find increasing utility. Mathematics can be a key technology in modern society.

The series aims to promote this trend by (1) providing comprehensive content on applications of mathematics, especially to industry technologies via various types of scientific research, (2) introducing basic, useful, necessary and crucial knowledge for several applications through concrete subjects, and (3) introducing new research results and developments for applications of mathematics in the real world. These points may provide the basis for opening a new mathematics-oriented technological world and even new research fields of mathematics.

More information about this series at http://www.springer.com/series/13254

Mutsuto Kawahara

Finite Element Methods in Incompressible, Adiabatic, and Compressible Flows

From Fundamental Concepts to Applications

Springer

Mutsuto Kawahara
Chuo University
Tokyo
Japan

ISSN 2198-350X ISSN 2198-3518 (electronic)
Mathematics for Industry
ISBN 978-4-431-56657-1 ISBN 978-4-431-55450-9 (eBook)
DOI 10.1007/978-4-431-55450-9

Printed on acid-free paper

This Springer imprint is published by Springer Nature
The registered company is Springer Japan KK

Preface

Professor John Tinsley Oden, a professor and vice-president of the University of Texas at Austin, in his 2013 speech in Tokyo suggested that computational mechanics is, or at least should be, the third pillar of science and engineering, in addition to theory and observation. The anatomy of computational mechanics consists of the mathematical model, the computational model, and the actual computation. This book covers a wide range of analyses of fluid flows, which make use of the mathematical and computational models of incompressible, adiabatic, and compressible flows. The book also touches on some actual computations. The mathematical models in the fluid flow field are established based on the conservation principles of mass, momentum, and energy. In the compressible flow analyses, the governing equation system can be solved directly with field variables of density, velocity, and energy. In the incompressible flow analyses, the field variables are velocity and pressure. By introducing the concept of the adiabatic flows, a wide range of fluid flows without heat conduction can be solved, for which the field variables are density and velocity, or velocity and pressure.

The finite element method is one of the most widely used computational methods in engineering. The central aim of this book is to introduce mathematical foundations and comprehensive theories of the finite element method in the field of incompressible, adiabatic, and compressible flows. Because flow characteristics are considerably different from those of solid materials, the computational techniques are also different. The finite element method is presented in its precise form based on the bubble function method, the improved balancing tensor diffusivity (IBTD) method, stream-line upwind Petrov–Galerkin (SUPG) method, and the characteristic method in addition to simple Euler and two-step methods. In particular, one key feature of this book is to provide the analysis of adiabatic flows that can solve incompressible flows using non-mixed interpolation.

The book does not assume that the readers are already familiar with the finite element methods. By reading through this book, readers should be able to gain a comprehensive understanding from introductory techniques to applications. The book consists of two parts: Part I introduces the finite element methods; Part II discusses the latest theories and applications, covering civil, mechanical, and aero engineering.

Acknowledgments

The author would like to acknowledge his gratefulness to the contributors of the numerical studies and applications adopted in this book. They include: Dr. Jun-ichi Mutsuhito (National Institute of Advanced Industrial Science and Technology), Prof. Takahiko Kurahashi (Nagaoka Institute of Technology), Mr. Shigeru Watanabe (Mitsui Engineering Co.), Prof. Hiroshi Okumura (University of Toyama), Dr. Taro Arikawa (Port and Airport Research Institute), and Prof. Akira Maruoka (Hachinohe College of Technology) for studies discussed in Chap. 8; Dr. Ichiro Uchiyama (OCJPN), Mr. Shuichi Nasu (Chuo University), and Prof. Akira Maruoka (Hachinohe College of Technology) in Chap. 9; Dr. Kazuya Nojima (Chuo University) in Chap. 10; and Mr. Kazuki Sawanobori (Chuo University) in Chap. 11. Mr. Shigeru Watanabe (Mitsui Engineering Co.) offered much needed help in preparing figures, texts and references. The English was edited by Prof. Shigeto Kawahara (Keio University). TEX writing was carried out by Mmes. Chiyoko Sato and Yukiko Nagamine, Uls & Co. The author is especially indebted to Mr. Takeyuki Yonezawa, Editorial Director and Dr. Yuko Sumino, former editor, Springer Japan. The author would like to express his gratitude to Sumiko Kawahara, his wife, for her continued support, without which this book would have not been completed.

Contents

Chapter 1
Introduction

1.1 Purpose and Contents of This Book

In general, natural phenomena can be described by analog differential equations. Since natural phenomena are complex, we must resort to complex nonlinear differential equations, which are hard to solve. Thus, realistically speaking, digital computing should be introduced instead. Then, we need a tool to convert the analog equations into the digital equations. The computational mechanics are created for this purpose. We develop in this book the finite element method in fluid flows. The finite element method is one of the most flexible computational methods to approximate boundary configurations and boundary conditions based on triangular or quadrilateral finite elements.

This book consists of two parts: Part I introduces the finite element method; Part II discusses the latest theories and applications of the finite element method. Readers will gain a comprehensive understanding including introductory techniques, precise theories, and applications of the finite element method in this one book. There are some other books dealing with the finite element method in fluid flows, e.g., Carey and Oden (1986), Pironneau (1988), Lewis et al. (2004), Löhner (2008), Zienkiewicz et al. (2014), etc. Those books, however, all assume that the readers have enough prior knowledge of the methodology of the finite element method. This book, on the other hand, assumes that the readers are not necessary familiar with the finite element methods.

Part I is devoted to introducing the concepts of the method explained using the elementary calculus of differentiation and matrix multiplication. Chapter 2 deals with the spring structure, because it is the most suitable topic to illustrate the understanding of the superposition, which is the fundamental concept of the finite element method. The fundamental discretization techniques are presented in Chap. 3 using pipeline systems. The basic concepts of the finite element method are described in detail using potential flows in Chap. 4. Those are the weighted residual equation, interpolation and weighting functions, superimposition over the flow field, and treatment of the boundary conditions. In Chap. 5, formulations of the time-dependent problems are developed for advection–diffusion problems. The bubble function method, the SUPG method, and the characteristic method are discussed in addition to the simple

M. Kawahara, *Finite Element Methods in Incompressible, Adiabatic, and Compressible Flows*, Mathematics for Industry 19,
DOI 10.1007/978-4-431-55450-9_1

Euler method, the two-step explicit method, Crank–Nicolson method, etc. Especially, CFL conditions of the stability are discussed. In Chap. 6, the mixed interpolation is explained using creeping flow.

Part II illustrates the theories and applications of the finite element methods in fluid flows. The indicial notation and summation convention are used to express precise forms of equations. In Chap. 7, the continuum mechanics, which we use in later chapters, are briefly presented. The analyses of fluid flows can be implemented, either assuming that fluids are incompressible or compressible. Chapter 8 deals with the finite element methods of the incompressible flows. A large number of research works have been conducted in this field. The governing equations are conservation equations of mass and momentum. Normally, conservation equation of mass is transformed into equation of continuity, if the incompressibility assumption is introduced. The working variables are velocity and pressure. We use the stabilization methods, such as the SUPG method, the IBTD method, the mixed interpolation method based on the bubble function. We will describe these methods which have almost the same stabilization effects.

The main cause of the instability for the computation of the incompressible flows stems from the fact that equation of continuity is a function of velocity only, but not a function of pressure. The incompressibility is a rather strong constraint. We can weaken the constraint of incompressibility by physical consideration. To do this, we can employ the adiabatic fluid flow concept. In Chap. 9, the computational methods and applications of the adiabatic flows are discussed. Assuming that pressure is a function of density only in equation of state, we can derive the governing equations of adiabatic fluid flows. Namely, the governing equations of adiabatic flows are conservation equations of mass and momentum and equation of state. The working variables are density, velocity, and pressure. Using appropriate variable transformation, we can derive the equations of momentum and continuity including acoustic velocity. The working variables are velocity and pressure if we subscribe to the constant acoustic velocity. The equation system is derived as the extension of that of the incompressible flows adding the terms of velocity and pressure to the equation of continuity by the physical consideration. The terms added play important roles of the stabilization of computation. Moreover, if we let the acoustic velocity approach infinity, we obtain the normal equation of continuity of the incompressible fluid flows. The computational methods we employed are the Euler method, the two-step explicit method, the characteristic method, and the SUPG method. Chapter 10 is concerned with the computational methods and applications of the compressible fluid flows. The governing equations are conservation equations of mass, momentum, and energy and equation of state. The working variables are density, velocity, pressure, and energy. The mixed interpolation method based on the bubble function and the SUPG method are used as the computational method. There are many actual problems, in which the finite element meshes do not always coincide with the moving fluids, e.g., problems of free surface, moving rigid body located in fluid flows, fluid flows in solid containers, etc. For those problems, ALE formulation is effective, which is described in Chap. 11. The differential equations in the reference domain are derived.

Although the mesh generation is important in the finite element method, we do not deal with this issue due to space limitation. See Cheng et al. (2012), Lo (2015), Shewchuck (2014). We do not intend to cite all of the relevant references, but selected references helpful for the understanding of the subjects are cited. The author believes and hopes that this book is valuable to get knowledge of the finite element method in fluid flows for the readers in mathematical science and civil, mechanical, and aero engineerings.

1.2 Notation and Terminology

In Part I, we mainly use the matrix notation. Zero element is not shown explicitly, and the blank in the matrix represents zero. In some parts of Part I and in almost all parts of Part II, we use the indicial notation. The indicial notation is extension of the componential notation.

There are two notation systems to express vectors and tensors: one is vectorial notation and the other is componential notation. A vector can be expressed by **a**, for example. If we know that the component of the vector is a_i, we can write

$$\mathbf{a} = a_i \mathbf{g}_i \tag{1.1}$$

where $\mathbf{g}_i$ is the base vector. Equation (1.1) is equivalent to

$$\mathbf{a} = a_1 \mathbf{g}_1 + a_2 \mathbf{g}_2 + a_3 \mathbf{g}_3 \tag{1.2}$$

in the case of three dimensional vector. We usually use the summation convention, which means that the summation is carried out until the maximum number of the index after multiplication for the repeated indices in the same term. Thus, we can denote the vector **a** or a_i; the former is the vectorial notation and the latter is the componential notation. If we know the base vector, we can simply convert the componential notation into the vectorial notation. Generally, the vectorial notation is used, because it is simpler to express the equations. On the other hand, to express precise equations of the finite element method, the componential notation is more convenient, although the expression can become rather complex. In the indicial notations, the special functions are effectively used; for example, the Kronecker delta function δ_{ij} and the permutation function ε_{ijk}, are defined as

$$\delta_{ij} = \begin{cases} 1 & (i = j) \\ 0 & (i \neq j) \end{cases} \tag{1.3}$$

$$\varepsilon_{ijk} = \begin{cases} 1 & (i,\ j,\ k,\ \text{even permutation}) \\ -1 & (i,\ j,\ k,\ \text{odd permutation}) \\ 0 & \text{otherwise} \end{cases} \tag{1.4}$$

We can express effective expressions using Eqs. (1.3) and (1.4) combined with the summation convention. Precise introduction of the indicial notation is found in Hopman (2002). Throughout Part I, we use matrix multiplication to express equations. To express the detailed expression of the matrix multiplication, the indicial notation with summation convention is employed. We use the indicial notation with summation convention in Part II to show the detailed expressions not only in the governing equations but also in the finite element method. To express the differentiation with respect to coordinate i, we use $,i$. For the differentiation with respect to time, in Chap. 7, we use over dot $\dot{}$, which is the material derivative with respect to time. In other chapters over dot $\dot{}$ means partial differentiation with respect to time.

For terminological usage, we follow the conventions in the current field as much as possible. However, there are some exceptions. We use the terminology volume V and surface S even in two-dimensional flow. This is because we can consider area A is the volume with unit thickness. We use "edge" in the sense of the end point of a triangular element. However, we use mid points or barycentric points as well in the finite element analysis. The term "node" represents those points. However, these two terms are sometimes used interchangeably. We sometime employ the mixed use of "scheme" and "method". If we like to show simply time integration, we use "scheme". If we like to express the scheme combined with the stabilization method, we employ "method". For the numerical integration in time, we use the "fractional step method". This is also called the "pressure Poisson equation method". We use the terminology "acoustic velocity," which is sometimes referred to as "sound speed".

Part I
Introduction to Finite Element Methods in Fluid Flows

Chapter 2
Basic Concepts of Finite Element Method

2.1 Introduction

This chapter provides an overview of basic concepts and mathematical foundations of finite element method. Understanding the fundamental concepts of the finite element method is useful not only in structures but also in fluid flows. The most basic concepts of the finite element method can be best explained using a structure composed of springs, which we are going to discuss in this chapter. This spring is one dimensional and it is thus simple to understand the important concepts for the finite element method. The physical aspects of springs are also comprehensive at the intuitive level.

In general, the finite element analysis amounts to solving a structure—which means to determine displacements and spring forces. To do so, we first decompose the structure into pieces of springs and nodal points, to which the springs are connected. Then, we build up the final equation, using three basic equations and boundary conditions. The basic equations consist of three elements: (1) deformation equation, (2) equilibrium equation, and (3) compatibility equation. In the build-up phase of the analysis, systematic algorithm is necessary from the perspective of computational coding. The precise formulation and build-up methods, which we call "superposing procedures," will also be discussed in this chapter. We describe both the general formulation and specific equations, using a simple structure as an example.

2.2 Spring Structure: Definitions of the Conventions Used

Consider the spring structure given in Fig. 2.1. This structure consists of springs and nodal points which connect the springs. We will start with some notations that are deployed mainly in Part I in this book. Here and throughout this book "nodal points" will be simply referred to as "nodes," for the sake of brevity. We number nodes consecutively as 1, 2, 3, …, N, where N is the total number of nodes.

M. Kawahara, *Finite Element Methods in Incompressible, Adiabatic, and Compressible Flows*, Mathematics for Industry 19,
DOI 10.1007/978-4-431-55450-9_2

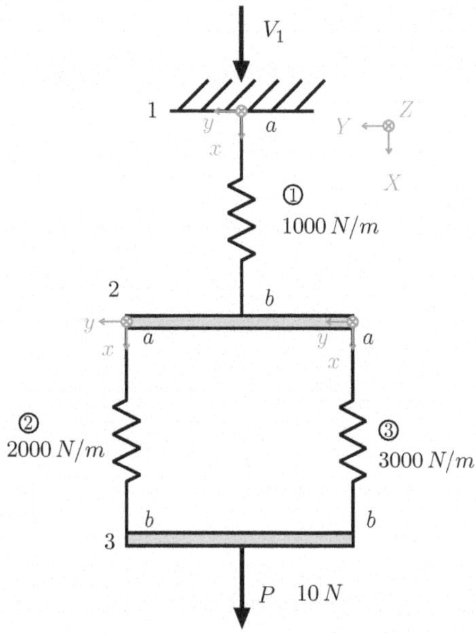

Fig. 2.1 Springs structure

In Fig. 2.1, for example, two springs are connected to on nodes 2 and 3. To represent these connections, nodes 2 and 3 are expressed using rigid body. We also number springs consecutively, as ①, ②, …, Ⓜ, where Ⓜ is the total number of springs. For each spring, local coordinate x–y–z is assigned, which is called "member coordinate system." For the structure as a whole, one global coordinate X–Y–Z is assigned. For those coordinates, we use right-hand system. The quantities associated with nodal points are denoted by subscripts 1, 2, 3, …N, and those associated with springs by circled subscripts ①, ②, …, Ⓜ.

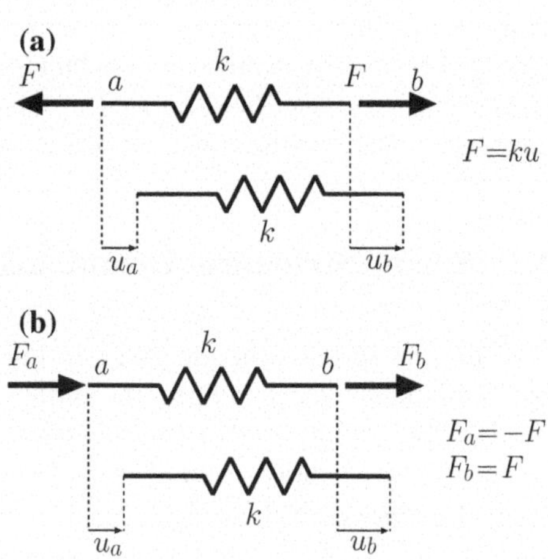

Fig. 2.2 **a** Spring force and displacement. **b** Edge force and displacement

To distinguish between the two edges of a spring, we name them "edge a" and "edge b," as shown in Fig. 2.2. The choice of a and b is arbitrary here; however, we will use these labels consistently. We introduce the member coordinate x–y–z of which x-axis is directed from edge a to edge b. The quantity associated with edge a or b is expressed with the subscript a or b.

2.3 Deformation Equation

It is well known that a spring force is proportional to a deformation of spring, as in

$$F = ku \tag{2.1}$$

where F is the force applied to the spring, u is the deformation, and k is a spring constant. As shown in Fig. 2.2a, the tension is taken to be positive. The deformation is expressed as

$$u = u_b - u_a \tag{2.2}$$

where u_a and u_b are displacements at edge a and edge b, respectively. In Fig. 2.2a, we observe that only the force at edge a shows a negative direction of the member coordinate x–y–z. Another force at edge b and both displacements are directed to a positive direction of the coordinate. It would thus be more convenient to use the forces and displacements that are directed to the same direction, as shown in Fig. 2.2b. Define

$$F_a = -F \tag{2.3}$$

$$F_b = F \tag{2.4}$$

This postulation allows us to rewrite Eq. (2.1) as follows:

$$\begin{bmatrix} F_a \\ \hline F_b \end{bmatrix} = \begin{bmatrix} -F \\ \hline F \end{bmatrix} = \begin{bmatrix} -ku \\ \hline ku \end{bmatrix} = \left[\begin{array}{c|c} k & -k \\ \hline -k & k \end{array}\right] \begin{bmatrix} u_a \\ \hline u_b \end{bmatrix} \tag{2.5}$$

Those forces F_a and F_b are referred to as the edge forces at edge a and edge b. The advantage of using those forces is clarified in Eq. (2.5): the coefficient matrix in the last term is symmetric, and in the computational mechanics, treating symmetric matrices is much easier than treating nonsymmetric matrices.

Recall now that the spring should keep the balance. Thus, the following equation needs to hold:

$$F_a + F_b = 0 \tag{2.6}$$

If we sum the first and second equations of Eq. (2.5), we find Eq. (2.6) is satisfied. Equation (2.5) can be rewritten as

$$\begin{bmatrix} F_a \\ \hline F_b \end{bmatrix} = \left[\begin{array}{c|c} k_{aa} & k_{ab} \\ \hline k_{ba} & k_{bb} \end{array}\right] \begin{bmatrix} u_a \\ \hline u_b \end{bmatrix} \tag{2.7}$$

where

$$\begin{aligned} k_{aa} &= k \\ k_{ab} &= k_{ba} = -k \\ k_{bb} &= k \end{aligned} \tag{2.8}$$

The coefficients k_{aa}, k_{ab}, k_{ba}, and k_{bb} represent the relations between the edge forces and edge displacements in such a way that they express the following relations:

k_{aa}: edge force F_a and edge displacement u_a
k_{ab}: edge force F_a and edge displacement u_b
k_{ba}: edge force F_b and edge displacement u_a
k_{bb}: edge force F_b and edge displacement u_b

Equation (2.7) holds for all springs ①, ②, and ③ in case of the structure shown in Fig. 2.1, which can be derived as follows:

$$\begin{bmatrix} F_a^{①} \\ F_a^{②} \\ F_a^{③} \\ \hline F_b^{①} \\ F_b^{②} \\ F_b^{③} \end{bmatrix} = \left[\begin{array}{ccc|ccc} k_{aa}^{①} & & & k_{ab}^{①} & & \\ & k_{aa}^{②} & & & k_{ab}^{②} & \\ & & k_{aa}^{③} & & & k_{ab}^{③} \\ \hline k_{ba}^{①} & & & k_{bb}^{①} & & \\ & k_{ba}^{②} & & & k_{bb}^{②} & \\ & & k_{ba}^{③} & & & k_{bb}^{③} \end{array}\right] \begin{bmatrix} u_a^{①} \\ u_a^{②} \\ u_a^{③} \\ \hline u_b^{①} \\ u_b^{②} \\ u_b^{③} \end{bmatrix} \tag{2.9}$$

where $F_a^{ⓜ}$ and $F_b^{ⓜ}$ denote edge forces at edge a and edge b of spring ⓜ, whereas $u_a^{ⓜ}$ and $u_b^{ⓜ}$ are edge displacements at edge a and edge b.

Now if we posit the following,

$$[\mathbf{F_a}] = \begin{bmatrix} F_a^{①} \\ F_a^{②} \\ F_a^{③} \end{bmatrix}, \quad [\mathbf{F_b}] = \begin{bmatrix} F_b^{①} \\ F_b^{②} \\ F_b^{③} \end{bmatrix}$$

$$
[\mathbf{u_a}] = \begin{bmatrix} u_a^{①} \\ u_a^{②} \\ u_a^{③} \end{bmatrix}, \quad [\mathbf{u_b}] = \begin{bmatrix} u_b^{①} \\ u_b^{②} \\ u_b^{③} \end{bmatrix}
$$

and

$$
[\mathbf{k_{aa}}] = \begin{bmatrix} k_{aa}^{①} & & \\ & k_{aa}^{②} & \\ & & k_{aa}^{③} \end{bmatrix}, \quad [\mathbf{k_{ab}}] = \begin{bmatrix} k_{ab}^{①} & & \\ & k_{ab}^{②} & \\ & & k_{ab}^{③} \end{bmatrix}
$$

$$
[\mathbf{k_{ba}}] = \begin{bmatrix} k_{ba}^{①} & & \\ & k_{ba}^{②} & \\ & & k_{ba}^{③} \end{bmatrix}, \quad [\mathbf{k_{bb}}] = \begin{bmatrix} k_{bb}^{①} & & \\ & k_{bb}^{②} & \\ & & k_{bb}^{③} \end{bmatrix}
$$

then Eq. (2.9) can be expressed as follows:

$$
\begin{bmatrix} \mathbf{F}_a \\ \hline \mathbf{F}_b \end{bmatrix} = \left[\begin{array}{c|c} \mathbf{k}_{aa} & \mathbf{k}_{ab} \\ \hline \mathbf{k}_{ba} & \mathbf{k}_{bb} \end{array}\right] \begin{bmatrix} \mathbf{u}_a \\ \hline \mathbf{u}_b \end{bmatrix} \tag{2.10}
$$

Comparing Eqs. (2.10) and (2.7), they are essentially of the same form. Equation (2.7) is referred to as the "deformation equation of each member," whereas Eq. (2.10) is referred to as the "deformation equation." In Sect. 2.6, we will show that the deformation equation plays an important role in the finite element method. For this reason, the deformation equation is sometimes called the "finite element equation."

2.4 Equilibrium Equation

Now moving on to Fig. 2.3, let us first decompose the whole structure into nodes and springs, and then put together the structures again. One important equation to understand is the equilibrium equation, which expresses the condition that the spring forces assembled at a node should be summed to zero. Considering the direction, the equilibrium condition at nodes 1, 2, and 3 can be derived as follows:

$$
\begin{aligned}
&\text{node 1:} \quad & -F_a^{①} + V_1 = 0 \\
&\text{node 2:} \quad & -F_b^{①} - F_a^{②} - F_a^{③} = 0 \\
&\text{node 3:} \quad & -F_b^{②} - F_b^{③} + P = 0
\end{aligned} \tag{2.11}
$$

where V_1 is a reaction force at node 1, and P is the external load applied at node 3. Equation (2.11) can be expressed in the form of a matrix:

Fig. 2.3 Equilibrium of spring structures

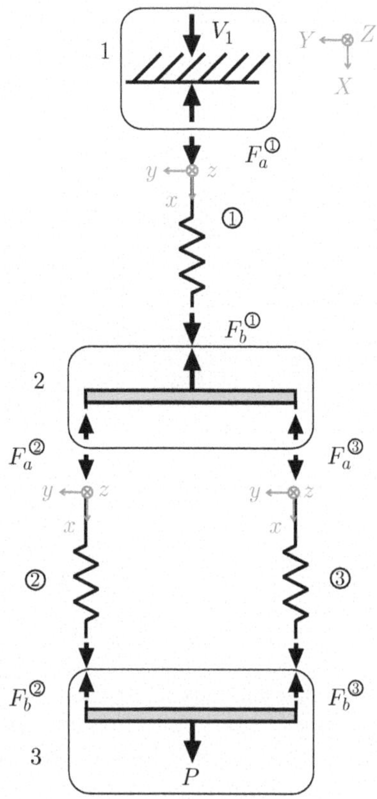

$$\begin{array}{c} \\ 1 \\ 2 \\ 3 \end{array}\begin{array}{c} \begin{array}{cccccc} ① & ② & ③ & ① & ② & ③ \end{array} \\ \left[\begin{array}{ccc|ccc} 1 & & & & & \\ & 1 & 1 & 1 & & \\ & & & & 1 & 1 \end{array}\right] \end{array} \left[\begin{array}{c} F_a^{①} \\ F_a^{②} \\ F_a^{③} \\ \hline F_b^{①} \\ F_b^{②} \\ F_b^{③} \end{array}\right] = \left[\begin{array}{c} \\ \\ P \end{array}\right] + \left[\begin{array}{c} V_1 \\ \\ \\ \end{array}\right] \tag{2.12}$$

Equation (2.12) is the three equations for nodes 1, 2, and 3. On the right side of Eq. (2.12) is the external force and the reaction force terms.

Now let us examine the coefficient matrix of Eq. (2.12). The first part of the coefficient matrix is

$$\begin{array}{c} \\ 1 \\ 2 \\ 3 \end{array} \begin{array}{c} \begin{array}{ccc} ① & ② & ③ \end{array} \\ \left[\begin{array}{ccc} 1 & & \\ & 1 & 1 \\ & & \end{array}\right. \end{array}$$

The row numbers correspond to node numbers and column numbers represent spring numbers.

Since edge *a*
of spring ① is 1, then element 1-① is 1,
of spring ② is 2, then element 2-② is 1, and
of spring ③ is 2, then element 2-③ is 1.

To put it differently, this matrix expresses the connection between nodes and edge *a* of spring ⓜ (where ⓜ = 1, 2, 3). Let this matrix be denoted by $[\alpha]$.

The second part of the coefficient matrix in Eq. (2.12) is

$$\begin{array}{c} \begin{array}{ccc} ① & ② & ③ \end{array} \\ \left.\begin{array}{ccc} & & \\ 1 & & \\ & 1 & 1 \end{array}\right] \end{array} \begin{array}{c} \\ 1 \\ 2 \\ 3 \end{array}$$

The row numbers correspond to node numbers and column numbers correspond to spring numbers.

Since edge *b*
of spring ① is 2, then element 2-① is 1,
of spring ② is 3, then element 3-② is 1, and
of spring ③ is 3, then element 3-③ is 1.

Differently put, this matrix expresses the connection between nodes and edge *b* of spring ⓜ (where ⓜ = 1,2,3). This matrix is denoted by $[\beta]$. For the sake of convenience, we use $[\alpha]$ and $[\beta]$ to refer to the connectivity matrix. Using the connectivity matrix, Eq. (2.12) can be denoted as follows:

$$\left[\alpha \,\vdots\, \beta\right] \begin{bmatrix} \mathbf{F_a} \\ \hdashline \mathbf{F_b} \end{bmatrix} = \left[\mathbf{P}\right] + \left[\mathbf{V}\right] \tag{2.13}$$

The external load matrix [**P**] and reaction force matrix [**V**] are

$$\left[\mathbf{P}\right]=\begin{matrix}1\\2\\3\end{matrix}\begin{bmatrix} \\ \\ P\end{bmatrix},\quad \left[\mathbf{V}\right]=\begin{matrix}1\\2\\3\end{matrix}\begin{bmatrix}V_1\\ \\ \end{bmatrix} \tag{2.14), (2.15}$$

Equations (2.12) and (2.13) are the "equilibrium equations."

2.5 Compatibility Equation

The displacement of the edges of a spring should be the same as that of nodes connected to the spring. Looking at Fig. 2.4, the relation between nodes and edges of springs can be determined as follows:

$$\begin{aligned} u_a^{①} &= u_1, \quad & u_b^{①} &= u_2 \\ u_a^{②} &= u_2, \quad & u_b^{②} &= u_3 \\ u_a^{③} &= u_2, \quad & u_b^{③} &= u_3 \end{aligned} \tag{2.16}$$

These equations can be summarized in the matrix form:

$$\begin{bmatrix} u_a^{①} \\ u_a^{②} \\ u_a^{③} \\ \hdashline u_b^{①} \\ u_b^{②} \\ u_b^{③} \end{bmatrix} = \begin{matrix} ① \\ ② \\ ③ \\ ① \\ ② \\ ③ \end{matrix} \overset{\begin{matrix}1 & 2 & 3\end{matrix}}{\begin{bmatrix} 1 & & \\ & 1 & \\ & 1 & \\ \hdashline & 1 & \\ & & 1 \\ & & 1 \end{bmatrix}} \begin{bmatrix} u_1 \\ u_2 \\ u_3 \end{bmatrix} \tag{2.17}$$

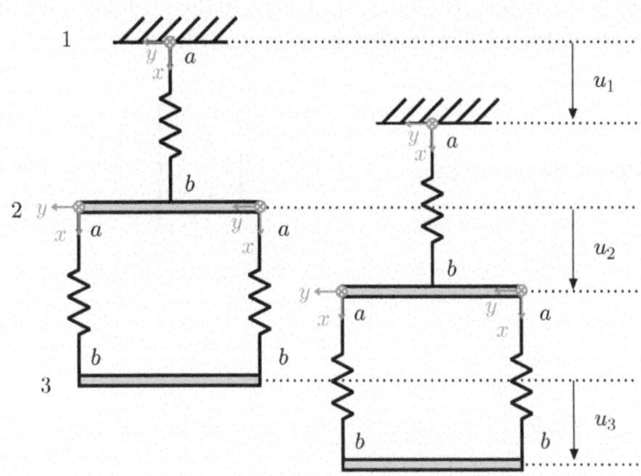

Fig. 2.4 Compatibility of string structures

It is interesting that the coefficient matrix in Eq. (2.17) is the transposed matrix in Eq. (2.13), i.e., $[\alpha^T]$ and $[\beta^T]$. Using the connectivity matrix, Eq. (2.17) can then be expressed as

$$\begin{bmatrix} \mathbf{u}_a \\ \hdashline \mathbf{u}_b \end{bmatrix} = \begin{bmatrix} \alpha^T \\ \hdashline \beta^T \end{bmatrix} \begin{bmatrix} \mathbf{U} \end{bmatrix} \tag{2.18}$$

where $[\mathbf{u}_a]$ and $[\mathbf{u}_b]$ are displacements at nodes a and b as in Eq. (2.17) and

$$\begin{bmatrix} \mathbf{U} \end{bmatrix} = \begin{bmatrix} u_1 \\ u_2 \\ u_3 \end{bmatrix} \tag{2.19}$$

which means the displacement of the nodes. Equations (2.17) and (2.18) are the "compatibility equation."

2.6 Finite Element Equation

The finite element equation consists of main three equations and boundary conditions. The main system of the whole spring structure is deformation equation, Eq. (2.10), equilibrium equation, Eq. (2.13), and compatibility equation, Eq. (2.18). By introducing Eq. (2.18) into Eq. (2.10) and rearranging the terms, we get

$$\begin{bmatrix} \mathbf{F}_a \\ \hdashline \mathbf{F}_b \end{bmatrix} = \begin{bmatrix} \mathbf{k}_{aa} & \mathbf{k}_{ab} \\ \mathbf{k}_{ba} & \mathbf{k}_{bb} \end{bmatrix} \begin{bmatrix} \alpha^T \\ \hdashline \beta^T \end{bmatrix} \begin{bmatrix} \mathbf{U} \end{bmatrix} \tag{2.20}$$

Equation (2.20) is equation for the computation of spring forces $[\mathbf{F}_a]$ and $[\mathbf{F}_b]$, given the nodal displacements $[\mathbf{U}]$. We call Eq. (2.20) the "spring force equation." Substituting Eq. (2.20) into Eq. (2.13), and rearranging the terms, the "finite element equation" is obtained as follows:

$$\begin{bmatrix} \alpha & \beta \end{bmatrix} \begin{bmatrix} \mathbf{k}_{aa} & \mathbf{k}_{ab} \\ \mathbf{k}_{ba} & \mathbf{k}_{bb} \end{bmatrix} \begin{bmatrix} \alpha^T \\ \hdashline \beta^T \end{bmatrix} \begin{bmatrix} \mathbf{U} \end{bmatrix} = \begin{bmatrix} \mathbf{P} \end{bmatrix} + \begin{bmatrix} \mathbf{V} \end{bmatrix} \tag{2.21}$$

In case of spring structures shown in Fig. 2.1, the spring force equation is

$$
\begin{bmatrix} F_a^{(1)} \\ F_a^{(2)} \\ F_a^{(3)} \\ \hdashline F_b^{(1)} \\ F_b^{(2)} \\ F_b^{(3)} \end{bmatrix}
=
\left[\begin{array}{ccc:ccc}
k_{aa}^{(1)} & & & k_{ab}^{(1)} & & \\
& k_{aa}^{(2)} & & & k_{ab}^{(2)} & \\
& & k_{aa}^{(3)} & & & k_{ab}^{(3)} \\ \hdashline
k_{ba}^{(1)} & & & k_{bb}^{(1)} & & \\
& k_{ba}^{(2)} & & & k_{bb}^{(2)} & \\
& & k_{ba}^{(3)} & & & k_{bb}^{(3)}
\end{array}\right]
\left[\begin{array}{ccc}
1 & & \\
& 1 & \\
& 1 & \\ \hdashline
& 1 & \\
& & 1 \\
& & 1
\end{array}\right]
\begin{bmatrix} u_1 \\ u_2 \\ u_3 \end{bmatrix}
$$

$$
=
\left[\begin{array}{ccc}
k_{aa}^{(1)} & k_{ab}^{(1)} & \\
& k_{aa}^{(2)} & k_{ab}^{(2)} \\
& k_{aa}^{(3)} & k_{ab}^{(3)} \\ \hdashline
k_{ba}^{(1)} & k_{bb}^{(1)} & \\
& k_{ba}^{(2)} & k_{bb}^{(2)} \\
& k_{ba}^{(3)} & k_{bb}^{(3)}
\end{array}\right]
\begin{bmatrix} u_1 \\ u_2 \\ u_3 \end{bmatrix}
\tag{2.22}
$$

which can be obtained by introducing Eq. (2.17) into Eq. (2.9). Next, introducing Eq. (2.22) into Eq. (2.12), we get the following:

$$
\left[\begin{array}{ccc:ccc}
1 & & & & & \\
& 1 & 1 & 1 & & \\
& & & & 1 & 1
\end{array}\right]
\left[\begin{array}{ccc}
k_{aa}^{(1)} & k_{ab}^{(1)} & \\
& k_{aa}^{(2)} & k_{ab}^{(2)} \\
& k_{aa}^{(3)} & k_{ab}^{(3)} \\ \hdashline
k_{ba}^{(1)} & k_{bb}^{(1)} & \\
& k_{ba}^{(2)} & k_{bb}^{(2)} \\
& k_{ba}^{(3)} & k_{bb}^{(3)}
\end{array}\right]
\begin{bmatrix} u_1 \\ u_2 \\ u_3 \end{bmatrix}
$$

$$
= \begin{bmatrix} \\ \\ P \end{bmatrix} + \begin{bmatrix} V_1 \\ \\ \\ \end{bmatrix}
\tag{2.23}
$$

Performing the matrix multiplication of Eq. (2.23), we get the final form of the finite element equation:

$$
\begin{bmatrix}
k_{aa}^{(1)} & k_{ab}^{(1)} & \\
k_{ba}^{(1)} & k_{bb}^{(1)} + k_{aa}^{(2)} + k_{aa}^{(3)} & k_{ab}^{(2)} + k_{ab}^{(3)} \\
& k_{ba}^{(2)} + k_{ba}^{(3)} & k_{bb}^{(2)} + k_{bb}^{(3)}
\end{bmatrix}
\begin{bmatrix} u_1 \\ u_2 \\ u_3 \end{bmatrix}
= \begin{bmatrix} \\ \\ P \end{bmatrix} + \begin{bmatrix} V_1 \\ \\ \\ \end{bmatrix}
\tag{2.24}
$$

Using the spring constant of each spring shown in Eq. (2.8), Eq. (2.24) can be transformed as follows:

$$\begin{bmatrix} k_{\textcircled{1}} & -k_{\textcircled{1}} & \\ -k_{\textcircled{1}} & k_{\textcircled{1}}+k_{\textcircled{2}}+k_{\textcircled{3}} & -k_{\textcircled{2}}-k_{\textcircled{3}} \\ & -k_{\textcircled{2}}-k_{\textcircled{3}} & k_{\textcircled{2}}+k_{\textcircled{3}} \end{bmatrix} \begin{bmatrix} u_1 \\ u_2 \\ u_3 \end{bmatrix} = \begin{bmatrix} \\ \\ P \end{bmatrix} + \begin{bmatrix} V_1 \\ \\ \\ \end{bmatrix} \tag{2.25}$$

Equation (2.25) is the finite element equation in the global form. Let us examine Eqs. (2.21) and (2.24) in further detail. Performing the matrix multiplication of coefficient matrices, Eq. (2.21) leads to the following equation:

$$\left[\mathbf{K}\right]\left[\mathbf{u}\right] = \left[\mathbf{P}\right] + \left[\mathbf{V}\right] \tag{2.26}$$

where

$$\left[\mathbf{K}\right] = \left(\left[\alpha \mathbf{k}_{aa} \alpha^T\right] + \left[\alpha \mathbf{k}_{ab} \beta^T\right] + \left[\beta \mathbf{k}_{ba} \alpha^T\right] + \left[\beta \mathbf{k}_{bb} \beta^T\right]\right)$$

Equation (2.26) means that the coefficients of the finite element equation consist of four constituents, which are given below:

$$\begin{aligned} &\left(\begin{bmatrix} k_{aa}^{\textcircled{1}} & & \\ & k_{aa}^{\textcircled{2}}+k_{aa}^{\textcircled{3}} & \\ & & \end{bmatrix} + \begin{bmatrix} & k_{ab}^{\textcircled{1}} & \\ & & k_{ab}^{\textcircled{2}}+k_{ab}^{\textcircled{3}} \\ & & \end{bmatrix} \right. \\ &\left. + \begin{bmatrix} & & \\ k_{ba}^{\textcircled{1}} & & \\ & k_{ba}^{\textcircled{2}}+k_{ba}^{\textcircled{3}} & \end{bmatrix} + \begin{bmatrix} & & \\ & k_{bb}^{\textcircled{1}} & \\ & & k_{bb}^{\textcircled{2}}+k_{bb}^{\textcircled{3}} \end{bmatrix} \right) \begin{bmatrix} u_1 \\ u_2 \\ u_3 \end{bmatrix} \\ &= \begin{bmatrix} \\ \\ P \end{bmatrix} + \begin{bmatrix} V_1 \\ \\ \\ \end{bmatrix} \end{aligned} \tag{2.27}$$

Equation (2.27) is derived from performing the multiplication of each constituent matrix of Eq. (2.26). Equation (2.27) can also be rewritten as

$$\begin{aligned} &\left(\begin{bmatrix} k_{aa}^{\textcircled{1}} & k_{ab}^{\textcircled{1}} & \\ k_{ba}^{\textcircled{1}} & k_{bb}^{\textcircled{1}} & \\ & & \end{bmatrix} + \begin{bmatrix} & & \\ & k_{aa}^{\textcircled{2}} & k_{ab}^{\textcircled{2}} \\ & k_{ba}^{\textcircled{2}} & k_{bb}^{\textcircled{2}} \end{bmatrix} + \begin{bmatrix} & & \\ & k_{aa}^{\textcircled{3}} & k_{ab}^{\textcircled{3}} \\ & k_{ba}^{\textcircled{3}} & k_{bb}^{\textcircled{3}} \end{bmatrix} \right) \begin{bmatrix} u_1 \\ u_2 \\ u_3 \end{bmatrix} \\ &= \begin{bmatrix} \\ \\ P \end{bmatrix} + \begin{bmatrix} V_1 \\ \\ \\ \end{bmatrix} \end{aligned} \tag{2.28}$$

The finite element equation in the global form Eq. (2.24) can easily be derived from Eq. (2.27). The finite element equation is derived by superimposing the deformation equation into the whole equation. In this sense, the deformation equation, Eq. (2.10), is sometimes called the "local finite element equation," whereas Eq. (2.21) is called the "global finite element equation." This is the basic concept of the finite element method. This concept is in fact not limited to spring structures, but extends to all analyses of structures and fluids, as we will see in the rest of this book.

2.7 Boundary Condition

Adding all rows of Eq. (2.21) or Eq. (2.25), we find that the sum is zero. This means that the coefficient matrix of Eq. (2.21) or Eq. (2.25) is singular, and we thus cannot solve the simultaneous equation, because the boundary conditions are not imposed. In case of the structure shown in Fig. 2.1, the boundary condition is

$$u_1 = 0 \tag{2.29}$$

where u_1 is the displacement on node 1 and at the node reaction force V_1 is unknown. The displacements of nodes of a whole structure can be divided into two groups: unknown and known displacements. To collectively analyze the unknown displacements, the following matrix $[\boldsymbol{\xi}^T]$ will be defined. The row number corresponds to node number, and column number corresponds to number of unknown displacement, which is noted by $\overline{2}$ and $\overline{3}$,

$$\left[\boldsymbol{\xi}^T\right] = \begin{matrix} \\ 1 \\ 2 \\ 3 \end{matrix}\begin{matrix} \overline{2} & \overline{3} \\ \left[\begin{matrix} & \\ 1 & \\ & 1 \end{matrix}\right. & \left.\right] \end{matrix} \tag{2.30}$$

and corresponding diagonal element is 1. Then the following expression should hold:

$$\left[\mathbf{u}\right] = \left[\boldsymbol{\xi}^T\right]\left[\overline{\mathbf{u}}\right] \tag{2.31}$$

where $[\overline{\mathbf{u}}]$ is the unknown displacement, which is

$$\left[\overline{\mathbf{u}}\right] = \begin{bmatrix} u_2 \\ u_3 \end{bmatrix} \tag{2.32}$$

Thus, Eq. (2.31) can be written as

$$\begin{bmatrix} u_1 \\ u_2 \\ u_3 \end{bmatrix} = \begin{bmatrix} & \\ 1 & \\ & 1 \end{bmatrix}\begin{bmatrix} u_2 \\ u_3 \end{bmatrix} \tag{2.33}$$

Equations (2.31) and (2.33) mean that the known displacements are subtracted from the whole displacements. Multiplying to [**P**] by [$\boldsymbol{\xi}$], the following relations can be obtained:

$$\left[\boldsymbol{\xi}\right]\left[\mathbf{P}\right]=\left[\overline{\mathbf{P}}\right] \tag{2.34}$$

which is

$$\begin{bmatrix} & 1 & \\ & & 1 \end{bmatrix}\begin{bmatrix} \\ \\ P \end{bmatrix}=\begin{bmatrix} \\ P \end{bmatrix} \tag{2.35}$$

where [$\overline{\mathbf{P}}$] is the applied external force matrix. Subtracting the nodes on which the boundary condition is imposed, and the following relation holds

$$\left[\boldsymbol{\xi}\right]\left[\mathbf{V}\right]=\left[\mathbf{0}\right] \tag{2.36}$$

which is

$$\begin{bmatrix} & 1 & \\ & & 1 \end{bmatrix}\begin{bmatrix} V_1 \\ \\ \\ \end{bmatrix}=\begin{bmatrix} \\ \\ \end{bmatrix} \tag{2.37}$$

This derivation shows that reaction force can be subtracted from the whole system. Further, introducing Eq. (2.31) into Eq. (2.26), we obtain the following:

$$\left[\mathbf{K}\right]\left[\boldsymbol{\xi}^T\right]\left[\bar{\mathbf{u}}\right]=\left[\mathbf{P}\right]+\left[\mathbf{V}\right] \tag{2.38}$$

which can be decomposed as

$$\begin{bmatrix} k_{\textcircled{1}} & -k_{\textcircled{1}} & \\ -k_{\textcircled{1}} & (k_{\textcircled{1}}+k_{\textcircled{2}}+k_{\textcircled{3}}) & -(k_{\textcircled{2}}+k_{\textcircled{3}}) \\ & -(k_{\textcircled{2}}+k_{\textcircled{3}}) & (k_{\textcircled{2}}+k_{\textcircled{3}}) \end{bmatrix}\begin{bmatrix} & \\ 1 & \\ & 1 \end{bmatrix}\begin{bmatrix} u_2 \\ u_3 \end{bmatrix}$$
$$=\begin{bmatrix} \\ \\ P \end{bmatrix}+\begin{bmatrix} V_1 \\ \\ \\ \end{bmatrix} \tag{2.39}$$

By rearranging the terms, we get

$$\begin{bmatrix} -k_{\textcircled{1}} & \\ (k_{\textcircled{1}}+k_{\textcircled{2}}+k_{\textcircled{3}}) & -(k_{\textcircled{2}}+k_{\textcircled{3}}) \\ -(k_{\textcircled{2}}+k_{\textcircled{3}}) & (k_{\textcircled{2}}+k_{\textcircled{3}}) \end{bmatrix}\begin{bmatrix} u_2 \\ u_3 \end{bmatrix}=\begin{bmatrix} \\ \\ P \end{bmatrix}+\begin{bmatrix} V_1 \\ \\ \\ \end{bmatrix} \tag{2.40}$$

Multiplying $[\boldsymbol{\xi}]$ from the left to Eq. (2.38) and using Eq. (2.36), the following equation can be obtained:

$$\left[\overline{\mathbf{K}}\right]\left[\overline{\mathbf{u}}\right]=\left[\overline{\mathbf{P}}\right] \tag{2.41}$$

where $[\overline{\mathbf{K}}]=[\boldsymbol{\xi}][\mathbf{K}][\boldsymbol{\xi}^T]$ can be expanded as

$$\begin{bmatrix} (k_{①}+k_{②}+k_{③}) & -(k_{②}+k_{③}) \\ -(k_{②}+k_{③}) & (k_{②}+k_{③}) \end{bmatrix}\begin{bmatrix} u_2 \\ u_3 \end{bmatrix}=\begin{bmatrix} \\ P \end{bmatrix} \tag{2.42}$$

Equation (2.41) or Eq. (2.42) is the final form of the finite element equation including the boundary condition. Specifying the external force P, the displacements of structure can be obtained by Eq. (2.41) or Eq. (2.42). Subtracting Eq. (2.42) from Eq. (2.40), the remainder equation is

$$\begin{bmatrix} -k_{①} & \end{bmatrix}\begin{bmatrix} u_2 \\ u_3 \end{bmatrix}=\begin{bmatrix} V_1 \end{bmatrix} \tag{2.43}$$

from which the reaction force can be obtained. Thus, this equation is called the reaction equation.

2.8 Solution

In addition to the basic machinery that we have introduced so far in this chapter, it is necessary to deploy some other products to solve the finite element equation Eq. (2.42). We can solve Eq. (2.42) for u_2 and u_3 and get

$$u_2=\frac{(k_{②}+k_{③})}{k_{①}(k_{②}+k_{③})}P=\frac{1}{k_{①}}P \tag{2.44}$$

$$u_3=\frac{(k_{①}+k_{②}+k_{③})}{k_{①}(k_{②}+k_{③})}P=\frac{1}{k_{②}+k_{③}}P+\frac{P}{k_{①}} \tag{2.45}$$

Substituting Eqs. (2.44) and (2.45) into Eq. (2.43), the reaction force V_1 can be obtained:

$$V_1=-k_{①}u_2=-P \tag{2.46}$$

Equation (2.46) means that the magnitude of V_1 is P and direction of V_1 is anti-direction of P. This is a natural consequence of the equilibrium of the whole spring structure.

To compute the spring forces, Eq. (2.22) is reformulated as follows, using Eqs. (2.8) and (2.29):

$$\begin{bmatrix} F_a^{①} \\ F_a^{②} \\ F_a^{③} \end{bmatrix} = \begin{bmatrix} -k_{①} & \\ k_{②} & -k_{②} \\ k_{③} & -k_{③} \end{bmatrix} \begin{bmatrix} u_2 \\ u_3 \end{bmatrix} \tag{2.47}$$

$$\begin{bmatrix} F_b^{①} \\ F_b^{②} \\ F_b^{③} \end{bmatrix} = \begin{bmatrix} k_{①} & \\ -k_{②} & k_{②} \\ -k_{③} & k_{③} \end{bmatrix} \begin{bmatrix} u_2 \\ u_3 \end{bmatrix} \tag{2.48}$$

Introducing Eqs. (2.44) and (2.45) into Eqs. (2.47) and (2.48), we obtain

$$\begin{aligned} F_a^{①} &= -k_{①} u_2 = -k_{①} \frac{P}{k_{①}} = -P \\ F_a^{②} &= k_{②}(u_2 - u_3) = -k_{②} \frac{P}{k_{②} + k_{③}} = -\frac{k_{②}}{k_{②} + k_{③}} P \\ F_a^{③} &= k_{③}(u_2 - u_3) = -k_{③} \frac{P}{k_{②} + k_{③}} = -\frac{k_{③}}{k_{②} + k_{③}} P \\ F_b^{①} &= k_{①} u_2 = k_{①} \frac{P}{k_{①}} = P \\ F_b^{②} &= k_{②}(-u_2 + u_3) = k_{②} \frac{P}{k_{②} + k_{③}} = \frac{k_{②}}{k_{②} + k_{③}} P \\ F_b^{③} &= k_{③}(-u_2 + u_3) = k_{③} \frac{P}{k_{②} + k_{③}} = \frac{k_{③}}{k_{②} + k_{③}} P \end{aligned} \tag{2.49}$$

Recall Eqs. (2.3) and (2.4), then $F_b^{①}$, $F_b^{②}$, and $F_b^{③}$ can be adopted as the spring forces. Thus, we have solved the spring structure shown in Fig. 2.1.

2.9 Systematic Formulation of the Finite Element Matrix

Now we take a look at the global finite element equation Eq. (2.24) from a different perspective. The coefficient matrix of Eq. (2.24) can be systematically formulated using the coefficient matrix of the deformation equation, which is

$$\left[\begin{array}{c|c} k_{aa}^{ⓜ} & k_{ab}^{ⓜ} \\ \hline k_{ba}^{ⓜ} & k_{bb}^{ⓜ} \end{array}\right] \tag{2.50}$$

To highlight the fact that Eq. (2.50) is for the spring ⓜ, ⓜ is superscripted on the elements of the matrix to which it is associated. Table 2.1 summarizes the connectivity between springs and nodes. The first column represents the number of spring ⓜ,

Table 2.1 Connectivity relation

ⓜ	a	b
①	1	2
②	2	3
③	2	3

whereas the second and third columns show the number of nodes which correspond to the edges a and b, respectively.

Building on Table 2.1, the coefficient matrix of the global finite element equation Eq. (2.24) can be formulated as follows. The global matrix is a matrix whose row and column numbers correspond to nodes. Given this matrix, apply the following procedures:

Superimpose $k_{aa}^{\textcircled{m}}$ onto row a and column a;
Superimpose $k_{ab}^{\textcircled{m}}$ onto row a and column b;
Superimpose $k_{ba}^{\textcircled{m}}$ onto row b and column a;
Superimpose $k_{bb}^{\textcircled{m}}$ onto row b and column b.

Let us now consider this line of approach in detail with a specific example. For spring ①, the local coefficient matrix is

$$\begin{bmatrix} k_{\textcircled{1}} & -k_{\textcircled{1}} \\ -k_{\textcircled{1}} & k_{\textcircled{1}} \end{bmatrix} \tag{2.51}$$

which we can superimpose onto the global coefficient matrix, and we get

$$\begin{array}{c} \\ 1 \\ 2 \\ 3 \end{array} \begin{array}{c} \begin{array}{ccc} \quad 1 \quad & \quad 2 \quad & \quad 3 \quad \end{array} \\ \begin{bmatrix} k_{\textcircled{1}} & -k_{\textcircled{1}} & \\ -k_{\textcircled{1}} & k_{\textcircled{1}} & \\ & & \end{bmatrix} \end{array} \tag{2.52}$$

To take another example, for spring ②, the local coefficient matrix is

$$\begin{bmatrix} k_{\textcircled{2}} & -k_{\textcircled{2}} \\ -k_{\textcircled{2}} & k_{\textcircled{2}} \end{bmatrix} \tag{2.53}$$

We can again superimpose this coefficient onto the global coefficient matrix, and we get

$$
\begin{array}{c}
\\
1 \\
2 \\
3
\end{array}
\begin{array}{c}
\begin{array}{ccc} \quad 1 \quad & \quad 2 \quad & \quad 3 \quad \end{array} \\
\begin{bmatrix}
k_{①} & -k_{①} & \\
-k_{①} & k_{①} + k_{②} & -k_{②} \\
 & -k_{②} & k_{②}
\end{bmatrix}
\end{array}
\tag{2.54}
$$

Finally, for spring ③, recall that the local coefficient matrix is

$$
\begin{bmatrix}
k_{③} & -k_{③} \\
-k_{③} & k_{③}
\end{bmatrix}
\tag{2.55}
$$

and by incorporating the coefficient matrix into the global coefficient matrix:

$$
\begin{array}{c}
\\
1 \\
2 \\
3
\end{array}
\begin{array}{c}
\begin{array}{ccc} \quad 1 \quad & \quad 2 \quad & \quad 3 \quad \end{array} \\
\begin{bmatrix}
k_{①} & -k_{①} & \\
-k_{①} & k_{①} + k_{②} + k_{③} & -k_{②} - k_{③} \\
 & -k_{②} - k_{②} & k_{②} + k_{③}
\end{bmatrix}
\end{array}
\tag{2.56}
$$

The final matrix obtained in this way is identical to the coefficient matrix of Eq. (2.25). The algorithm demonstrated in this section is systematic and efficient for computational coding.

2.10 Special Cases of Spring Structures

The algorithm described in Sect. 2.9 allows us to deal with special cases of spring structures. The first example of this kind is parallel springs, as exemplified in Fig. 2.5. We follow the previous notational conventions and mark node numbers with 1 and 2,

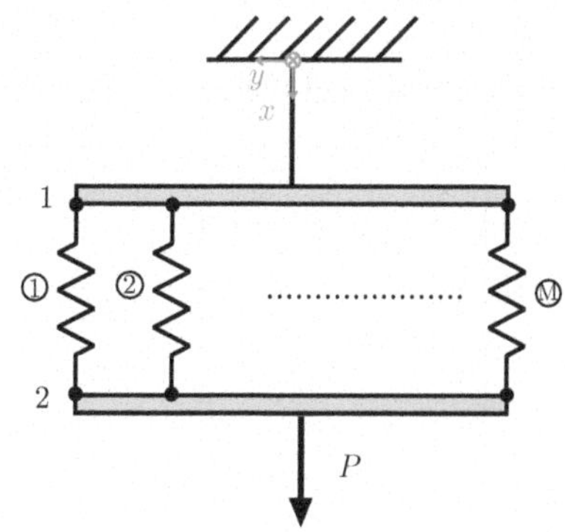

Fig. 2.5 Parallel springs

Table 2.2 Connectivity relations of parallel springs

ⓜ	a	b
①	1	2
②	1	2
⋮	⋮	⋮
Ⓜ	1	2

spring numbers with ①, ②, ..., Ⓜ, and spring constants with $k_{①}, k_{②}, \ldots, k_{Ⓜ}$. Table 2.2 summarizes the connectivity relations.

Based on Table 2.2, we can express the coefficient matrix of the global finite element equation as

$$\begin{bmatrix} k_{①}+k_{②}+\cdots+k_{Ⓜ} & -k_{①}-k_{②}-\cdots-k_{Ⓜ} \\ -k_{①}-k_{②}-\cdots-k_{Ⓜ} & k_{①}+k_{②}+\cdots+k_{Ⓜ} \end{bmatrix} \tag{2.57}$$

Denoting displacements at each node u_1 and u_2, and imposing the following boundary condition,

$$u_1 = 0 \tag{2.58}$$

The global finite element equation can be directly obtained as

$$\begin{bmatrix} k_{①}+k_{②}+\cdots+k_{Ⓜ} \end{bmatrix}\begin{bmatrix} u_2 \end{bmatrix} = \begin{bmatrix} P \end{bmatrix} \tag{2.59}$$

where P is the external force. This procedure allows us to formulate Eq. (2.59) more easily, and to get displacement u_2. It also enables us to compute the spring forces in an efficient manner.

The second example is a series of springs, as illustrated in Fig. 2.6. Again we represent node numbers with 0, 1, 2, ..., n, spring numbers with ①, ②, ..., ⓝ, and spring constants with $k_{①}, k_{②}, \ldots, k_{ⓝ}$. We can summarize connectivity relation as in Table 2.3.

The coefficient matrix of the global finite element equation is

$$\begin{bmatrix} k_{①} & -k_{①} & & & & & & \\ -k_{①} & k_{①}+k_{②} & -k_{②} & & & & & \\ & -k_{②} & k_{②}+k_{③} & \ddots & & & & \\ & & \ddots & \ddots & \ddots & & & \\ & & & -k_{\textcircled{i-1}} & k_{\textcircled{i-1}}+k_{ⓘ} & -k_{ⓘ} & & \\ & & & & \ddots & \ddots & \ddots & \\ & & & & & -k_{\textcircled{n-1}} & k_{\textcircled{n-1}}+k_{ⓝ} & -k_{ⓝ} \\ & & & & & & -k_{ⓝ} & k_{ⓝ} \end{bmatrix} \tag{2.60}$$

Fig. 2.6 Series of springs

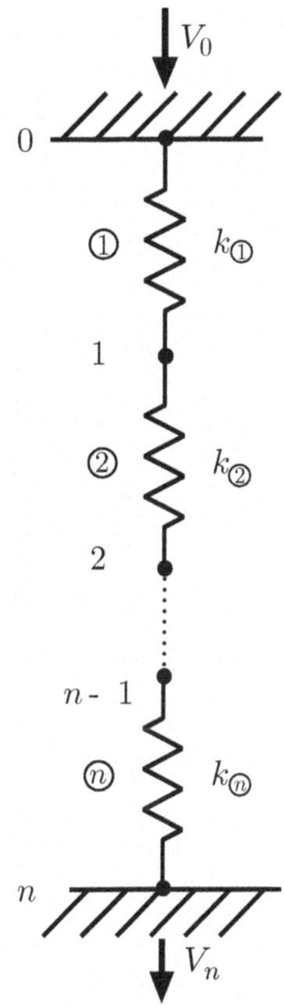

The global finite element equation can be described as follows:

$$\begin{bmatrix} k_{\textcircled{1}} & -k_{\textcircled{1}} & & & & & & \\ -k_{\textcircled{1}} & k_{\textcircled{1}}+k_{\textcircled{2}} & -k_{\textcircled{2}} & & & & & \\ & -k_{\textcircled{2}} & k_{\textcircled{2}}+k_{\textcircled{3}} & \ddots & & & & \\ & & \ddots & \ddots & \ddots & & & \\ & & & -k_{\textcircled{i-1}} & k_{\textcircled{i-1}}+k_{\textcircled{i}} & -k_{\textcircled{i}} & & \\ & & & & \ddots & \ddots & \ddots & \\ & & & & & -k_{\textcircled{n-1}} & k_{\textcircled{n-1}}+k_{\textcircled{n}} & -k_{\textcircled{n}} \\ & & & & & & -k_{\textcircled{n}} & k_{\textcircled{n}} \end{bmatrix} \begin{bmatrix} u_0 \\ u_1 \\ u_2 \\ \vdots \\ u_i \\ \vdots \\ u_{n-1} \\ u_n \end{bmatrix}$$

Table 2.3 Connectivity relations of a series of springs. See Fig. 2.6

ⓜ	a	b
①	0	1
②	1	2
⋮	⋮	⋮
ⓝ	$n-1$	n

$$= \begin{bmatrix} \\ P_1 \\ P_2 \\ \vdots \\ P_i \\ \vdots \\ P_{n-1} \\ \\ \end{bmatrix} + \begin{bmatrix} V_0 \\ \\ \\ \\ \\ \\ \\ V_n \end{bmatrix} \tag{2.61}$$

where $u_0, u_1, \ldots, u_n$ represent displacements, $P_1, P_2, \ldots, P_{n-1}$ are external forces, and V_0 and V_n are reaction forces at nodes 0 and n. The boundary conditions are

$$u_0 = u_n = 0 \tag{2.62}$$

We first divide Eq. (2.61) into two parts:

$$\begin{bmatrix} -k_{①} & & \\ & & -k_{ⓝ} \end{bmatrix} \begin{bmatrix} u_1 \\ u_2 \\ \vdots \\ \vdots \\ \vdots \\ u_{n-1} \end{bmatrix} = \begin{bmatrix} V_0 \\ V_n \end{bmatrix} \tag{2.63}$$

and

$$
\begin{bmatrix}
k_{\textcircled{1}}+k_{\textcircled{2}} & -k_{\textcircled{2}} & & & & & \\
-k_{\textcircled{2}} & k_{\textcircled{2}}+k_{\textcircled{3}} & -k_{\textcircled{3}} & & & & \\
 & \ddots & \ddots & \ddots & & & \\
 & & -k_{\textcircled{i-1}} & k_{\textcircled{i-1}}+k_{\textcircled{i}} & -k_{\textcircled{i}} & & \\
 & & & \ddots & \ddots & \ddots & \\
 & & & & -k_{\textcircled{n-1}} & k_{\textcircled{n-1}}+k_{\textcircled{n}} &
\end{bmatrix}
\begin{bmatrix} u_1 \\ u_2 \\ \vdots \\ u_i \\ \vdots \\ u_{n-1} \end{bmatrix}
=
\begin{bmatrix} P_1 \\ P_2 \\ \vdots \\ P_i \\ \vdots \\ P_{n-1} \end{bmatrix}
\tag{2.64}
$$

When the following condition holds,

$$
k_{\textcircled{1}} = k_{\textcircled{2}} = \cdots k_{\textcircled{n}} \cdots = \boldsymbol{k} \tag{2.65}
$$

$$
\boldsymbol{k}
\begin{bmatrix}
2 & -1 & & & & \\
-1 & 2 & -1 & & & \\
 & \ddots & \ddots & \ddots & & \\
 & & -1 & 2 & -1 & \\
 & & & \ddots & \ddots & \ddots \\
 & & & & -1 & 2
\end{bmatrix}
\begin{bmatrix} u_1 \\ u_2 \\ \vdots \\ u_i \\ \vdots \\ u_{n-1} \end{bmatrix}
=
\begin{bmatrix} P_1 \\ P_2 \\ \vdots \\ P_i \\ \vdots \\ P_{n-1} \end{bmatrix}
\tag{2.66}
$$

Equation (2.66) is similar to the finite difference approximation of one-dimensional second-order ordinary differential equation. By solving Eq. (2.66), we obtain displacements $u_1, \ldots, u_{n-1}$, as well as spring forces. Although it is very hard—if not impossible—to solve the global finite element equation in general, the analytical form of the inverse matrix of Eq. (2.66) can be derived. Suppose that we posit

$$
A_n =
\begin{bmatrix}
2 & -1 & & & & & \\
-1 & 2 & -1 & & & & \\
 & \ddots & \ddots & \ddots & & & \\
 & & -1 & 2 & -1 & & \\
 & & & \ddots & \ddots & \ddots & \\
 & & & & -1 & 2 & -1 \\
 & & & & & -1 & 2
\end{bmatrix}
\tag{2.67}
$$

then we get

$$A_n^{-1} = \frac{1}{n+1}\begin{bmatrix} n & (n-1) & (n-2) & \dots & (n+1-k) & \dots & (n+1-l) & \dots & 1 \\ (n-1) & 2(n-1) & 2(n-2) & \dots & 2(n+1-k) & \dots & 2(n+1-l) & \dots & 2 \\ (n-2) & (n-2)2 & 3(n-2) & \dots & 3(n+1-k) & \dots & 3(n+1-l) & \dots & 3 \\ \vdots & \vdots & \vdots & \ddots & \vdots & & \vdots & & \vdots \\ (n+1-k) & (n+1-k)2 & (n+1-k)3 & \dots & k(n+1-k) & \dots & k(n+1-l) & \dots & k \\ \vdots & \vdots & \vdots & & \vdots & \ddots & \vdots & \dots & \vdots \\ (n+1-k) & (n+1-k)2 & (n+1-k)3 & \dots & (n+1-k)l & \dots & l(n+1-l) & \dots & l \\ \vdots & \vdots & \vdots & & \vdots & & \vdots & \ddots & \vdots \\ 1 & 2 & 3 & \dots & k & \dots & l & \dots & n \end{bmatrix} \tag{2.68}$$

Matrix elements in Eq. (2.68) are as follows.

$$\begin{array}{ll} \dfrac{k(n+1-k)}{n+1} & l \geqq k \\ \dfrac{(n+1-k)l}{n+1} & l < k \end{array} \tag{2.69}$$

In general, the numerical methods are employed to solve the global finite element equation.

2.11 Summary and Conclusion

In this chapter we have observed that the finite element equation is derived from the theorems based on the three main equations—deformation, equilibrium, and compatibility equations—in addition to boundary conditions. The unknown degrees of freedom can be determined by solving those equations. We have presented the precise formulation of spring structures in order to illustrate the fundamentals of finite element method. To recapitulate, the main procedure consists of the superimposition of the local finite element equation onto the whole system to obtain the global finite element equation based on the connectivity conditions. Thus, the formulation is simple and adaptable to explain to any arbitrary shaped structures, as well as to fluids with arbitrary shaped boundaries.

We will observe in the following chapters that in fact, the connectivity procedures extend to many other cases. Therefore, if we know the local finite element equation in any structures or fluids, then it means that we can derive the global finite element equation. Therefore, the main theme of this book focuses on deriving the local finite element equation for the analyses of fluid flows. This is also the main reason why almost if not all of the literature on the finite element method provides the derivation of the local finite element equation.

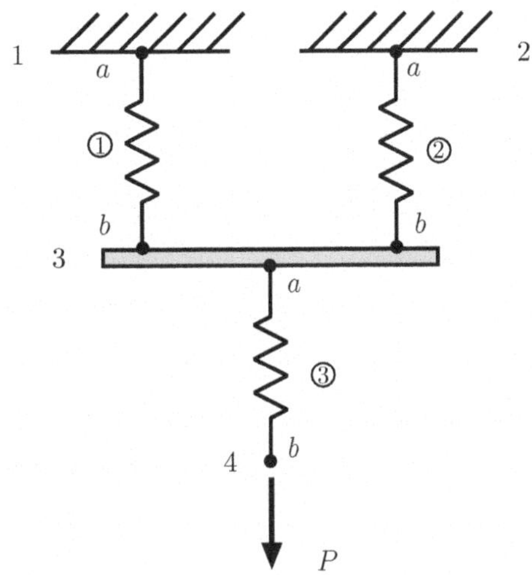

Fig. 2.7 Spring structure

The finite element equation of structures consisting of series of the same springs can be also derived. In this case the final equation is the same as obtained by the finite difference method. Moreover, it is also possible to use different sets of coefficients of springs for the whole structure in the finite element method. Therefore, the finite element method is sometimes said to include the finite difference method.

2.12 Exercises

1. Compute displacements and forces of spring structures shown in Fig. 2.1 using spring constants in the figure.
2. Derive the global finite element equation of spring structures shown in Fig. 2.7.
3. Consider the meaning of the symmetry of the finite element matrix, i.e., what is the implication of the symmetry ($k_{ab} = k_{ba}$) is.
4. Prove that the matrix (2.67) is the inverse matrix of (2.66).
5. Solve the spring structure shown in Fig. 2.7.

Chapter 3
Pipeline Systems

3.1 Introduction

Chapter 2 covered the fundamentals of the finite element method, illustrating the derivation of the global finite element method using a simple spring structure. How to drive the global finite element equation from the local finite element equation is more or less identical across all finite element analyses that apply to structures and fluids. Therefore, it is all the more critical to understand the derivation of the local finite element equation. In essence, then, the bulk of the literature on the finite element method addresses how to derive the local finite element equation. This chapter extends the previous chapter by presenting a case study using a simple pipeline system. We will go through the method to derive its local finite element equation. The formulation we are going to introduce in this chapter is called the Galerkin method. This chapter forms a basis for the next chapter, which together deals with computational methods for the finite element method. An introductory computer program is also included, which allows readers to understand the basic ideas developed in Chaps. 2 and 3.

3.2 A Pipeline System

Suppose we are solving the flow though pipeline shown in Fig. 3.1. Following and extending on the notational conventions that we deployed in Chap. 2, we number nodes numerically as 1, 2, 3, 4, and pipe components as ①, ②, ③, ④. For the sake of illustration, we set aside the effect of bending. The two edges of the pipe component, abbreviated as pipe, are distinguished by edge a and edge b, as shown in Fig. 3.2. The length of the pipe is represented as l, and the section area as A. The local coordinate system x–y–z is the right-hand system with x-axis directed from edge a to edge b and with origin located on edge a. The global coordinate system X–Y–Z is also used, but its origin can be arbitrary.

M. Kawahara, *Finite Element Methods in Incompressible, Adiabatic, and Compressible Flows*, Mathematics for Industry 19,
DOI 10.1007/978-4-431-55450-9_3

Fig. 3.1 Pipeline system

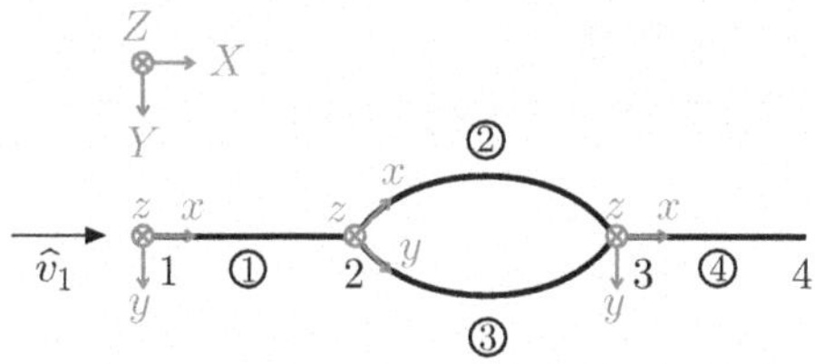

Fig. 3.2 Pipe component

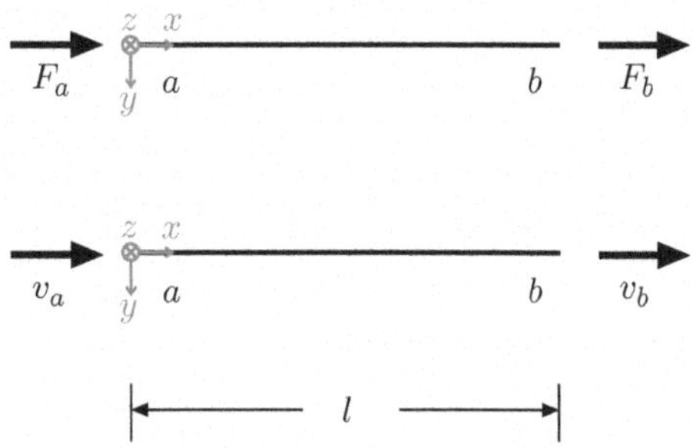

Assume that the flow is very slow and has high viscosity. The relation between the flux of flow F and the deformation rate d is linear in such a way that

$$F = \mu A d = B d \tag{3.1}$$

where μ represents viscosity. For the sake of exposition, we put B as a place holder for μA, which functions as a proportional constant. The flux F is the deriving force to generate the flow, and the deformation rate is

$$d = \frac{dv}{dx} \tag{3.2}$$

where v is the velocity of the flow in the pipe. The flux should consistently be in an equilibrium state, which means that

$$\frac{dF}{dx} = 0 \tag{3.3}$$

From Eqs. (3.1) and (3.2), the flux can thus be expressed as

$$F = B\frac{dv}{dx} \tag{3.4}$$

Introducing Eq. (3.4) into Eq. (3.3), we can derive the governing equation as given below:

$$B\frac{d^2v}{dx^2} = 0 \tag{3.5}$$

We now denote the velocities at the two edges of the pipe a and b as follows:

$$\begin{aligned} v_a &= (v)_{x=0} \\ v_b &= (v)_{x=l} \end{aligned} \tag{3.6}$$

Likewise, the fluxes at the two edges are

$$\begin{aligned} F_a &= -(F)_{x=0} \\ F_b &= (F)_{x=l} \end{aligned} \tag{3.7}$$

It is easy to solve Eq. (3.5) under the two conditions, Eqs. (3.6) and (3.7). Then the local finite element equation can be derived. However, unfortunately, the general governing equation cannot be solved in this manner. Therefore, we need the finite element method, which we turn to now.

3.3 Local Finite Element Equation

The Galerkin method allows us to derive the local finite element equation here. Multiplying both sides of Eq. (3.5) by weighting function v^*, and integrating over the pipe, we get

$$\int_0^l \left(v^* B \frac{d^2 v}{dx^2} \right) dx = 0 \tag{3.8}$$

The weighting function v^* is an arbitrary function, and we can in fact choose any function that is continuous. The naming and the meaning are found in Sect. 4.9. Indeed, insofar as Eq. (3.5) is valid, Eq. (3.8) is also valid, for any function v^*. Applying the equality

$$\frac{d}{dx} \left(v^* B \frac{dv}{dx} \right) = B \frac{dv^*}{dx} \frac{dv}{dx} + v^* B \frac{d^2 v}{dx^2} \tag{3.9}$$

to Eq. (3.8), it can be transformed into the following form:

$$\int_0^l \frac{d}{dx} \left(v^* B \frac{dv}{dx} \right) dx - \int_0^l \left(B \frac{dv^*}{dx} \frac{dv}{dx} \right) dx = 0 \tag{3.10}$$

Integrating the first term of Eq. (3.10), we obtain

$$\left[v^* B \frac{dv}{dx} \right]_0^l - \int_0^l \left(B \frac{dv^*}{dx} \frac{dv}{dx} \right) dx = 0 \tag{3.11}$$

Using Eqs. (3.4) and (3.7), Eq. (3.11) can be rewritten as

$$F_a v_a^* + F_b v_b^* = \int_0^l \left(B \frac{dv^*}{dx} \frac{dv}{dx} \right) dx \tag{3.12}$$

where

$$\begin{aligned} v_a^* &= (v^*)_{x=0} \\ v_b^* &= (v^*)_{x=l} \end{aligned} \tag{3.13}$$

Recall that v^* is an arbitrary function, and hence v_a^* and v_b^* are also arbitrary. Consider the interpolation for the velocity as

$$v = \left(1 - \frac{x}{l}\right) v_a + \left(\frac{x}{l}\right) v_b \tag{3.14}$$

Equation (3.14) is linear, and Eq. (3.5) is satisfied. Thus, v_a and v_b are velocities of the edges a and b. Equation (3.14) is referred to as the interpolation function. For the weighting function of velocity, the following equation can be used:

$$v^* = \left(1 - \frac{x}{l}\right) v_a^* + \left(\frac{x}{l}\right) v_b^* \tag{3.15}$$

In this method, the weighting function is of the same type of interpolation function—this sort of method is referred to as the Galerkin method. In this method, note that Eq. (3.15) satisfies Eq. (3.13). Let me reiterate here that v_a^* and v_b^* are arbitrary constants, not velocities. For substituting Eqs. (3.14) and (3.15) into Eq. (3.12), we can compute the first derivatives:

$$\frac{dv}{dx} = \left(-\frac{1}{l}\right) v_a + \left(\frac{1}{l}\right) v_b \tag{3.16}$$

$$\frac{dv^*}{dx} = \left(-\frac{1}{l}\right) v_a^* + \left(\frac{1}{l}\right) v_b^* \tag{3.17}$$

Incorporating Eqs. (3.16) and (3.17) into Eq. (3.12), and rearranging the terms, we obtain

$$\begin{aligned} &v_a^* \left[F_a - \int_0^l \left(-\frac{1}{l}\right) B \left(-\frac{1}{l}\right) dx\, v_a - \int_0^l \left(-\frac{1}{l}\right) B \left(\frac{1}{l}\right) dx\, v_b \right] \\ &+ v_b^* \left[F_b - \int_0^l \left(\frac{1}{l}\right) B \left(-\frac{1}{l}\right) dx\, v_a - \int_0^l \left(\frac{1}{l}\right) B \left(\frac{1}{l}\right) dx\, v_b \right] = 0 \end{aligned} \tag{3.18}$$

We can rewrite Eq. (3.18) by rearranging the terms:

$$\begin{aligned} & v_a^* \left[F_a - \left(\frac{B}{l}\right) v_a - \left(-\frac{B}{l}\right) v_b \right] \\ & \quad + v_b^* \left[F_b - \left(-\frac{B}{l}\right) v_a - \left(\frac{B}{l}\right) v_b \right] = 0 \end{aligned} \tag{3.19}$$

Recall again that v_a^* and v_b^* are arbitrary constants; thus, the local form of the finite element equation can now be derived as follows:

$$\begin{bmatrix} F_a \\ \hdashline F_b \end{bmatrix} = \left[\begin{array}{c:c} \left(\frac{B}{l}\right) & \left(-\frac{B}{l}\right) \\ \hdashline \left(-\frac{B}{l}\right) & \left(\frac{B}{l}\right) \end{array}\right] \begin{bmatrix} v_a \\ \hdashline v_b \end{bmatrix} \tag{3.20}$$

Suppose we posit $v_a^* = 1$ and $v_b^* = 0$, we get the first equation of Eq. (3.20); suppose on the other hand that $v_a^* = 0$ and $v_b^* = 1$, then we can obtain the second equation of Eq. (3.20). Equation (3.20) is the same type of Eq. (2.5), and Eq. (3.20) can be rewritten as

$$\begin{bmatrix} F_a \\ \hdashline F_b \end{bmatrix} = \left[\begin{array}{c:c} h & -h \\ \hdashline -h & h \end{array}\right] \begin{bmatrix} v_a \\ \hdashline v_b \end{bmatrix} \tag{3.21}$$

or

$$\begin{bmatrix} F_a \\ \hdashline F_b \end{bmatrix} = \left[\begin{array}{c:c} h_{aa} & h_{ab} \\ \hdashline h_{ba} & h_{bb} \end{array}\right] \begin{bmatrix} v_a \\ \hdashline v_b \end{bmatrix} \tag{3.22}$$

where

$$\begin{aligned} h_{aa} &= h \\ h_{ab} &= -h \\ h_{ba} &= -h \\ h_{bb} &= h \end{aligned}$$

To specifically signal the fact that Eq. (3.21) is for the pipe Ⓜ, we note equation terms with the superscript Ⓜ, as in

$$\begin{bmatrix} F_a^{ⓜ} \\ \hdashline F_b^{ⓜ} \end{bmatrix} = \begin{bmatrix} h_{aa}^{ⓜ} & h_{ab}^{ⓜ} \\ \hdashline h_{ba}^{ⓜ} & h_{bb}^{ⓜ} \end{bmatrix} \begin{bmatrix} v_a^{ⓜ} \\ \hdashline v_b^{ⓜ} \end{bmatrix} \tag{3.23}$$

where

$$\begin{aligned} h_{aa}^{ⓜ} &= h^{ⓜ} \\ h_{ab}^{ⓜ} &= -h^{ⓜ} \\ h_{ba}^{ⓜ} &= -h^{ⓜ} \\ h_{bb}^{ⓜ} &= h^{ⓜ} \end{aligned}$$

The local finite element equations are Eqs. (3.21) and (3.22), or Eq. (3.23).

3.4 Global Finite Element Equation

Using Eq. (3.23), the global form of the member finite element equation can be derived as follows:

$$\begin{bmatrix} \mathbf{F}_a \\ \hdashline \mathbf{F}_b \end{bmatrix} = \begin{bmatrix} \mathbf{h}_{aa} & \mathbf{h}_{ab} \\ \hdashline \mathbf{h}_{ba} & \mathbf{h}_{bb} \end{bmatrix} \begin{bmatrix} \mathbf{v}_a \\ \hdashline \mathbf{v}_b \end{bmatrix} \tag{3.24}$$

The derivation procedure is identical to what we have gone through in Sect. 2.3—Eq. (3.24) corresponds to Eq. (2.10). Let us posit the connectivity matrices $[\boldsymbol{\alpha}]$ and $[\boldsymbol{\beta}]$, which are obtained by the same procedure illustrated in Sect. 2.4. The matrices are based on the connectivity relations, which are summarized in Table 3.1.

Using the procedures illustrated in Sects. 2.3–2.5, we can obtain the global finite element equation:

$$\begin{bmatrix} \boldsymbol{\alpha} & \boldsymbol{\beta} \end{bmatrix} \begin{bmatrix} \mathbf{h}_{aa} & \mathbf{h}_{ab} \\ \hdashline \mathbf{h}_{ba} & \mathbf{h}_{bb} \end{bmatrix} \begin{bmatrix} \boldsymbol{\alpha}^T \\ \hdashline \boldsymbol{\beta}^T \end{bmatrix} \begin{bmatrix} \mathbf{v} \end{bmatrix} = \begin{bmatrix} \mathbf{V} \end{bmatrix} \tag{3.25}$$

Table 3.1 Connectivity relations

ⓜ	a	b
①	1	2
②	2	3
③	2	3
④	3	4

where [**v**] represents velocity at each node, which can be specifically represented as

$$[\mathbf{v}] = \begin{bmatrix} v_1 \\ v_2 \\ v_3 \\ v_4 \end{bmatrix} \tag{3.26}$$

and [**V**] is the flux on which the velocity is specified, which is

$$[\mathbf{V}] = \begin{bmatrix} V_1 \\ \\ \\ \end{bmatrix} \tag{3.27}$$

where V_1 is the flux on node 1, which is the deriving force to determine the velocity to be specific values $\hat{v}_1$. The velocity $\hat{v}_1$ is known, but the flux V_1 is not. The precise form of Eq. (3.25) can be described as follows:

$$\begin{bmatrix} h_{①} & -h_{①} & & \\ -h_{①} & h_{①} + h_{②} + h_{③} & -h_{②} - h_{③} & \\ & -h_{②} - h_{③} & h_{②} + h_{③} + h_{④} & -h_{④} \\ & & -h_{④} & h_{④} \end{bmatrix} \begin{bmatrix} v_1 \\ v_2 \\ v_3 \\ v_4 \end{bmatrix} = \begin{bmatrix} V_1 \\ \\ \\ \\ \end{bmatrix} \tag{3.28}$$

Equation (3.25), or its precise form Eq. (3.28), is the global finite element equation. Equation (3.28) can be rewritten as in the following form:

$$[\mathbf{H}][\mathbf{v}] = [\mathbf{V}] \tag{3.29}$$

where

$$[\mathbf{H}] = \left([\boldsymbol{\alpha}\mathbf{h}_{aa}\boldsymbol{\alpha}^T] + [\boldsymbol{\alpha}\mathbf{h}_{ab}\boldsymbol{\beta}^T] + [\boldsymbol{\beta}\mathbf{h}_{ba}\boldsymbol{\alpha}^T] + [\boldsymbol{\beta}\mathbf{h}_{bb}\boldsymbol{\beta}^T] \right)$$

3.5 Boundary Conditions and the Final Form of the Global Finite Element Method

The boundary condition of the pipeline system shown in Fig. 3.1 is that the velocity on node 1 is known in such a way that

$$v_1 = \hat{v}_1 \tag{3.30}$$

where $\hat{v}_1$ is a specified value. The flux V_1 is unknown. To impose the boundary condition of Eq. (3.30), the velocity can be separated into the known portions and unknown portions. More specifically,

$$\begin{bmatrix} v_1 \\ v_2 \\ v_3 \\ v_4 \end{bmatrix} = \begin{bmatrix} & & \\ 1 & & \\ & 1 & \\ & & 1 \end{bmatrix} \begin{bmatrix} v_2 \\ v_3 \\ v_4 \end{bmatrix} + \begin{bmatrix} 1 \\ \\ \\ \end{bmatrix} \begin{bmatrix} \hat{v}_1 \end{bmatrix} \tag{3.31}$$

We can rewrite Eq. (3.31) simply as

$$\left[\mathbf{v}\right] = \left[\boldsymbol{\xi}^T\right]\left[\overline{\mathbf{v}}\right] + \left[\boldsymbol{\eta}^T\right]\left[\hat{\mathbf{v}}\right] \tag{3.32}$$

where $[\overline{\mathbf{v}}]$ represents unknown variables, whereas $[\hat{\mathbf{v}}]$ represents known variables. The matrix $[\boldsymbol{\xi}^T]$ is a matrix that is similar to Eq. (2.30). The matrix $[\boldsymbol{\eta}^T]$ stands for the correspondence of the node on which the velocity is specified with respect to the whole nodal velocity. Incorporating Eq. (3.32) into Eq. (3.29), we get

$$\left[\mathbf{H}\right]\left[\boldsymbol{\xi}^T\right]\left[\overline{\mathbf{v}}\right] + \left[\mathbf{H}\right]\left[\boldsymbol{\eta}^T\right]\left[\hat{\mathbf{v}}\right] = \left[\mathbf{V}\right] \tag{3.33}$$

Multiplying $[\boldsymbol{\xi}]$ from the left to both sides of Eq. (3.33), and by making using of the following relation

$$\left[\boldsymbol{\xi}\right]\left[\mathbf{V}\right] = \left[\mathbf{0}\right] \tag{3.34}$$

the final form of the global finite element equation, including the boundary condition, can be obtained:

$$\left[\overline{\mathbf{H}}\right]\left[\overline{\mathbf{v}}\right] = \left[\overline{\mathbf{Q}}\right] \tag{3.35}$$

where

$$\left[\overline{\mathbf{H}}\right] = \left[\boldsymbol{\xi}\right]\left[\mathbf{H}\right]\left[\boldsymbol{\xi}^T\right]$$

and

$$\left[\overline{\mathbf{Q}}\right] = -\left[\boldsymbol{\xi}\right]\left[\mathbf{H}\right]\left[\boldsymbol{\eta}^T\right]\left[\hat{\mathbf{v}}\right]$$

A final form of Eq. (3.35) is as follows:

$$\begin{bmatrix} (h_{①} + h_{②} + h_{③}) & -(h_{②} + h_{③}) & \\ -(h_{②} + h_{③}) & (h_{②} + h_{③} + h_{④}) & h_{④} \\ & -h_{④} & h_{④} \end{bmatrix} \begin{bmatrix} v_2 \\ v_3 \\ v_4 \end{bmatrix} = \begin{bmatrix} \overline{Q}_2 \\ \overline{Q}_3 \\ \overline{Q}_4 \end{bmatrix} \tag{3.36}$$

$$
\begin{bmatrix} \overline{Q}_2 \\ \overline{Q}_3 \\ \overline{Q}_4 \end{bmatrix} = - \begin{bmatrix} -h_{①} \\ \\ \\ \end{bmatrix} \begin{bmatrix} \hat{v}_1 \end{bmatrix} \tag{3.37}
$$

Equation (3.36) can be derived from Eq. (3.28) in a straightforward manner; i.e., the rows and columns corresponding to the known velocities are subtracted from Eq. (3.28). The right-side terms are obtained by transforming the left-side terms of corresponding columns of the coefficient matrix times the known variables. This algorithm is simple and thus suitable for computational coding.

Multiplying $[\eta]$ from the left to Eq. (3.29) and using the relation

$$
\begin{bmatrix} \eta \end{bmatrix} \begin{bmatrix} \mathbf{V} \end{bmatrix} = \begin{bmatrix} \overline{\mathbf{V}} \end{bmatrix} \tag{3.38}
$$

we obtain the flux equation as follows:

$$
\begin{bmatrix} \eta \end{bmatrix} \begin{bmatrix} \mathbf{H} \end{bmatrix} \begin{bmatrix} \mathbf{v} \end{bmatrix} = \begin{bmatrix} \overline{\mathbf{V}} \end{bmatrix} \tag{3.39}
$$

whose componential form is as follows:

$$
\begin{bmatrix} h_{①} & -h_{①} & & \end{bmatrix} \begin{bmatrix} v_1 \\ v_2 \\ v_3 \\ v_4 \end{bmatrix} = \begin{bmatrix} V_1 \end{bmatrix} \tag{3.40}
$$

Introducing the solution of Eq. (3.36) into Eq. (3.40), the driving flux at node 1 can be computed.

3.6 Computer Program: pipeline_system

This section describes a computer program, which is named as pipeline_system and coded by FORTRAN 90. The aim of this section is to give readers an idea of how it works, but it is not suited for actual computational implementation. The readers can take the program to be a concrete implementation of the ideas written in Sects. 3.2–3.5.

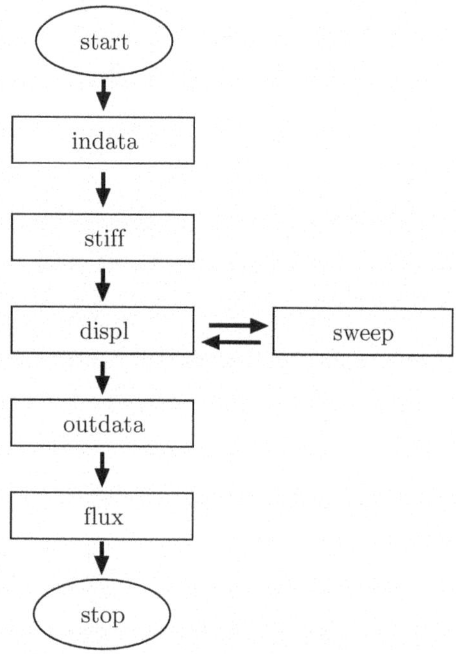

Fig. 3.3 Flow chart

3.6.1 Flow Chart

To get an overall image of the computer program: pileline_system, we consider the flow chart of the program, which is shown in Fig. 3.3. Main program pipeline_system consists of five subroutines, one of which has one subroutine in itself. The flow of the computer program is simple. In the subroutine indata, data in this program is fed in. In subroutine stiff, the coefficient matrix of the global finite element equation is made. The known term is computed in the subroutine displ, in which the subroutine sweep solves the simultaneous equation using the coefficient matrix and known term. The computed velocity is output-fed in subroutine outdata. The flux is computed in subroutine flux.

3.6.2 Subroutine Indata

Let a total numbers of nodes be n, of pipes be m, and of nodes on which the velocities are specified be ix. In the case that the pipeline system shown in Fig. 3.1, $n = 4$, $m = 4$, and $ix = 1$. The connectivity relation that is expressed by the arrays $ia(i)$ and $ib(i)$, which $i = 1$ to m, i.e., i means the number of pipe. In the array $ia(i)$, the

Table 3.2 Connectivity relations

i	1	2	3	4
$ia(i)$	1	2	2	3
$ib(i)$	2	3	3	4

Table 3.3 Specified values array

i	1
$ifx(i)$	1
$pp(i)$	1.0

Table 3.4 Coefficients array

i	1	2	3	4
$b(i)$	1.0	0.5	0.5	1.0
$sl(i)$	1.0	1.0	1.0	1.0

node number to which the edge a of pipe i is connected is stored. In the array $ib(i)$, the node number to which the edge b of pipe i is connected is stored. In the case of the pipeline system shown in Fig. 3.1, stored numbers in $ia(i)$ and $ib(i)$ are listed in Table 3.2.

The nodal number of nodes on which the velocity is specified is stored in $ifx(i)$ and specified values are in $pp(i)$, where $i = 1$ to ix. In the case shown in Fig. 3.1, $ix = 1$, and specified value is assumed as 1.0. Those are listed in Table 3.3.

The coefficient values of the local finite element equation $B_{ⓜ}$ and $l_{ⓜ}$ are in the array $b(i)$, $sl(i)$, $i = 1$ to m. Those are listed in Table 3.4.

3.6.3 Subroutine Stiff

The main idea explained in Sect. 3.4 is programmed in this subroutine, i.e., the coefficient matrix of the global finite element equation is programed. We do not need to program the coefficient matrix of the local finite element matrix. The algorithm written in Sect. 2.9 is illustrated again in Fig. 3.4.

Fig. 3.4 Algorithm for global coefficient matrix

$$\begin{array}{ccc} & ia(ⓜ) & ib(ⓜ) \\ & \uparrow & \uparrow \\ & a & b \\ ia(ⓜ) \leftarrow a & h_{ⓜ} & -h_{ⓜ} \\ ib(ⓜ) \leftarrow b & -h_{ⓜ} & h_{ⓜ} \end{array} \tag{3.41}$$

Let the global finite element matrix be $sa(i, j)$ where $i = 1$ to n, $j = 1$ to n. The program is as follows:

$$
\begin{aligned}
& do\ \ i = 1,\ m \\
& \quad ii = ia(i) \\
& \quad jj = ib(i) \\
& \quad sa(ii, ii) = sa(ii, ii) + h(i) \\
& \quad sa(ii, jj) = sa(ii, jj) - h(i) \\
& \quad sa(jj, ii) = sa(jj, ii) - h(i) \\
& \quad sa(jj, jj) = sa(jj, jj) + h(i) \\
& end\ \ do\ \ i
\end{aligned}
\tag{3.42}
$$

The index i is the number of pipe. The index ii is the number of node to which the edge a of pipe i is attached. The index jj is the number of node to which the edge b of pipe i is attached. Let us work through the schematic change of $sa(i)$ according to i. Readers should make sure that the final matrix in Eq. (3.46) is the coefficient matrix of Eq. (3.28) (Fig. 3.5).

3.6.4 Subroutine Displ

To introduce the boundary condition stated in Sect. 3.5 into the program, we must treat the matrix, whose corresponding rows and columns are erased from the global finite element matrix. Rather than introducing an additional array, we use the array $sa(i, j)$ itself by erasing the corresponding rows and columns, and replacing diagonal terms from zero to 1. This algorithm is schematically shown as follows (Fig. 3.6):

The program works as follows:

$$
\begin{aligned}
& do\ \ i = 1,\ ix \\
& \quad ii = ifx(i) \\
& \quad do\ \ j = 1,\ n \\
& \quad\quad sa(ii, j) = 0.0d0 \\
& \quad\quad sa(j, ii) = 0.0d0 \\
& \quad end\ \ do\ \ j \\
& \quad sa(ii, ii) = 1.0d0 \\
& end\ \ do\ \ i
\end{aligned}
$$

$i = 1$

$$sa(i,j) = \begin{array}{c} \\ 1 \\ 2 \\ 3 \\ 4 \end{array} \begin{array}{c} \begin{array}{cccc} 1 & 2 & 3 & 4 \end{array} \\ \left[\begin{array}{cccc} h_{\textcircled{1}} & -h_{\textcircled{1}} & & \\ -h_{\textcircled{1}} & h_{\textcircled{1}} & & \\ & & & \\ & & & \end{array} \right] \end{array} \tag{3.43}$$

$i = 2$

$$sa(i,j) = \begin{array}{c} \\ 1 \\ 2 \\ 3 \\ 4 \end{array} \begin{array}{c} \begin{array}{cccc} 1 & 2 & 3 & 4 \end{array} \\ \left[\begin{array}{cccc} h_{\textcircled{1}} & -h_{\textcircled{1}} & & \\ -h_{\textcircled{1}} & h_{\textcircled{1}}+h_{\textcircled{2}} & -h_{\textcircled{2}} & \\ & -h_{\textcircled{2}} & h_{\textcircled{2}} & \\ & & & \end{array} \right] \end{array} \tag{3.44}$$

$i = 3$

$$sa(i,j) = \begin{array}{c} \\ 1 \\ 2 \\ 3 \\ 4 \end{array} \begin{array}{c} \begin{array}{cccc} 1 & 2 & 3 & 4 \end{array} \\ \left[\begin{array}{cccc} h_{\textcircled{1}} & -h_{\textcircled{1}} & & \\ -h_{\textcircled{1}} & h_{\textcircled{1}}+h_{\textcircled{2}}+h_{\textcircled{3}} & -h_{\textcircled{2}}-h_{\textcircled{3}} & \\ & -h_{\textcircled{2}}-h_{\textcircled{3}} & h_{\textcircled{2}}+h_{\textcircled{3}} & \\ & & & \end{array} \right] \end{array} \tag{3.45}$$

$i = 4$

$$sa(i,j) = \begin{array}{c} \\ 1 \\ 2 \\ 3 \\ 4 \end{array} \begin{array}{c} \begin{array}{cccc} 1 & 2 & 3 & 4 \end{array} \\ \left[\begin{array}{cccc} h_{\textcircled{1}} & -h_{\textcircled{1}} & & \\ -h_{\textcircled{1}} & h_{\textcircled{1}}+h_{\textcircled{2}}+h_{\textcircled{3}} & -h_{\textcircled{2}}-h_{\textcircled{3}} & \\ & -h_{\textcircled{2}}-h_{\textcircled{3}} & h_{\textcircled{2}}+h_{\textcircled{3}}+h_{\textcircled{4}} & -h_{\textcircled{4}} \\ & & -h_{\textcircled{4}} & h_{\textcircled{4}} \end{array} \right] \end{array} \tag{3.46}$$

Fig. 3.5 Change of $sa(i, j)$ according to number of pipe $i = 1$ to 4

Fig. 3.6 Boundary conditions

$$\left[\begin{array}{cccc} h_{\textcircled{1}} & -h_{\textcircled{1}} & & \\ -h_{\textcircled{1}} & h_{\textcircled{1}}+h_{\textcircled{2}}+h_{\textcircled{3}} & -h_{\textcircled{2}}-h_{\textcircled{3}} & \\ & -h_{\textcircled{2}}-h_{\textcircled{3}} & h_{\textcircled{2}}+h_{\textcircled{3}}+h_{\textcircled{4}} & -h_{\textcircled{4}} \\ & & -h_{\textcircled{4}} & h_{\textcircled{4}} \end{array} \right]$$

$$= \left[\begin{array}{cccc} 1 & & & \\ & h_{\textcircled{1}}+h_{\textcircled{2}}+h_{\textcircled{3}} & -h_{\textcircled{2}}-h_{\textcircled{3}} & \\ & -h_{\textcircled{2}}-h_{\textcircled{3}} & h_{\textcircled{2}}+h_{\textcircled{3}}+h_{\textcircled{4}} & -h_{\textcircled{4}} \\ & & -h_{\textcircled{4}} & h_{\textcircled{4}} \end{array} \right] \tag{3.47}$$

Fig. 3.7 Known term $pq(i)$

$$\begin{array}{c} \\ 1 \\ 2 \\ 3 \\ 4 \end{array} \quad \begin{array}{c} 1 \\ \left[\begin{array}{c} \\ -h_{①} \\ \\ \\ \end{array}\right] \end{array} \tag{3.48}$$

The index i represents a set of sequential numbers of nodes on which the velocities are specified and the index ii is actual number of nodes. Inner do loop is the routine to erase the corresponding rows and columns. The last statement puts 1 on the corresponding diagonal terms of $sa(i, j)$.

The known terms $pq(i)$ of Eq. (3.37) are programed as follows:

```
do i = 1, ix
  ii = ifx(i)
  do j = 1, nx
    pq(j) = pq(j) - sa(j, ii) * pp(i)
  end do j
  pq(ii) = 0.0d0
end do i
```

The index i is the sequential number of nodes on which the velocities are specified. The index ii is the actual number of nodes on which the velocities are specified. Inner do loop computes the known terms. Of importance is the last statement, which means corresponding rows of $pq(i)$ are erased from the computation. By this procedure, the known term $pq(i)$ can be obtained as shown in Fig. 3.7.

Using the arrays $sa(i, j)$ and $pq(i)$, subroutine sweep can compute the solution of linear simultaneous equation. The values on the positions corresponding to the known term are zero. The computed velocities are obtained and stored in array $pq(i)$. The algorithm is the standard sweep out method.

3.6.5 Subroutine Outdata

Subroutine outdata outputs the velocities on the whole nodes, which are stored in the array $pq(i)$. Using the input data of the pipeline system shown in Fig. 3.1, the output data listed in Tables 3.5 and 3.6.

Table 3.5 Velocity

Node	Velocity
1	1.0
2	1.0
3	1.0
4	1.0

Table 3.6 Flux of pipe

Pipe	Flux
1	0.0
2	0.0
3	0.0
4	0.0

3.6.6 *Subroutine Flux*

The flux of each pipe can be computed using Eq. (3.20). As stated in Sect. 3.2, the F_b is output considering the plus sign.

3.6.7 *Program List: pipeline_system*

The computer program list is shown below. For the input data, pipeline system shown in Fig. 3.1 is used (Fig. 3.8).

3.6.8 *Output List*

The output results are shown in Fig. 3.9. Those include input data, output data, velocity values, and flux values.

3.7 Alternatives to Main Algorithm

What is given above is just an example: there are a plenty of alternatives. The program we introduced in Sects. 3.6.1–3.6.6 was intended to show how to program the finite element method for beginners.

```
!**************************************************************
!**************************************************************
!**********                            ***********************
!**********     pipeline system        ***********************
!**********     coded by Wakui         ***********************
!**********                            ***********************
!**************************************************************
!**************************************************************
!**                                                          **
!**  n--------total number of nodes                          **
!**  m--------tatal number of pipes                          **
!**  ix-------total number of nodes on which velocity is     **
!**           specified                                      **
!**  ia,ib----conectivities of edges a and b                 **
!**  ipx------number of nodes on which velocities are        **
!**           specified                                      **
!**  pp-------specified velocities                           **
!**                                                          **
!**************************************************************
!**************************************************************


!**************************************************************
!**                  *******************************************
!**  main program    *******************************************
!**                  *******************************************
!**************************************************************
     program pipeline_system

     implicit real(8)  (a-h,o-z)
     integer,parameter :: MN=20
     integer :: ia(MN),ib(MN),ipx(MN)
     real(8) :: sh(MN),pp(MN),b(MN),sl(MN)
     real(8) :: sa(MN,MN),pq(MN)

     open(10,file='input.txt')
     open(30,file='output.txt')

     call indata  (n,m,ix,ia,ib,b,sl,ipx,pp)
     call stiff   (n,m,ia,ib,sa,b,sl,sh,MN)
     call displ   (n,ix,ipx,sa,pq,pp,MN)
     call outdata (n,pq,ix,ipx,pp)
     call flux    (m,ia,ib,pq,b,sl,sh)

     close(10)
     close(30)

     stop
     end program pipeline_system
!**************************************************************
```

Fig. 3.8 Program list: pipeline_system

```
!****************************************************************
!**                 *********************************************
!**  sub program    *********************************************
!**                 *********************************************
!****************************************************************
     subroutine indata (n,m,ix,ia,ib,b,sl,ipx,pp)

     implicit real(8)(a-h,o-z)
     integer :: ia(*),ib(*),ipx(*)
     real(8) :: pp(*),b(*),sl(*)
!****************************************************************
!**                 *********************************************
!**  file input     *********************************************
!**                 *********************************************
!****************************************************************
     read(10,*) n,m,ix
     read(10,*) (j,ia(j),ib(j),b(j),sl(j), i=1,m)
     read(10,*) (ipx(i),pp(i),i=1,ix)
!****************************************************************
!**                     *****************************************
!**  input data output  *****************************************
!**                     *****************************************
!****************************************************************
      write(30,*) '*******  pipeline system program  *****'
      write(30,*)
      write(30,*) '*******  input data  *****************'
      write(30,*)
      write(30,*) '  num.ele     edge.a     edge.b'
      write(30,*)
      do i=1,m
        write(30,300)i,ia(i),ib(i)
      end do

      write(30,*)
 200   format(5x,i5,f11.3)
 300  format(5x,i5,2i11)

      return
      end subroutine indata
!****************************************************************


!****************************************************************
!****                    ****************************************
!****  subroutine stiff  ****************************************
!****                    ****************************************
!****************************************************************
      subroutine stiff(n,m,ia,ib,sa,b,sl,sh,MN)
```

Fig. 3.8 (continued)

```
      implicit real(8)(a-h,o-z)
      integer :: ia(*),ib(*)
      real(8):: sa(MN,*),b(*),sl(*),sh(*)

      do i = 1,n
        do j = 1,n
          sa(i,j) = 0.0d0
        end do
      end do

      write(3,*) '  num.ele  coeff.b  coeff.l  coeff.h'
      write(30,*)'  num.ele  coeff.b  coeff.l  coeff.h'
      write(30,*)

      do i = 1,m
        ii = ia(i)
        jj = ib(i)
        sh(i) = b(i) / sl(i)
        sa(ii,ii) = sa(ii,ii) + sh(i)
        sa(ii,jj) = sa(ii,jj) - sh(i)
        sa(jj,ii) = sa(jj,ii) - sh(i)
        sa(jj,jj) = sa(jj,jj) + sh(i)
      end do

      do i = 1,m
        write(30,301) i,b(i),sl(i),sh(i)
      end do

      write(30,*)
  301 format(5x,i5,3f11.3)

      return
      end subroutine stiff
!**************************************************************


!**************************************************************
!****         ********************************************
!****  displ  ********************************************
!****         ********************************************
!**************************************************************
      subroutine displ(n,ix,ipx,sa,pq,pp,MN)

      implicit real(8)(a-h,o-z)
      integer :: ipx(*)
      real(8) :: sa(MN,*),pq(*),pp(*)

      do i = 1,n
        pq(i) = 0.0d0
      end do
```

Fig. 3.8 (continued)

```
       do i = 1,ix
         ii = ipx(i)
         do j = 1,n
           pq(j) = pq(j) - sa(j,ii) * pp(i)
         end do
       end do

       do i = 1,ix
         ii = ipx(i)
         do j = 1,n
           sa(ii,j) = 0.0d0
           sa(j,ii) = 0.0d0
         end do
         sa(ii,ii) = 1.0d0
       end do
!*************************************************************


!*************************************************************
!****          ***********************************************
!****   sweep  ***********************************************
!****          ***********************************************
!*************************************************************
       call sweep(n,pq,sa,MN)

       return
       end subroutine displ
!*************************************************************


!*************************************************************
!****            *********************************************
!****   outdata  *********************************************
!****            *********************************************
!*************************************************************
       subroutine outdata(n,pq,ix,ipx,pp)

       implicit real(8)(a-h,o-z)
       integer :: ipx(*)
       real(8) :: pq(*),pp(*)

       do i = 1,ix
         ii = ipx(i)
         pq(ii) = pp(i)
       end do
       write(30,*) '**********  output results  *************'
       write(30,*)
       write(30,*) '  num.ele   velocity'
       write(30,*)
       do i=1,n
         write(30,302)i,pq(i)
       end do
```

Fig. 3.8 (continued)

```
      write(30,*)
 302  format(5x,i5,f11.3)

      return
      end subroutine outdata
!************************************************************

!************************************************************
!****         **********************************************
!****  flux   **********************************************
!****         **********************************************
!************************************************************
      subroutine flux(m,ia,ib,pq,b,sl,sh)

      implicit real(8)(a-h,o-z)
      integer :: ia(*),ib(*)
      real(8) :: pq(*),b(*),sl(*),sh(*)

      write(30,*) '  num.ele       flux'
      write(30,*)
      do i = 1,m
        ii = ia(i)
        jj = ib(i)
        va = pq(ii)
        vb = pq(jj)
        sh(i) = b(i) / sl(i)
        fb = sh(i) * (vb - va)
        write(30,303)i,fb
      end do

 303  format(5x,i5,1f11.3)

      return
      end subroutine flux
!************************************************************

!************************************************************
!****          *********************************************
!****  sweep   *********************************************
!****          *********************************************
!************************************************************
      subroutine sweep(n,pq,sa,MN)

      implicit real(8)(a-h,o-z)
      real(8) :: pq(*),sa(MN,*)

      do i = 1,n
        cc = 1.0d0 / sa(i,i)
        do j = i , n
          sa(i,j) = sa(i,j) * cc
        end do
```

Fig. 3.8 (continued)

```
        pq(i) = pq(i) * cc
        do k = i+1,n
          bb = sa(k,i)
          do j = 1,n
            sa(k,j) = sa(k,j) - sa(i,j) * bb
          end do
          pq(k) = pq(k) -pq(i) * bb
        end do
      end do

      do i = n-1,1,-1
        do j = i+1,n
          pq(i) = pq(i) - pq(j) * sa(i,j)
        end do
      end do

      return
      end subroutine sweep
!*************************************************************
```

Fig. 3.8 (continued)

```
************  pipeline system program  ***********

************  input data  ***********************

  num.ele      edge.a      edge.b

        1           1           2
        2           2           3
        3           2           3
        4           3           4

  num.ele     coeff.b     coeff.l     coeff.h

        1       1.000       1.000       1.000
        2       0.500       1.000       0.500
        3       0.500       1.000       0.500
        4       1.000       1.000       1.000

*************  output results  *******************

  num.ele   velocity

        1       1.000
        2       1.000
        3       1.000
        4       1.000

  num.ele        flux

        1       0.000
        2       0.000
        3       0.000
        4       0.000
```

Fig. 3.9 Output results

Table 3.7 Connectivity relations

i	1	2	3	4
$ie(1,i)$	1	2	2	3
$ie(2,i)$	2	3	3	4

Fig. 3.10 Algorithm for global coefficient matrix

$$\begin{array}{c} \\ ie(1,Ⓜ) \leftarrow 1 \\ ie(2,Ⓜ) \leftarrow 2 \end{array} \begin{array}{c} \begin{array}{cc} ie(1,Ⓜ) & ie(2,Ⓜ) \\ \uparrow & \uparrow \\ 1 & 2 \end{array} \\ \left[\begin{array}{c|c} h_{Ⓜ} & -h_{Ⓜ} \\ \hline -h_{Ⓜ} & h_{Ⓜ} \end{array}\right] \end{array} \equiv sm(i,i) \tag{3.49}$$

An another algorithm is shown in this section, which is the basis of the programs in the following chapters. In the subroutine indata, the connectivity relation can be stored as in Table 3.7.

Namely, $ia(i)$ and $ib(i)$ can be replaced by one array $ie(i, j)$ where $i = 1$ and 2, and $j = 1$ to m. The index $i = 1$ means edge a and index $i = 2$ is edge b. The array $sm(i, j)$ where $i = 1$ to 2 and $j = 1$ to 2 is used for the local coefficient matrix.

The superposition procedure shown in Fig. 3.10 can be changed to the algorithm shown in Fig. 3.4. The program is as follows:

```
do im = 1, m
  sm(1, 1) =  h(im)
  sm(1, 2) = -h(im)
  sm(2, 1) = -h(im)
  sm(2, 2) =  h(im)
  do i = 1, 2
    ii = ie(i, im)
    do j = 1, 2
      jj = ie(j, im)
      sa(ii, jj) = sa(ii, jj) + sm(i, j)
    end do j
  end do i
end do im
```

The index im represents the number of pipe, $sm(i, j)$ is the array for local matrix, and $sa(i, j)$ is the global matrix. The index i denotes edge a and edge b, ii is the nodal number to which edge i is connected, and the index j also denotes edge a and edge b, jj is the nodal number to which edge j is connected.

For the algorithm described in the previous chapters, we need a large size of the computer storage for the main array $sa(i, j)$ whose size is $n \times n$, where n is the total numbers of nodes. Among the actual existing computer programs of the finite element method, many of them do not employ the array $sa(i, j)$. In those programs, the solution of simultaneous equation can be obtained without making the main matrix $sa(i, j)$, and using the iterative methods, such as AD method, CG method, etc.

Another approach is to use only nonzero element in the solution of the simultaneous equation. For those solution procedures, see the specific literatures on them (e.g., Coleman and Van Loan 1988; Watkins 2010; Golub and Van Loan 2013, etc.).

3.8 Summary and Conclusion

In Chap. 2, we have provided an overview of the important procedure of superposition. Meanwhile, we have shown that the finite element equation is essential, i.e., if we get the local finite element equation, then we can easily obtain the global finite element equation using the connectivity relation. In this chapter, using the simplest example, i.e., pipeline system, we have shown the derivation of the finite element equation. The Galerkin method which is introduced in this chapter serves as fundamental concepts in the later chapters, and therefore readers are expected to familiarize themselves with what is presented in this chapter before proceeding.

There are a plenty of variations when it comes to actual implementation techniques of the Galerkin method, including the nonlinear problem. We have explained the superposition procedures and treatment of boundary condition. The computer program coded by FORTRAN 90 has been presented to illustrate the basic computational concept of the finite element method. Although other procedures do exist, which are related to solve the simultaneous equation system, the simplest sweep out procedure is used in this computer program. Let us reiterate, however, that the program is presented for the sake of illustration, not necessarily for the sake of actual implementation.

3.9 Exercises

1. Confirm that the following equation of equilibrium is satisfied:

$$\begin{bmatrix} \boldsymbol{\alpha} & \boldsymbol{\beta} \end{bmatrix} \begin{bmatrix} \mathbf{F}_a \\ \mathbf{F}_b \end{bmatrix} = \begin{bmatrix} \mathbf{V} \end{bmatrix} \tag{3.50}$$

or

$$\begin{bmatrix} 1 & & & & & & & \\ & 1 & 1 & & 1 & & & \\ & & & 1 & & 1 & 1 & \\ & & & & & & & 1 \end{bmatrix} \begin{bmatrix} F_a^{①} \\ F_a^{②} \\ F_a^{③} \\ F_a^{④} \\ F_b^{①} \\ F_b^{②} \\ F_b^{③} \\ F_b^{④} \end{bmatrix} = \begin{bmatrix} V_1 \\ \\ \\ \\ \end{bmatrix} \tag{3.51}$$

2. Consider the following differential equation:

$$\frac{du}{dx} = 0 \tag{3.52}$$

$$u(0) = u_0 \tag{3.53}$$

Based on the interpolation function

$$u = \left(1 - \frac{x}{l}\right) u_a + \left(\frac{x}{l}\right) u_b \tag{3.54}$$

where u_a and u_b are function values on edges a and b, the length of interval is l (see Fig. 3.11), and the weighting function is

$$u^* = \left(1 - \frac{x}{l}\right) u_a^* + \left(\frac{x}{l}\right) u_b^* \tag{3.55}$$

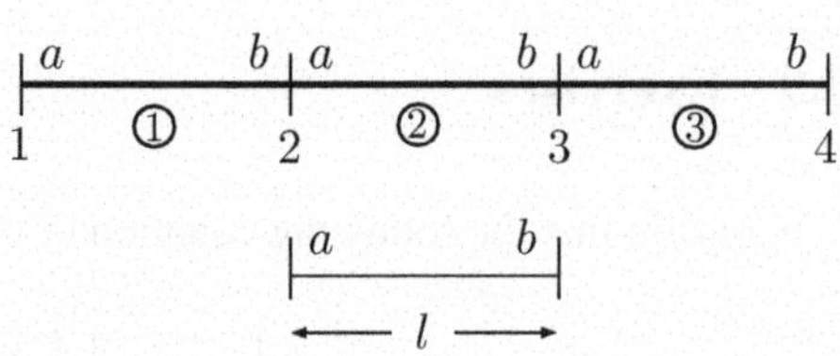

Fig. 3.11 One-dimensional mesh

where u_a^*, u_b^* are arbitrary constants, show that the global finite element equation is

$$\begin{bmatrix} -\frac{1}{2} & \frac{1}{2} & & \\ -\frac{1}{2} & & -\frac{1}{2} & \\ & -\frac{1}{2} & & \frac{1}{2} \\ & & -\frac{1}{2} & \frac{1}{2} \end{bmatrix} \begin{bmatrix} u_1 \\ u_2 \\ u_3 \\ u_4 \end{bmatrix} = \begin{bmatrix} \\ \\ \\ \\ \end{bmatrix} \tag{3.56}$$

and the global finite element equation including boundary condition is

$$\begin{bmatrix} & -\frac{1}{2} & \\ -\frac{1}{2} & & \frac{1}{2} \\ & -\frac{1}{2} & \frac{1}{2} \end{bmatrix} \begin{bmatrix} u_2 \\ u_3 \\ u_4 \end{bmatrix} = - \begin{bmatrix} -\frac{1}{2} \\ \\ \\ \end{bmatrix} \begin{bmatrix} u_1 \end{bmatrix}. \tag{3.57}$$

Chapter 4
Potential Flow

4.1 Introduction

The previous chapters have used one-dimensional second-order differential equation to explain the basic concepts of the finite element method. As we have observed, it is possible to obtain the exact solution in the case of one-dimensional problem in general. Therefore, the advantage of deploying the finite element method may not appear to be so substantial given one-dimensional problems, although it does indeed have an advantage of allowing us to deal with the nonuniform length distributions.

This chapter extends on the previous discussion and deals with two-dimensional simple nonviscous flows, i.e., potential flows, analyzing them using the finite element method. The analysis of the potential flow is one of the most useful flow analyses in the domain of engineering and science in general. This analysis therefore introduces many fundamental concepts of the finite element method, although we will use steady nonviscous flow.

4.2 Potential Flow in a Channel

Given Fig. 4.1, consider the two-dimensional steady potential flow in a nonuniform channel. Fluids flow from boundaries A–B to C–D—the potentials for A–B and C–D are given. Assume that velocities u and v are given by the gradient of the potential as

$$u = -\frac{\partial \phi}{\partial x} \tag{4.1}$$

$$v = -\frac{\partial \phi}{\partial y} \tag{4.2}$$

M. Kawahara, *Finite Element Methods in Incompressible, Adiabatic, and Compressible Flows*, Mathematics for Industry 19,
DOI 10.1007/978-4-431-55450-9_4

Fig. 4.1 Potential flow

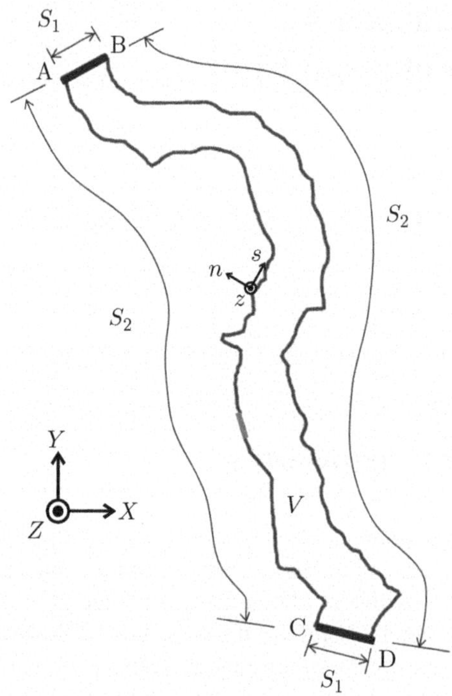

The minus signs indicate the fact that the fluid flows from the place of high potential to that of low potential. The continuity of the flow can be shown below:

$$\frac{\partial u}{\partial x} + \frac{\partial v}{\partial y} = 0 \tag{4.3}$$

Substituting Eqs. (4.1) and (4.2) into Eq. (4.3) and rearranging the terms, we can derive the Laplace equation, as shown in Eq. (4.4):

$$-\left(\frac{\partial^2 \phi}{\partial x^2} + \frac{\partial^2 \phi}{\partial y^2}\right) = 0 \qquad in \quad V \tag{4.4}$$

where V is the flow field. We will denote the boundary of flow field by S. As boundary conditions, we introduce two types, which are essential and natural boundary conditions. There must be some differences between the potentials on boundaries A–B and those on C–D of the flow field shown in Fig. 4.1, because there is a flow. To be more specific, some values of the potential are specified on boundaries A–B and those on B–C. Those boundary conditions are referred to as "essential boundary conditions," and in general, those are shown as

$$\phi = \hat{\phi} \qquad on \quad S_1 \tag{4.5}$$

where S_1 is the essential boundary and $\hat{\phi}$ is the potential value given on the boundary. Consider the boundary condition that should hold on the two sidewalls—A–C and B–D—of the channel. That is the velocity normal to the boundary, which must be zero. Let q be the velocity normal to the boundary

$$q = ul + vm = \frac{\partial \phi}{\partial x} l + \frac{\partial \phi}{\partial y} m = \hat{q} \qquad on \quad S_2 \tag{4.6}$$

where $\hat{q}$ is the velocity value given on the boundary, and l and m are the direction cosines of the unit normal to the boundary:

$$l = \frac{dy}{ds} \tag{4.7}$$

$$m = \frac{dx}{ds} \tag{4.8}$$

where s is the coordinate along the boundary. Boundary condition expressed in Eq. (4.6) is the condition that the gradient of the potential is given—and this sort of condition is called the "natural boundary condition." On the channel wall, we can take

$$\hat{q} = 0 \tag{4.9}$$

Now against this background, we move on to describe the finite element method of the flow expressed by Eq. (4.4) under the boundary condition Eqs. (4.5) and (4.6). The following two conditions should hold for the flow field shown in Fig. 4.1:

$$\begin{aligned} S_1 \cup S_2 &= S \\ S_1 \cap S_2 &= \mathbf{0} \end{aligned} \tag{4.10}$$

where $\mathbf{0}$ is a null set. Simply, the boundary S consists of two boundaries—S_1 and S_2, and S_1 and S_2 do not overlap.

4.3 Weighted Residual Equation

Let us start with the generation of the finite element idealization of flow field. Finite elements of various shapes—triangular, rectangular, etc.—can be potentially deployed to illustrate this. However, because triangular elements are the simplest to deal with, we use the triangular finite elements in the analysis. Using the triangular elements, we can approximate any complex shape of the boundary. An example is given in Fig. 4.2.

We start our discussion by focusing on the weighted residual equation on an arbitrary volume V, through which fluid flows. The naming and meaning of the weighted residual equation will be explained later in Sect. 4.9. For now, choose

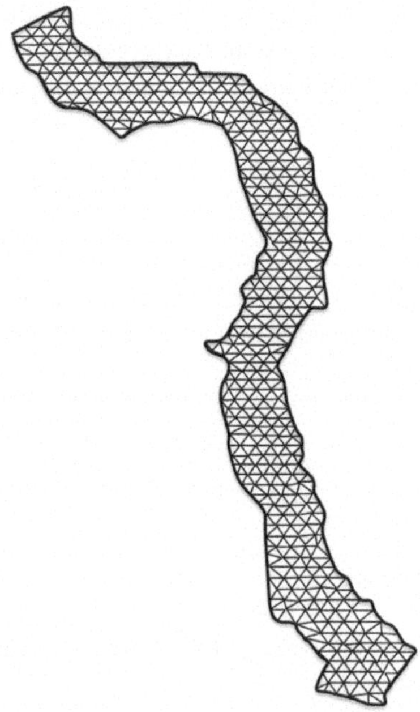

Fig. 4.2 Finite element mesh

an arbitrary function ϕ^*, which will be henceforth referred to as the "weighting function." Multiplying both sides of Eq. (4.4) by ϕ^*, and integrating over the volume V, we obtain

$$\int_V \phi^* \left(\frac{\partial^2 \phi}{\partial x^2} + \frac{\partial^2 \phi}{\partial y^2} \right) dxdy = 0 \tag{4.11}$$

The left side of Eq. (4.11) can be transformed using the following equalities:

$$\int_V \left(\phi^* \frac{\partial^2 \phi}{\partial x^2} \right) dxdy = \int_V \frac{\partial}{\partial x} \left(\phi^* \frac{\partial \phi}{\partial x} \right) dxdy - \int_V \left(\frac{\partial \phi^*}{\partial x} \frac{\partial \phi}{\partial x} \right) dxdy \tag{4.12}$$

$$\int_V \left(\phi^* \frac{\partial^2 \phi}{\partial y^2} \right) dxdy = \int_V \frac{\partial}{\partial y} \left(\phi^* \frac{\partial \phi}{\partial y} \right) dxdy - \int_V \left(\frac{\partial \phi^*}{\partial y} \frac{\partial \phi}{\partial y} \right) dxdy \tag{4.13}$$

The first terms on the right side of Eqs. (4.12) and (4.13) can be transformed into the boundary integrals as

$$\int_V \frac{\partial}{\partial x} \left(\phi^* \frac{\partial \phi}{\partial x} \right) dxdy = \int_S \phi^* \frac{\partial \phi}{\partial x} \frac{dy}{ds} ds \tag{4.14}$$

$$\int_V \frac{\partial}{\partial y} \left(\phi^* \frac{\partial \phi}{\partial y} \right) dxdy = \int_S \phi^* \frac{\partial \phi}{\partial y} \frac{dx}{ds} ds \tag{4.15}$$

Introducing Eqs. (4.14) and (4.15) into Eqs. (4.12) and (4.13), respectively, and further introducing the resulting equations into Eq. (4.11), we get

$$\int_V \left(\frac{\partial \phi^*}{\partial x}\frac{\partial \phi}{\partial y} + \frac{\partial \phi^*}{\partial y}\frac{\partial \phi}{\partial y} \right) dxdy = \int_S \phi^* \left(\frac{\partial \phi}{\partial x} l + \frac{\partial \phi}{\partial y} m \right) ds \tag{4.16}$$

Based on Eqs. (4.6) and (4.16) it can be expressed as

$$\int_V \left(\frac{\partial \phi^*}{\partial x}\frac{\partial \phi}{\partial x} + \frac{\partial \phi^*}{\partial y}\frac{\partial \phi}{\partial y} \right) dxdy = \int_S (\phi^* q)\, ds \tag{4.17}$$

Equation (4.17) is valid over any volume V and its boundary S; for instance, over a finite element, a whole flow field, and even a part of flow field, to the extent that the boundary is closed. For the sake of simplicity, let us assume that ϕ^* is zero on S_1. Recall that all boundary S of the whole flow field V consists of nonoverlapping boundaries S_1 and S_2, as expressed in Eq. (4.10), then we can derive the following:

$$\int_V \left(\frac{\partial \phi^*}{\partial x}\frac{\partial \phi}{\partial x} + \frac{\partial \phi^*}{\partial y}\frac{\partial \phi}{\partial y} \right) dxdy = \int_{S_2} (\phi^* \hat{q}) ds \tag{4.18}$$

Notice that the right side of Eq. (4.18) consists of the known term. Equation (4.18) is derived from Eq. (4.4) without using the boundary condition Eq. (4.5). Therefore, we can regard the set of Eqs. (4.18) and (4.5) as basic equations of the potential flow.

4.4 Interpolation Function and Weighting Function

Equation (4.17) holds over a finite element, as shown in Fig. 4.3, and inside of it, denoted by V and its boundary S. Its three edges are labeled as a, b, and c. The global coordinate x-y-z is located on an arbitrary point, which is right-hand system.

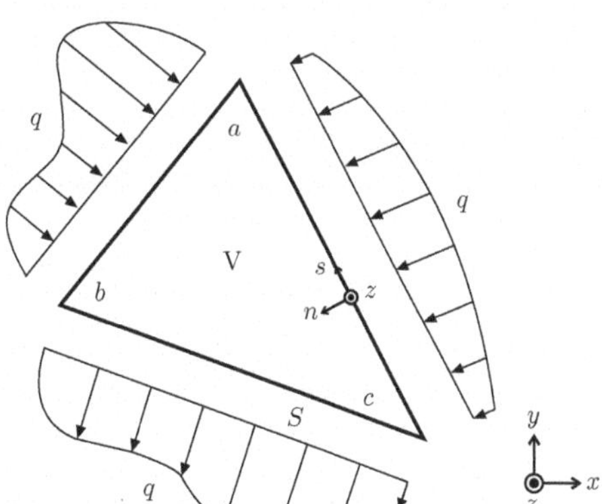

Fig. 4.3 A finite element

The coordinate system s-n-z is placed along the boundary S, and the direction of s is from a to b, b to c, and c to a. Through the three sides of an element, fluid flow q comes in and out. To the extent that the finite element mesh is satisfactorily refined, the linear interpolation for an element can be adapted. In this case, the potential inside an element can be expressed as

$$\phi = \alpha_0 + \alpha_1 x + \alpha_2 y \tag{4.19}$$

where $\alpha_0, \alpha_1, \alpha_2$ are unknown constants, which are to be determined. In Fig. 4.3, let the potential on edges a, b, c be ϕ_a, ϕ_b, ϕ_c, then the following relations hold:

$$\begin{aligned} \phi_a &= \alpha_0 + \alpha_1 x_a + \alpha_2 y_a \\ \phi_b &= \alpha_0 + \alpha_1 x_b + \alpha_2 y_b \\ \phi_c &= \alpha_0 + \alpha_1 x_c + \alpha_2 y_c \end{aligned} \tag{4.20}$$

where $(x_a, y_a), (x_b, y_b), (x_c, y_c)$ are coordinates of edges a, b, c, respectively. Inverting the relations in Eq. (4.20), we obtain

$$\begin{aligned} \alpha_0 &= a_a\phi_a + a_b\phi_b + a_c\phi_c \\ \alpha_1 &= b_a\phi_a + b_b\phi_b + b_c\phi_c \\ \alpha_2 &= c_a\phi_a + c_b\phi_b + c_c\phi_c \end{aligned} \tag{4.21}$$

where

$$\begin{aligned} a_i &= \frac{1}{2\mathbf{\Delta}}(x_j y_k - x_k y_j) \\ b_i &= \frac{1}{2\mathbf{\Delta}}(y_j - y_k) \\ c_i &= \frac{1}{2\mathbf{\Delta}}(x_k - x_j) \end{aligned} \tag{4.22}$$

and

$$\mathbf{\Delta} = \frac{1}{2}\{(x_j - x_i)(y_k - y_i) - (x_k - x_i)(y_j - y_i)\} \tag{4.23}$$

where i, j, k are permutations of a, b, c, and $\mathbf{\Delta}$ is an area of a triangle. Substituting Eq. (4.21) into Eq. (4.19), the interpolation function can be expressed as follows:

$$\phi = \Phi_a(x, y)\phi_a + \Phi_b(x, y)\phi_b + \Phi_c(x, y)\phi_c \tag{4.24}$$

where $\Phi_a(x, y)$, $\Phi_b(x, y)$, $\Phi_c(x, y)$ are called shape functions, whose meanings will be clarified in Sect. 4.10. But for now

$$
\begin{aligned}
\Phi_a(x, y) &= a_a + b_a x + c_a y \\
\Phi_b(x, y) &= a_b + b_b x + c_b y \\
\Phi_c(x, y) &= a_c + b_c x + c_c y
\end{aligned} \tag{4.25}
$$

Equation (4.25) is the linear interpolation function, and the values on edges a, b, c are ϕ_a, ϕ_b, ϕ_c. Expressed in a matrix form, Eq. (4.25) can be described as

$$
\phi = \begin{bmatrix} \Phi_a(x, y) \; \Phi_b(x, y) \; \Phi_c(x, y) \end{bmatrix} \begin{bmatrix} \phi_a \\ \phi_b \\ \phi_c \end{bmatrix} \tag{4.26}
$$

We emphasize that ϕ_a, ϕ_b, ϕ_c are the potential values on edges a, b, c, which are to be determined. A function of the type that is the same as Eq. (4.24) is used as the weighting function ϕ^*:

$$
\phi^* = \Phi_a(x, y)\phi_a^* + \Phi_b(x, y)\phi_b^* + \Phi_c(x, y)\phi_c^* \tag{4.27}
$$

where $\Phi_a(x, y)$, $\Phi_b(x, y)$, $\Phi_c(x, y)$ are the functions defined in Eq. (4.25). The constants ϕ_a^*, ϕ_b^*, ϕ_c^* are the values of the weighting function at the edges a, b, c of an element. Those are arbitrary and independent, because the function ϕ^* itself is an arbitrary function. Then Eq. (4.27) can be rewritten in the following matrix form:

$$
\phi^* = \begin{bmatrix} \Phi_a(x, y) \; \Phi_b(x, y) \; \Phi_c(x, y) \end{bmatrix} \begin{bmatrix} \phi_a^* \\ \phi_b^* \\ \phi_c^* \end{bmatrix} \tag{4.28}
$$

In general, the weighting function is taken to be the same type of the interpolation function. This method, recall from Chap. 3, is referred to as the "Galerkin method." It is possible to use different type functions for weight and interpolation. However, because we obtain the symmetric coefficient matrix of the final finite element equation, normally, we use the same type function for weight and interpolation.

4.5 Local Finite Element Equation

For an element of a kind that is shown in Fig. 4.3, Eq. (4.17) holds. Based on the weighted residual equation Eq. (4.17), the interpolation function Eq. (4.26), and the weighting function (4.28), the local finite element equation can be derived. The derivatives of both the functions are calculated as follows:

$$\begin{bmatrix} \dfrac{\partial \phi^*}{\partial x} \\ \dfrac{\partial \phi^*}{\partial y} \end{bmatrix} = \begin{bmatrix} \dfrac{\partial \Phi_a}{\partial x} & \dfrac{\partial \Phi_b}{\partial x} & \dfrac{\partial \Phi_c}{\partial x} \\ \dfrac{\partial \Phi_a}{\partial y} & \dfrac{\partial \Phi_b}{\partial y} & \dfrac{\partial \Phi_c}{\partial y} \end{bmatrix} \begin{bmatrix} \phi_a^* \\ \phi_b^* \\ \phi_c^* \end{bmatrix} \tag{4.29}$$

$$\begin{bmatrix} \dfrac{\partial \phi}{\partial x} \\ \dfrac{\partial \phi}{\partial y} \end{bmatrix} = \begin{bmatrix} \dfrac{\partial \Phi_a}{\partial x} & \dfrac{\partial \Phi_b}{\partial x} & \dfrac{\partial \Phi_c}{\partial x} \\ \dfrac{\partial \Phi_a}{\partial y} & \dfrac{\partial \Phi_b}{\partial y} & \dfrac{\partial \Phi_c}{\partial y} \end{bmatrix} \begin{bmatrix} \phi_a \\ \phi_b \\ \phi_c \end{bmatrix} \tag{4.30}$$

Equation (4.17) can then be expressed as

$$\int_V \begin{bmatrix} \dfrac{\partial \phi^*}{\partial x} & \dfrac{\partial \phi^*}{\partial y} \end{bmatrix} \begin{bmatrix} \dfrac{\partial \phi}{\partial x} \\ \dfrac{\partial \phi}{\partial y} \end{bmatrix} dxdy = \int_S (\phi^* q) ds \tag{4.31}$$

Introducing Eqs. (4.29) and (4.30) into Eq. (4.31) and rearranging the terms, we get the following equation:

$$\begin{bmatrix} \phi_a^* & \phi_b^* & \phi_c^* \end{bmatrix} \int_V \left(\begin{bmatrix} M_a & N_a \\ M_b & N_b \\ M_c & N_c \end{bmatrix} \begin{bmatrix} M_a & M_b & M_c \\ N_a & N_b & N_c \end{bmatrix} \right) dxdy \begin{bmatrix} \phi_a \\ \phi_b \\ \phi_c \end{bmatrix}$$

$$= \begin{bmatrix} \phi_a^* & \phi_b^* & \phi_c^* \end{bmatrix} \int_S \begin{bmatrix} Q_a \\ Q_b \\ Q_c \end{bmatrix} dS \tag{4.32}$$

where

$$M_a = \frac{\partial \Phi_a}{\partial x} \quad M_b = \frac{\partial \Phi_b}{\partial x} \quad M_c = \frac{\partial \Phi_c}{\partial x}$$

$$N_a = \frac{\partial \Phi_a}{\partial y} \quad N_b = \frac{\partial \Phi_b}{\partial y} \quad N_c = \frac{\partial \Phi_c}{\partial y}$$

$$Q_a = \Phi_a q \quad Q_b = \Phi_b q \quad Q_c = \Phi_c q$$

In Eq. (4.32), since ϕ_a^*, ϕ_b^*, ϕ_c^*, ϕ_a, ϕ_b, and ϕ_c are all independent of x and y, we can put them out of the integration symbols. With this, and rearranging the terms again, we get

$$\begin{bmatrix} \phi_a^* & \phi_b^* & \phi_c^* \end{bmatrix} \begin{bmatrix} K_{aa} & K_{ab} & K_{ac} \\ K_{ba} & K_{bb} & K_{bc} \\ K_{ca} & K_{cb} & K_{cc} \end{bmatrix} \begin{bmatrix} \phi_a \\ \phi_b \\ \phi_c \end{bmatrix} = \begin{bmatrix} \phi_a^* & \phi_b^* & \phi_c^* \end{bmatrix} \begin{bmatrix} F_a \\ F_b \\ F_c \end{bmatrix} \tag{4.33}$$

where

$$\begin{bmatrix} K_{aa} & K_{ab} & K_{ac} \\ K_{ba} & K_{bb} & K_{bc} \\ K_{ca} & K_{cb} & K_{cc} \end{bmatrix}$$

$$= \int_V \begin{bmatrix} (M_aM_a + N_aN_a) & (M_aM_b + N_aN_b) & (M_aM_c + N_aN_c) \\ (M_bM_a + N_bN_a) & (M_bM_b + N_bN_b) & (M_bM_c + N_bN_c) \\ (M_cM_a + N_cN_a) & (M_cM_b + N_cN_b) & (M_cM_c + N_cN_c) \end{bmatrix} dxdy \tag{4.34}$$

$$\begin{bmatrix} F_a \\ F_b \\ F_c \end{bmatrix} = \int_S \begin{bmatrix} Q_a \\ Q_b \\ Q_c \end{bmatrix} ds \tag{4.35}$$

A concrete rendition of Eq. (4.33) is

$$\begin{aligned} &\phi_a^*[K_{aa}\phi_a + K_{ab}\phi_b + K_{ac}\phi_c - F_a] \\ &\quad + \phi_a^*[K_{ba}\phi_a + K_{bb}\phi_b + K_{bc}\phi_c - F_b] \\ &\quad + \phi_a^*[K_{ca}\phi_a + K_{cb}\phi_b + K_{cc}\phi_c - F_c] = 0 \end{aligned} \tag{4.36}$$

Recall that ϕ_a^*, ϕ_b^*, and ϕ_c^* are arbitrary constants. Therefore, we put

$$\begin{aligned} \phi_a^* &= 1, \quad \phi_b^* = 0, \quad \phi_c^* = 0 \\ \phi_a^* &= 0, \quad \phi_b^* = 1, \quad \phi_c^* = 0 \\ \phi_a^* &= 0, \quad \phi_b^* = 0, \quad \phi_c^* = 1 \end{aligned}$$

then, we can derive the local finite element equation from Eq. (4.36), which is

$$\begin{bmatrix} F_a \\ F_b \\ F_c \end{bmatrix} = \begin{bmatrix} K_{aa} & K_{ab} & K_{ac} \\ K_{ba} & K_{bb} & K_{bc} \\ K_{ca} & K_{cb} & K_{cc} \end{bmatrix} \begin{bmatrix} \phi_a \\ \phi_b \\ \phi_c \end{bmatrix} \tag{4.37}$$

Equation (4.37) corresponds to Eqs. (2.7) and (3.22). Therefore, based on Eq. (4.37), the finite element method can be developed in the same manner as we did in Chaps. 2 and 3. The procedure to derive Eq. (4.37) is based on Eqs. (4.26) and (4.28), i.e., linear functions. The interpolation and weighting functions are not necessarily linear functions. The local finite element equation Eq. (4.37) in fact holds for more general interpolation functions.

4.6 Concrete Forms of the Local Finite Element Matrix

The coefficient matrix of the local finite element equation depends on its interpolation function. Therefore, this section will derive the finite element matrix using the linear interpolation function. Concretely, we can simply differentiate Eq. (4.25) and get

$$\begin{bmatrix} M_a \\ M_b \\ M_c \end{bmatrix} = \begin{bmatrix} \dfrac{\partial \Phi_a}{\partial x} \\ \dfrac{\partial \Phi_b}{\partial x} \\ \dfrac{\partial \Phi_c}{\partial x} \end{bmatrix} = \begin{bmatrix} b_a \\ b_b \\ b_c \end{bmatrix} \tag{4.38}$$

$$\begin{bmatrix} N_a \\ N_b \\ N_c \end{bmatrix} = \begin{bmatrix} \dfrac{\partial \Phi_a}{\partial y} \\ \dfrac{\partial \Phi_b}{\partial y} \\ \dfrac{\partial \Phi_c}{\partial y} \end{bmatrix} = \begin{bmatrix} c_a \\ c_b \\ c_c \end{bmatrix} \tag{4.39}$$

where b_a, b_b, b_c, c_a, c_b, and c_c are computed using the coordinate values of the nodal points, expressed in Eq. (4.22). By substituting Eqs. (4.38) and (4.39) into Eq. (4.34), we get the following coefficient matrix of the local finite element equation:

$$\begin{bmatrix} K_{aa} & K_{ab} & K_{ac} \\ K_{ba} & K_{bb} & K_{bc} \\ K_{ca} & K_{cb} & K_{cc} \end{bmatrix} = \mathbf{\Delta} \left(\begin{bmatrix} b_a \\ b_b \\ b_c \end{bmatrix} \begin{bmatrix} b_a & b_b & b_c \end{bmatrix} + \begin{bmatrix} c_a \\ c_b \\ c_c \end{bmatrix} \begin{bmatrix} c_a & c_b & c_c \end{bmatrix} \right) \tag{4.40}$$

For the integration, because the integrands in Eq. (4.34) are all constants, the integration can be achieved simply by multiplying the area of triangle $\mathbf{\Delta}$, which is the integration domain. Equation (4.40) can be computed if we know the coordinates of edges a, b, and c of a triangular element. In the case of linear interpolation functions, the integration is thus easy to implement. On the other hand, general interpolation functions require integral calculation. For this general purpose, we need to deploy the Gauss integration, area coordinates, and other methods.

To pick out the equation for a particular element $\textcircled{m}$, superscript $\textcircled{m}$ is attached in each term of Eq. (4.37):

$$\begin{bmatrix} F_a^{\textcircled{m}} \\ F_b^{\textcircled{m}} \\ F_c^{\textcircled{m}} \end{bmatrix} = \begin{bmatrix} K_{aa}^{\textcircled{m}} & K_{ab}^{\textcircled{m}} & K_{ac}^{\textcircled{m}} \\ K_{ba}^{\textcircled{m}} & K_{bb}^{\textcircled{m}} & K_{bb}^{\textcircled{m}} \\ K_{ca}^{\textcircled{m}} & K_{cb}^{\textcircled{m}} & K_{cc}^{\textcircled{m}} \end{bmatrix} \begin{bmatrix} \phi_a^{\textcircled{m}} \\ \phi_b^{\textcircled{m}} \\ \phi_c^{\textcircled{m}} \end{bmatrix} \tag{4.41}$$

Equation (4.41) is the local finite element equation for element $\textcircled{m}$.

4.7 Global Finite Element Equation

Based on the local finite element equation shown in Eq. (4.41), we can also derive the global finite element equation. The procedure is similar to the formulation that we developed in Sects. 3.4 and 3.5, which is also explained in more detail in Sects. 2.2–2.6. Because Eq. (4.41) holds over every element in the flow field, the finite element equation of the whole elements can be derived as follows:

$$\begin{bmatrix} \mathbf{F}_a \\ \hline \mathbf{F}_b \\ \hline \mathbf{F}_c \end{bmatrix} = \left[\begin{array}{c|c|c} \mathbf{K}_{aa} & \mathbf{K}_{ab} & \mathbf{K}_{ac} \\ \hline \mathbf{K}_{ba} & \mathbf{K}_{bb} & \mathbf{K}_{bc} \\ \hline \mathbf{K}_{ca} & \mathbf{K}_{cb} & \mathbf{K}_{cc} \end{array}\right] \begin{bmatrix} \phi_a \\ \hline \phi_b \\ \hline \phi_c \end{bmatrix} \tag{4.42}$$

where $[\phi_a]$, $[\phi_b]$, and $[\phi_c]$ are matrices expressing the potential values on edges a, b, and c of the whole finite elements, and $[\mathbf{F}_a]$, $[\mathbf{F}_b]$, and $[\mathbf{F}_c]$ are the matrices expressing the equivalent velocity values on edges a, b, and c of the whole finite elements. The coefficient matrices of $[\mathbf{K}_{aa}]$, $[\mathbf{K}_{ab}]$ to $[\mathbf{K}_{cc}]$ are diagonal matrices and can be obtained from the coefficient matrix in Eq. (4.41).

The equilibrium of the equivalent velocities can be expressed in the following formulation:

$$\begin{bmatrix} \alpha & \beta & \gamma \end{bmatrix} \begin{bmatrix} \mathbf{F}_a \\ \mathbf{F}_b \\ \mathbf{F}_c \end{bmatrix} = \begin{bmatrix} \mathbf{V} \end{bmatrix} \tag{4.43}$$

where $[\alpha]$, $[\beta]$, $[\gamma]$ are the connectivity matrices, which are used in the same algorithm discussed in Sect. 2.5. $[\mathbf{V}]$ is the matrix which expresses the unknown equivalent velocities on the whole nodes. Compatibility of nodal potentials in the whole flow field can be expressed as

$$\begin{bmatrix} \phi_a \\ \phi_b \\ \phi_c \end{bmatrix} = \begin{bmatrix} \alpha^T \\ \beta^T \\ \gamma^T \end{bmatrix} \begin{bmatrix} \phi \end{bmatrix} \tag{4.44}$$

where $[\phi]$ represents the matrix expressing the nodal potential values. Introducing Eq. (4.44) into Eq. (4.42), and furthermore, introducing again the resulting equation into Eq. (4.43), we get the following finite element equation:

$$\begin{bmatrix} \mathbf{K} \end{bmatrix} \begin{bmatrix} \phi \end{bmatrix} = \begin{bmatrix} \mathbf{V} \end{bmatrix} \tag{4.45}$$

where

$$\begin{aligned} \begin{bmatrix} \mathbf{K} \end{bmatrix} = & \begin{bmatrix} \alpha \mathbf{K_{aa}} \alpha^{\mathbf{T}} \end{bmatrix} + \begin{bmatrix} \alpha \mathbf{K_{ab}} \beta^{\mathbf{T}} \end{bmatrix} + \begin{bmatrix} \alpha \mathbf{K_{ac}} \gamma^{\mathbf{T}} \end{bmatrix} \\ & + \begin{bmatrix} \beta \mathbf{K_{ba}} \alpha^{\mathbf{T}} \end{bmatrix} + \begin{bmatrix} \beta \mathbf{K_{bb}} \beta^{\mathbf{T}} \end{bmatrix} + \begin{bmatrix} \beta \mathbf{K_{bc}} \gamma^{\mathbf{T}} \end{bmatrix} \\ & + \begin{bmatrix} \gamma\, \mathbf{K_{ca}} \alpha^{\mathbf{T}} \end{bmatrix} + \begin{bmatrix} \gamma\, \mathbf{K_{cb}} \beta^{\mathbf{T}} \end{bmatrix} + \begin{bmatrix} \gamma\, \mathbf{K_{cc}} \gamma^{\mathbf{T}} \end{bmatrix} \end{aligned} \tag{4.46}$$

The boundary conditions can be introduced using the same procedure that we described in Sect. 3.5. As we did in Sect. 3.5, the potential $[\phi]$ can be divided into two portions: known potential $[\hat{\phi}]$ and unknown potential $[\bar{\phi}]$.

$$\begin{bmatrix} \phi \end{bmatrix} = \begin{bmatrix} \xi^T \end{bmatrix} \begin{bmatrix} \bar{\phi} \end{bmatrix} + \begin{bmatrix} \eta^T \end{bmatrix} \begin{bmatrix} \hat{\phi} \end{bmatrix} \tag{4.47}$$

where $[\boldsymbol{\xi}]$ and $[\boldsymbol{\eta}]$ are the similar matrices expressing the correspondence relations stated in Sect. 3.5. Using the following relations:

$$\left[\xi\right]\left[\mathbf{V}\right]=\left[\mathbf{0}\right] \tag{4.48}$$

$$\left[\eta\right]\left[\mathbf{V}\right]=\left[\bar{\mathbf{V}}\right] \tag{4.49}$$

where $[\bar{\mathbf{V}}]$ represents a nonzero matrix of $[\mathbf{V}]$, the final form of the global finite element equation can be obtained:

$$\left[\bar{\mathbf{K}}\right]\left[\bar{\phi}\right]=\left[\hat{\boldsymbol{\Omega}}\right] \tag{4.50}$$

where

$$\left[\bar{\mathbf{K}}\right]=\left[\xi\right]\left[\mathbf{K}\right]\left[\xi^T\right]$$
$$\left[\hat{\boldsymbol{\Omega}}\right]=-\left[\xi\right]\left[\mathbf{K}\right]\left[\eta^T\right]\left[\hat{\phi}\right]$$

Solving Eq. (4.50) allows us to obtain the potentials for the whole nodes. Using the solution of Eq. (4.50), the unknown equivalent velocity on nodes, where the nodal potentials are specified, can be derived as follows:

$$\left[\bar{\mathbf{V}}\right]=\left[\eta\right]\left[\mathbf{K}\right]\left[\phi\right] \tag{4.51}$$

The next section will describe in detail the derivation of the global finite element equation using small flow field, as an illustrative example. We first obtain the potential values on the whole nodes to solve Eq. (4.50). Then the velocities in the whole flow field can be computed using Eqs. (4.1) and (4.2). Because the potential values of edges a, b, and c are known, the velocities at each element can be computed as

$$u=-\frac{\partial\phi}{\partial x}=-\left[M_a\ M_b\ M_c\right]\begin{bmatrix}\phi_a\\ \phi_b\\ \phi_c\end{bmatrix} \tag{4.52}$$

$$v=-\frac{\partial\phi}{\partial y}=-\left[N_a\ N_b\ N_c\right]\begin{bmatrix}\phi_a\\ \phi_b\\ \phi_c\end{bmatrix} \tag{4.53}$$

where M_a to M_c and N_a to N_c are defined in Eq. (4.32). If the potentials on nodes are known, velocities can then be computed in an element. Using the linear interpolation function, velocity inside an element is constant.

4.8 An Example of the Global Finite Element Equation

Consider a small finite element mesh of flow field shown in Fig. 4.4.

As the boundary condition, the values of potential nodes 1 and 5 are given as $\hat{\phi}_1 = 1$ and $\hat{\phi}_5 = 0$, and of equivalent velocities V_1 and V_5 are unknown.

Equation (4.41) holds for each element of ⓜ (ⓜ = ①, ②, ③), then we get

$$\begin{bmatrix} F_a^{①} \\ F_a^{②} \\ F_a^{③} \\ F_b^{①} \\ F_b^{②} \\ F_b^{③} \\ F_c^{①} \\ F_c^{②} \\ F_c^{③} \end{bmatrix} = \left[\begin{array}{ccc|ccc|ccc} K_{aa}^{①} & & & K_{ab}^{①} & & & K_{ac}^{①} & & \\ & K_{aa}^{②} & & & K_{ab}^{②} & & & K_{ac}^{②} & \\ & & K_{aa}^{③} & & & K_{ab}^{③} & & & K_{ac}^{③} \\ \hline K_{ba}^{①} & & & K_{bb}^{①} & & & K_{bc}^{①} & & \\ & K_{ba}^{②} & & & K_{bb}^{②} & & & K_{bc}^{②} & \\ & & K_{ba}^{③} & & & K_{bb}^{③} & & & K_{bc}^{③} \\ \hline K_{ca}^{①} & & & K_{cb}^{①} & & & K_{cc}^{①} & & \\ & K_{ca}^{②} & & & K_{cb}^{②} & & & K_{cc}^{②} & \\ & & K_{ca}^{③} & & & K_{cb}^{③} & & & K_{cc}^{③} \end{array}\right] \begin{bmatrix} \phi_a^{①} \\ \phi_a^{②} \\ \phi_a^{③} \\ \phi_b^{①} \\ \phi_b^{②} \\ \phi_b^{③} \\ \phi_c^{①} \\ \phi_c^{②} \\ \phi_c^{③} \end{bmatrix} \tag{4.54}$$

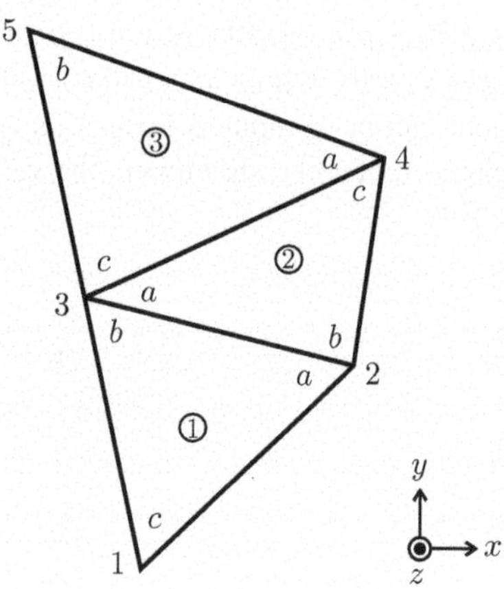

Fig. 4.4 Finite element mesh

Equilibrium of velocity gathered around each node should be zero; otherwise, the fluid would overflow at that node:

$$
\begin{aligned}
&\text{node 1: } -F_c^{\textcircled{1}} + V_1 = 0\\
&\text{node 2: } -F_a^{\textcircled{1}} - F_b^{\textcircled{2}} = 0\\
&\text{node 3: } -F_b^{\textcircled{1}} - F_a^{\textcircled{2}} - F_c^{\textcircled{3}} = 0\\
&\text{node 4: } -F_c^{\textcircled{2}} - F_a^{\textcircled{3}} = 0\\
&\text{node 5: } -F_b^{\textcircled{3}} + V_5 = 0
\end{aligned}
\tag{4.55}
$$

Equation (4.55) can be described in the following matrix form:

$$
\begin{array}{c}
\\ 1\\ 2\\ 3\\ 4\\ 5
\end{array}
\begin{array}{c}
\begin{array}{ccc:ccc:ccc}
\textcircled{1} & \textcircled{2} & \textcircled{3} & \textcircled{1} & \textcircled{2} & \textcircled{3} & \textcircled{1} & \textcircled{2} & \textcircled{3}
\end{array}\\
\left[\begin{array}{ccc:ccc:ccc}
 & & & & & & 1 & & \\
1 & & & & 1 & & & & \\
 & 1 & & 1 & & & & & 1\\
 & & 1 & & & & & 1 & \\
 & & & & & 1 & & &
\end{array}\right]
\end{array}
\begin{bmatrix}
F_a^{\textcircled{1}}\\ F_a^{\textcircled{2}}\\ F_a^{\textcircled{3}}\\ \hdashline F_b^{\textcircled{1}}\\ F_b^{\textcircled{2}}\\ F_b^{\textcircled{3}}\\ \hdashline F_c^{\textcircled{1}}\\ F_c^{\textcircled{2}}\\ F_c^{\textcircled{3}}
\end{bmatrix}
=
\begin{bmatrix}
V_1\\ \\ \\ \\ V_5
\end{bmatrix}
\tag{4.56}
$$

where the connectivity matrix can be obtained by the procedures described in Sect. 2.4, using Table 4.1. This equation corresponds to Eq. (4.43).

The compatibility at each node can be represented as follows:

Table 4.1 The connectivity relations of the mesh in Fig. 4.4

ⓜ	a	b	c
①	2	3	1
②	3	2	4
③	4	5	3

$$
\begin{aligned}
\phi_a^{\textcircled{1}} &= \phi_2 \\
\phi_b^{\textcircled{1}} &= \phi_3 \\
\phi_c^{\textcircled{1}} &= \phi_1 \\
\\
\phi_a^{\textcircled{2}} &= \phi_3 \\
\phi_b^{\textcircled{2}} &= \phi_2 \\
\phi_c^{\textcircled{2}} &= \phi_4 \\
\\
\phi_a^{\textcircled{3}} &= \phi_4 \\
\phi_b^{\textcircled{3}} &= \phi_5 \\
\phi_c^{\textcircled{3}} &= \phi_3
\end{aligned}
\tag{4.57}
$$

Equation (4.57) can be rewritten in a matrix format:

$$
\begin{array}{c}
\\
\left[\begin{array}{c}
\phi_a^{\textcircled{1}} \\ \phi_a^{\textcircled{2}} \\ \phi_a^{\textcircled{3}} \\ \hdashline
\phi_b^{\textcircled{1}} \\ \phi_b^{\textcircled{2}} \\ \phi_b^{\textcircled{3}} \\ \hdashline
\phi_c^{\textcircled{1}} \\ \phi_c^{\textcircled{2}} \\ \phi_c^{\textcircled{3}}
\end{array}\right]
\end{array}
=
\begin{array}{c}
\\
\begin{array}{c}
\textcircled{1} \\ \textcircled{2} \\ \textcircled{3} \\
\textcircled{1} \\ \textcircled{2} \\ \textcircled{3} \\
\textcircled{1} \\ \textcircled{2} \\ \textcircled{3}
\end{array}
\end{array}
\begin{array}{c}
\begin{array}{ccccc} 1 & 2 & 3 & 4 & 5 \end{array} \\
\left[\begin{array}{ccccc}
 & 1 & & & \\
 & & 1 & & \\
 & & & 1 & \\ \hdashline
 & & 1 & & \\
 & 1 & & & \\
 & & & & 1 \\ \hdashline
1 & & & & \\
 & & & 1 & \\
 & & 1 & &
\end{array}\right]
\end{array}
\begin{array}{c}
\\
\left[\begin{array}{c}
\phi_1 \\ \phi_2 \\ \phi_3 \\ \phi_4 \\ \phi_5
\end{array}\right]
\end{array}
\tag{4.58}
$$

This formulation corresponds to Eq. (4.44). Notice here that the coefficient matrices of Eqs. (4.56) and (4.58) are transpose matrices of each other. Substituting Eq. (4.58) into Eq. (4.54), we get the following equation:

$$
\left[\begin{array}{c}
F_a^{①} \\ F_a^{②} \\ F_a^{③} \\ \hdashline
F_b^{①} \\ F_b^{②} \\ F_b^{③} \\ \hdashline
F_c^{①} \\ F_c^{②} \\ F_c^{③}
\end{array}\right]
=
\left[\begin{array}{ccc:ccc:ccc}
K_{aa}^{①} & & & K_{ab}^{①} & & & K_{ac}^{①} & & \\
& K_{aa}^{②} & & & K_{ab}^{②} & & & K_{ac}^{②} & \\
& & K_{aa}^{③} & & & K_{ab}^{③} & & & K_{ac}^{③} \\ \hdashline
K_{ba}^{①} & & & K_{bb}^{①} & & & K_{bc}^{①} & & \\
& K_{ba}^{②} & & & K_{bb}^{②} & & & K_{bc}^{②} & \\
& & K_{ba}^{③} & & & K_{bb}^{③} & & & K_{bc}^{③} \\ \hdashline
K_{ca}^{①} & & & K_{cb}^{①} & & & K_{cc}^{①} & & \\
& K_{ca}^{②} & & & K_{cb}^{②} & & & K_{cc}^{②} & \\
& & K_{ca}^{③} & & & K_{cb}^{③} & & & K_{cc}^{③}
\end{array}\right]
$$

$$
\times
\left[\begin{array}{ccccc}
& 1 & & & \\
& & 1 & & \\
& & & 1 & \\ \hdashline
& & 1 & & \\
& 1 & & & \\
& & & & 1 \\ \hdashline
1 & & & & \\
& & & 1 & \\
& & 1 & &
\end{array}\right]
\left[\begin{array}{c}
\phi_1 \\ \phi_2 \\ \phi_3 \\ \phi_4 \\ \phi_5
\end{array}\right]
$$

$$
=
\left[\begin{array}{ccccc}
K_{ac}^{①} & K_{aa}^{①} & K_{ab}^{①} & & \\
& K_{ab}^{②} & K_{aa}^{②} & K_{ac}^{②} & \\
& & K_{ac}^{③} & K_{aa}^{③} & K_{ab}^{③} \\ \hdashline
K_{bc}^{①} & K_{ba}^{①} & K_{bb}^{①} & & \\
& K_{bb}^{②} & K_{ba}^{②} & K_{bc}^{②} & \\
& & K_{bc}^{③} & K_{ba}^{③} & K_{bb}^{③} \\ \hdashline
K_{cc}^{①} & K_{ca}^{①} & K_{cb}^{①} & & \\
& K_{cb}^{②} & K_{ca}^{②} & K_{cc}^{②} & \\
& & K_{cc}^{③} & K_{ca}^{③} & K_{cb}^{③}
\end{array}\right]
\left[\begin{array}{c}
\phi_1 \\ \phi_2 \\ \phi_3 \\ \phi_4 \\ \phi_5
\end{array}\right]
\tag{4.59}
$$

Substituting again Eq. (4.59) into Eq. (4.56) and rearranging the terms, the global finite element equation can be derived as

$$
\begin{bmatrix}
K_{cc}^{\textcircled{1}} & K_{ca}^{\textcircled{1}} & K_{cb}^{\textcircled{1}} & & \\
K_{ac}^{\textcircled{1}} & K_{aa}^{\textcircled{1}} + K_{bb}^{\textcircled{2}} & K_{ab}^{\textcircled{1}} + K_{ba}^{\textcircled{2}} & K_{bc}^{\textcircled{2}} & \\
K_{bc}^{\textcircled{1}} & K_{ba}^{\textcircled{1}} + K_{ab}^{\textcircled{2}} & K_{bb}^{\textcircled{1}} + K_{aa}^{\textcircled{2}} + K_{cc}^{\textcircled{3}} & K_{ac}^{\textcircled{2}} + K_{ca}^{\textcircled{3}} & K_{cb}^{\textcircled{3}} \\
 & K_{cb}^{\textcircled{2}} & K_{ca}^{\textcircled{2}} + K_{ac}^{\textcircled{3}} & K_{cc}^{\textcircled{2}} + K_{aa}^{\textcircled{3}} & K_{ab}^{\textcircled{3}} \\
 & & K_{bc}^{\textcircled{3}} & K_{ba}^{\textcircled{3}} & K_{bb}^{\textcircled{3}}
\end{bmatrix}
\begin{bmatrix} \phi_1 \\ \phi_2 \\ \phi_3 \\ \phi_4 \\ \phi_5 \end{bmatrix}
=
\begin{bmatrix} V_1 \\ \\ \\ \\ V_5 \end{bmatrix}
\tag{4.60}
$$

The result corresponds to Eq. (4.45). Notice that the coefficient matrix is symmetric. This nature of the equation stems from the fact that the coefficient matrices of equilibrium and compatibility are mutually transpose matrices, and the fact that the local finite element equation has the symmetric coefficient matrix. As we have seen in some detail in Sect. 2.9, the global finite element equation Eq. (4.60) can be systematically formulated.

The potential can be divided into known and unknown parts as follows:

$$
\begin{bmatrix} \phi_1 \\ \phi_2 \\ \phi_3 \\ \phi_4 \\ \phi_5 \end{bmatrix}
=
\begin{bmatrix} & & \\ 1 & & \\ & 1 & \\ & & 1 \\ & & \end{bmatrix}
\begin{bmatrix} \overline{\phi}_2 \\ \overline{\phi}_3 \\ \overline{\phi}_4 \end{bmatrix}
+
\begin{bmatrix} 1 & \\ & \\ & \\ & \\ & 1 \end{bmatrix}
\begin{bmatrix} \hat{\phi}_1 \\ \hat{\phi}_5 \end{bmatrix}
\tag{4.61}
$$

where $\overline{\phi}_2$, $\overline{\phi}_3$, $\overline{\phi}_4$ are unknown potentials and $\hat{\phi}_1$ and $\hat{\phi}_5$ are given as the boundary conditions. This equation corresponds to Eq. (4.47). Introducing Eq. (4.61) into Eq. (4.60) and using the following relation, which corresponds to Eq. (4.48):

$$
\begin{bmatrix} & 1 & & & \\ & & 1 & & \\ & & & 1 & \end{bmatrix}
\begin{bmatrix} V_1 \\ \\ \\ \\ V_5 \end{bmatrix}
=
\begin{bmatrix} \\ \\ \\ \end{bmatrix}
\tag{4.62}
$$

we obtain the final form of the global finite element equation including the boundary condition denoting $\overline{\phi}_2, \overline{\phi}_3, \overline{\phi}_4$ as ϕ_2, ϕ_3, ϕ_4:

$$
\begin{bmatrix}
K_{aa}^{\textcircled{1}} + K_{bb}^{\textcircled{2}} & K_{ab}^{\textcircled{1}} + K_{ba}^{\textcircled{2}} & K_{bc}^{\textcircled{2}} \\
K_{ba}^{\textcircled{1}} + K_{ab}^{\textcircled{2}} & K_{bb}^{\textcircled{1}} + K_{aa}^{\textcircled{2}} + K_{cc}^{\textcircled{3}} & K_{ac}^{\textcircled{2}} + K_{ca}^{\textcircled{3}} \\
K_{cb}^{\textcircled{2}} & K_{ca}^{\textcircled{2}} + K_{ac}^{\textcircled{3}} & K_{cc}^{\textcircled{2}} + K_{aa}^{\textcircled{3}}
\end{bmatrix}
\begin{bmatrix} \phi_2 \\ \phi_3 \\ \phi_4 \end{bmatrix}
= -
\begin{bmatrix}
K_{ac}^{\textcircled{1}} & \\
K_{bc}^{\textcircled{1}} & K_{cb}^{\textcircled{3}} \\
 & K_{ab}^{\textcircled{3}}
\end{bmatrix}
\begin{bmatrix} \hat{\phi}_1 \\ \hat{\phi}_5 \end{bmatrix}
\tag{4.63}
$$

Solving Eq. (4.63) will yield the potentials on the whole nodes in the flow field. Equation (4.63) corresponds to Eq. (4.50). Note also that the final form of the global finite element matrix is symmetric. The unknown velocities can be derived by applying the following relation to Eq. (4.60):

$$
\begin{bmatrix}
1 & & & & \\
 & & & & 1
\end{bmatrix}
\begin{bmatrix} V_1 \\ \\ \\ \\ V_5 \end{bmatrix}
=
\begin{bmatrix} V_1 \\ V_5 \end{bmatrix}
\tag{4.64}
$$

which corresponds to Eq. (4.49). The resulting equation for the unknown equivalent velocities on nodes 1 and 5 can be expressed as follows:

$$
\begin{bmatrix}
K_{cc}^{\textcircled{1}} & K_{ca}^{\textcircled{1}} & K_{cb}^{\textcircled{1}} & & \\
 & & K_{bc}^{\textcircled{3}} & K_{ba}^{\textcircled{3}} & K_{bb}^{\textcircled{3}}
\end{bmatrix}
\begin{bmatrix} \hat{\phi}_1 \\ \phi_2 \\ \phi_3 \\ \phi_4 \\ \hat{\phi}_5 \end{bmatrix}
=
\begin{bmatrix} V_1 \\ V_5 \end{bmatrix}
\tag{4.65}
$$

Equation (4.65) corresponds to Eq. (4.51). The velocities inside the flow field can be computed by the same procedure that we described in Sect. 4.7 using Eqs. (4.52) and (4.53).

4.9 Naming and Meaning of the Weighted Residual Equation

Now let us step back a bit and examine the meaning of Eq. (4.11) in further depth. If we employ the approximate function for the potential ϕ^h, and substitute it into Eq. (4.4), the approximate potential does not hold, and we get the residual R.

$$R = \frac{\partial^2 \phi^h}{\partial x^2} + \frac{\partial^2 \phi^h}{\partial y^2} \tag{4.66}$$

Multiplying the weighting function ϕ^* to Eq. (4.66) and integrating over an arbitrary volume V, we obtain

$$\int_V (\phi^* R) dxdy = \int_V \phi^* \left(\frac{\partial^2 \phi^h}{\partial x^2} + \frac{\partial^2 \phi^h}{\partial y^2} \right) dxdy \tag{4.67}$$

Now, we will sidestep a bit and introduce the notion of weighted average. We often use the concept of average for science and engineering data. When there are n data, $a_1, a_2, \ldots, a_n$, then the average a is

$$a = \frac{\sum_{i=1}^{n} a_i}{n} = \frac{\sum_{i=1}^{n} a_i}{\sum_{i=1}^{n} 1} \tag{4.68}$$

We can assign a weight r_i for each element a_i, then we can calculate its weighted average a_r:

$$a_r = \frac{\sum_{i=1}^{n} r_i a_i}{\sum_{i=1}^{n} r_i} \tag{4.69}$$

We can take arbitrary constants for the weight r_i. Equation (4.67) is the natural extension of Eq. (4.69), i.e., the weighted residual average. This should be summed to zero:

$$\frac{\int_V (\phi^* R) dxdy}{\int_V \phi^* dxdy} = 0 \tag{4.70}$$

Equation (4.70) leads to Eq. (4.11). We want to solve the exact solution of Eq. (4.4), but this task seems insurmountable. Thus instead, we pursue a different route, in which we attempt to solve weighted average of the differential equation, which is essentially the meaning of the weighted residual equation. The naming of the weighting function is also the natural extension of the weight r_i, and an arbitrary function can be employed. Thus we should make sure that we pursue the average value of phenomena by the finite element method, not the extreme values.

4.10 Shape Function

If we posit

$$\phi_a = 1, \quad \phi_b = 0, \quad \phi_c = 0 \tag{4.71}$$

then, Eq. (4.24) must be

$$\phi = \Phi_a(x, y) \tag{4.72}$$

What this represents is that $\Phi_a(x, y)$ expresses the distribution of potential with $\phi_a = 1, \phi_b = 0, \phi_c = 0$. The same relation holds for $\Phi_b(x, y)$ and $\Phi_c(x, y)$. Those relations are illustrated in Fig. 4.5.

The three functions —$\Phi_a(x, y)$, $\Phi_b(x, y)$, and $\Phi_c(x, y)$— are the natural extensions of the functions introduced in Eq. (3.14) in Chap. 3 to an triangular element and is called as shape function. Many shape functions, such as linear, quadratic, cubic, has been examined in the past. Let us now reconsider the meaning of Eq. (4.37). The left side terms of Eq. (4.37) is the integration over the three sides of an element. The integrand is velocity q on three sides of the element, shown in Fig. 4.3, times shape function. The shape function is a linear function, whose values on considering edge is 1, and whose values on two other edges are zero, as shown in Fig. 4.5. For instance, F_a is the equivalent velocity on the sides of ab and ac. The influence of velocity on edge a is 1, and on edges b and c are zero, and linearly changing between $a - b$ and $a - c$. The same holds for F_b and F_c. Therefore, F_a, F_b, and F_c are the equivalent velocities on edges a, b, and c to which the velocities on the three sides of the element are converted.

If we put in Eq. (4.37)

$$\phi_a = 1, \quad \phi_b = 0, \quad \phi_c = 0 \tag{4.73}$$

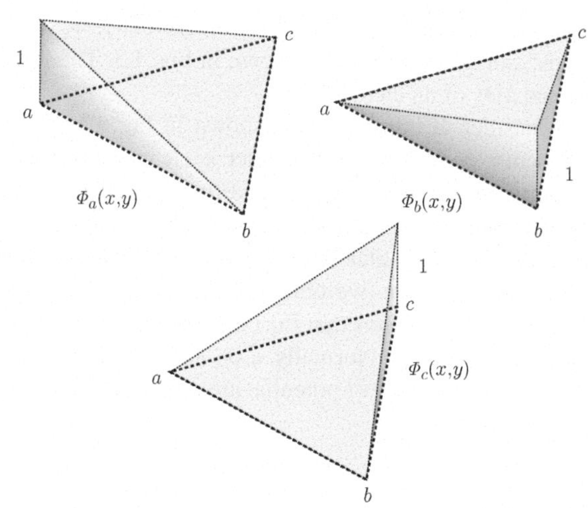

Fig. 4.5 Shape function

then we get

$$F_a = K_{aa}, \quad F_b = K_{ba}, \quad F_c = F_{ca} \tag{4.74}$$

Equation (4.74) means that the coefficients K_{aa}, K_{ba}, and K_{ca} are equivalent velocities in the case that the potential values on edge a is 1, and those on other edges are zero. The same holds for K_{ba}, K_{bb}, K_{bc}, K_{ca}, K_{cb}, and K_{cc}. Thus, K_{aa} to K_{cc} are the relational coefficients between the equivalent velocities and potential values. Moreover, we have discussed the wearing of the fact that the coefficient matrix is symmetric. For instance, $K_{ac} = K_{ca}$; this means that the equivalent velocity F_a when the potential is $\phi_a = \phi_b = 0$ and $\phi_c = 1$ is equal to the equivalent velocity F_c when the potential is $\phi_a = 1$ and $\phi_b = \phi_c = 0$. The same is true for $K_{ab} = K_{ba}$ and $K_{bc} = K_{cb}$. Those are the meanings of the symmetry of finite element equation.

4.11 Meaning of Equilibrium of Equivalent Velocity

We have formulated the equivalent velocity in Sect. 4.5. We also formulated the equilibrium of equivalent velocity in Eqs. (4.43) and (4.56). This section reconsiders the meaning of the equilibrium. We defined the equivalent velocity in Eq. (4.35), which is repeated here:

$$\begin{aligned} F_a &= \int_S (\Phi_a q) ds \\ F_b &= \int_S (\Phi_b q) ds \\ F_c &= \int_S (\Phi_c q) ds \end{aligned} \tag{4.75}$$

where F_a, F_b, and F_c are equivalent velocity, Φ_a, Φ_b, and Φ_c are the shape functions, and q is the velocity shown in Fig. 4.3. The integration is carried out over all boundaries of an element S.

Consider an inner node i, shown in Fig. 4.6. The node i is surrounded by the elements ⓚ, ⓛ, ⓜ, ⓝ. Consider each inner side of integration, the direction of integration are mutually opposed to each other between two adjacent elements. If we summarize the equivalent velocities gathered around i, we obtain zero, and only quantities on the outer sides are nonzero. After reiterating the same operation over all of the elements, we obtain the equivalent velocities on the whole sides of the flow field. This means that Eq. (4.43) or Eq. (4.56) is the equilibrium equation of the equivalent velocity. Normally, $\hat{q}$ on S_2 is equal to zero, and thus, except on boundary S_1, where the values of potential are all given, the summation value of the equivalent velocity is zero.

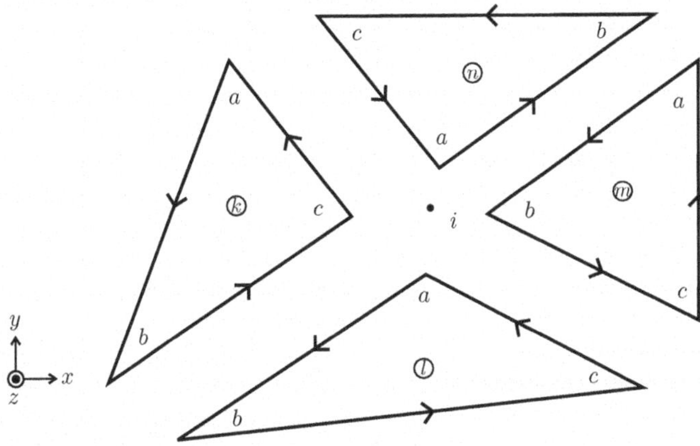

Fig. 4.6 Superposition at node i

4.12 Systematic Formulation of the Finite Element Matrix

The coefficient matrix of Eq. (4.60) can be systematically formulated using the coefficient matrix of the local finite element equation, which is in Eq. (4.41). The connectivity relations between edges a, b, and c and nodes are summarized in Table 4.1. The global matrix is a square matrix, whose row and column numbers correspond to nodes. Given this matrix, apply the following procedures:

Superimpose $K_{aa}^{ⓜ}$ onto row a and column a;

Superimpose $K_{ab}^{ⓜ}$ onto row a and column b;

Superimpose $K_{ac}^{ⓜ}$ onto row a and column c;

Superimpose $K_{ba}^{ⓜ}$ onto row b and column a;

Superimpose $K_{bb}^{ⓜ}$ onto row b and column b;

Superimpose $K_{bc}^{ⓜ}$ onto row b and column c;

Superimpose $K_{ca}^{ⓜ}$ onto row c and column a;

Superimpose $K_{cb}^{ⓜ}$ onto row c and column b;

Superimpose $K_{cc}^{ⓜ}$ onto row c and column c.

Let us consider the specific example of the flow field, shown in Fig. 4.4. For element ①, the relevant edges are $a = 2,\ b = 3,\ c = 1$. Thus, the coefficient matrix of Eq. (4.41) can be rearranged in the following way:

$$\begin{array}{c} \\ 1\\2\\3\\4\\5 \end{array}\begin{array}{c} \begin{array}{ccccc} 1 & 2 & 3 & 4 & 5 \end{array} \\ \begin{bmatrix} K^{①}_{cc} & K^{①}_{ca} & K^{①}_{cb} & & \\ K^{①}_{ac} & K^{①}_{aa} & K^{①}_{ab} & & \\ K^{①}_{bc} & K^{①}_{ba} & K^{①}_{bb} & & \\ & & & & \\ & & & & \end{bmatrix} \end{array} \tag{4.76}$$

For element ②, its edges are $a = 3,\ b = 2,\ c = 4$. Therefore, the coefficient matrix of Eq. (4.41) can be rearranged as

$$\begin{array}{c} \\ 1\\2\\3\\4\\5 \end{array}\begin{array}{c} \begin{array}{ccccc} 1 & 2 & 3 & 4 & 5 \end{array} \\ \begin{bmatrix} & & & & \\ & K^{②}_{bb} & K^{②}_{ba} & K^{②}_{bc} & \\ & K^{②}_{ab} & K^{②}_{aa} & K^{②}_{ac} & \\ & K^{②}_{cb} & K^{②}_{ca} & K^{②}_{cc} & \\ & & & & \end{bmatrix} \end{array} \tag{4.77}$$

For element ③, its edges are $a = 4,\ b = 5,\ c = 3$. Hence, its coefficient matrix is

$$\begin{array}{c} \\ 1\\2\\3\\4\\5 \end{array}\begin{array}{c} \begin{array}{ccccc} 1 & 2 & 3 & 4 & 5 \end{array} \\ \begin{bmatrix} & & & & \\ & & & & \\ & & K^{③}_{cc} & K^{③}_{ca} & K^{③}_{cb} \\ & & K^{③}_{ac} & K^{③}_{aa} & K^{③}_{ab} \\ & & K^{③}_{bc} & K^{③}_{ba} & K^{③}_{bb} \end{bmatrix} \end{array} \tag{4.78}$$

Equations (4.76)–(4.78) show components of global coefficient matrix, according to number of element $i = 1$ to 3. Finally, we sum up all the three matrices listed above, which will give us the coefficient matrix of Eq. (4.60). The algorithm demonstrated in this section is a natural extension of the algorithm discussed in Sect. 2.9.

4.13 Computer Program: potential_flow

This section presents a simple computer program called potential_flow. Generally speaking, when readers write a computer program, they have to make each and every step of computation explicit; such an exercise allows us to make concrete each step that is discussed in this chapter.

The primary goal of this program is to formulate and solve Eq. (4.63). The coefficient matrix of Eq. (4.60) consists of the coefficient matrix of Eq. (4.41), which can be shown by a systematic algorithm. However, the algorithm depends on the solution algorithm to solve the simultaneous equation as well. For potential_flow, we use the simplest procedure to solve simultaneous equation, because the aim of the program is to understand the fundamental ideas of the finite element method, and the computational efficiency is not the top priority.

4.13.1 Uniform Channel Flow

The first step is to create an appropriate finite element mesh. It is critical to choose the right kind of mesh in general. As a rule of thumb, for a field where a phenomenon is changing slowly, a relatively coarse mesh can be used, whereas for a field where the change is rather rapid, finer mesh should be used. Therefore, the optimal mesh can be best found after reiteration of actual analyses. When solving important problems, it is recommended that many iterations should be tried out to find the optimal solution.

As an illustrative example, Fig. 4.7 shows a uniform channel flow. The mesh used for this figure consists of triangular elements, whose shapes are all identical. Numberings of the nodes and elements are shown in figure. The numbering convention here is more or less arbitrary; however, the differences among the node numbers in an element should be as small as possible, as it would be convenient for solving the finial simultaneous equations by sweep-out method.

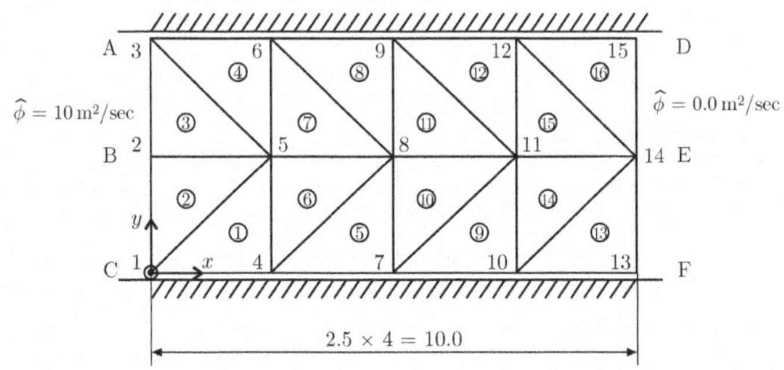

Fig. 4.7 Uniform channel flow

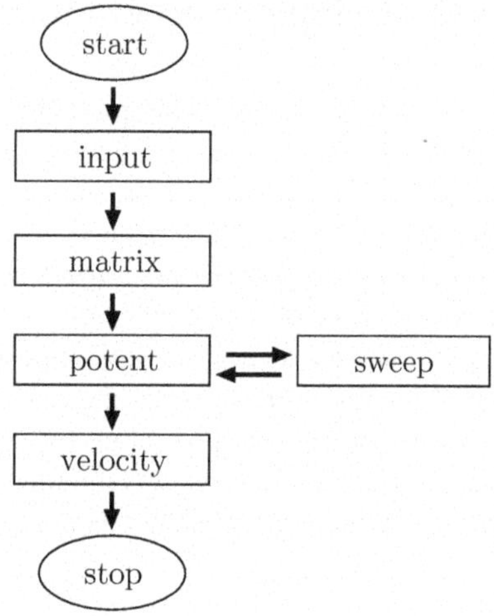

Fig. 4.8 Flow chart of the program: potential_flow

Specifying the potentials on A–B–C as $\hat{\phi} = 10\,\mathrm{m^2/s}$ and on D–E–F as $\hat{\phi} = 0.0\,\mathrm{m^2/s}$, the velocities in the channel should be all $u = 1.0\,\mathrm{m/s}$ and $v = 0.0\,\mathrm{m/s}$.

4.13.2 A Flow Chart

Figure 4.8 provides a flow chart. The main program consists of four subroutines, one of which is on its own sub-subroutine. The flow of the computer program is simple. In the subroutine input, necessary data is read into the program. In subroutine matrix, the coefficient matrix of the global finite element equation is created. In subroutine potent, the known term and the potential are computed. Subroutine sweep solves the simultaneous equation, using the coefficient matrix and known term. The velocity is computed in the subroutine velocity, and the computed potential and velocity are output.

4.13.3 Subroutine Input

The input data are shown in Table 4.2. For the example shown in Fig. 4.7, the total number of nodes nx is 15, and the total number of elements mx is 16. The total number of nodes with potential specifications ix is 6. The coordinates of the nodes are tabulated in Table 4.3 with the origin being on node 1. Coordinate x is $xy(1, i)$, and coordinate y is $xy(2, i)$. The connectivity relation is shown in Table 4.4. Edges a, b, c are shown as 1, 2, 3. Therefore, $ie(1, i)$ is edge a, $ie(2, i)$ is edge b, and $ie(3, i)$ is edge c of element i. Naming of edges a, b, c is arbitrary. However, the rotation

Table 4.2 Input data

Identifier	Maximum	Explanation
nx	1	Total number of nodes
mx	1	Total number of elements
ix	1	Total number of nodes with potential specification
$xy(1, i)$	nx	Coordinate x of node i
$xy(2, i)$	nx	Coordinate y of node i
$ie(i, j)$	$3, mx$	Connectivity relation of edges a, b, c of element j
$itx(i)$	ix	Number of nodes with potential specification
$tfx(i)$	ix	Specified value of the potential

Table 4.3 Coordinate of nodes

i	$xy(1, i)$	$xy(2, i)$	i	$xy(1, i)$	$xy(2, i)$
1	0.0	0.0	9	5.0	5.0
2	0.0	2.5	10	7.5	0.0
3	0.0	5.0	11	7.5	2.5
4	2.5	0.0	12	7.5	5.0
5	2.5	2.5	13	10.0	0.0
6	2.5	5.0	14	10.0	2.5
7	5.0	0.0	15	10.0	5.0
8	5.0	2.5			

Table 4.4 Conectivity of elements

im	$ie(1, im)$	$ie(2, im)$	$ie(3, im)$	im	$ie(1, im)$	$ie(2, im)$	$ie(3, im)$
1	4	5	1	9	10	11	7
2	2	1	5	10	8	7	11
3	2	5	3	11	8	11	9
4	6	3	5	12	12	9	11
5	7	8	4	13	13	14	10
6	5	4	8	14	11	10	14
7	5	8	6	15	11	14	12
8	9	6	8	16	15	12	14

around a-b-c should coincide with the rotation of z-axis. Boundary conditions of potential are given on the nodes 1, 2, 3 (A–B–C), and 13, 14, 15 (F–E–D). Those numbers are listed in $ifx(i)$. On nodes 1, 2, 3, the potential is given as 10.0d0, and on nodes 13, 14, 15, the potential is as 0.0d0. Those numbers are listed in $tfx(i)$ (Table 4.5).

Table 4.5 Boundary condition

i	$ifx(i)$	$tfx(i)$	i	$ifx(i)$	$tfx(i)$
1	1	10.0	4	13	0.0
2	2	10.0	5	14	0.0
3	3	10.0	6	15	0.0

4.13.4 Subroutine Matrix

The main idea developed in Sect. 4.12 is implemented in this subroutine, which corresponds to a natural extension of the idea discussed in Sect. 3.6.3.

The purpose of the subroutine matrix is to make the global finite element matrix $sk(i, j),\ i = 1$ to nx and $j = 1$ to nx, expressed in Eq. (4.45). This matrix obtained by the superposition procedure of $sm(i, j),\ i = 1$ to 3 and $j = 1$ to 3; the local finite element matrix in Eq. (4.41). The superposition procedure is represented in Sect. 4.12. The main index in this subroutine is im, $im = 1$ to mx. The im means ⓜ, i.e., the number of elements under consideration. In this computer program, three edges of an element $a,\ b$, and c are denoted as 1, 2, and 3. This is because numerical variables are much easier to handle in the FORTRAN program. Therefore, the node number connected with edge a of element ⓜ is stored in $ie(1, im)$, edge b in $ie(2, im)$, and edge c in $ie(3, im)$; see Table 4.4.

Therefore, the node numbers whose edges a, b, and c of element ⓜ are connected can be retrieved by

$$ia = ie(1, im)$$
$$ib = ie(2, im)$$
$$ic = ie(3, im)$$

where ia, ib, ic represent the node numbers to which the edges a, b, and c of element im are connected. Then, the coordinates (x_a, y_a), (x_b, y_b), (x_c, y_c) of element ⓜ are computed as

$$xa = xy(1, ia)$$
$$ya = xy(2, ia)$$
$$xb = xy(1, ib)$$
$$yb = xy(2, ib)$$
$$xc = xy(1, ic)$$
$$yc = xy(2, ic)$$

$$
\begin{array}{ccc}
 & & \begin{array}{ccc} ie(1,im) & ie(2,im) & ie(3,im) \\ \uparrow & \uparrow & \uparrow \\ 1 & 2 & 3 \end{array} \\
\begin{array}{c} ie(1,im) \\ ie(2,im) \\ ie(3,im) \end{array} & \begin{array}{c} \leftarrow 1 \\ \leftarrow 2 \\ \leftarrow 3 \end{array} & \left[\begin{array}{ccc} K^{\textcircled{m}}_{aa} & K^{\textcircled{m}}_{ab} & K^{\textcircled{m}}_{ac} \\ K^{\textcircled{m}}_{ba} & K^{\textcircled{m}}_{bb} & K^{\textcircled{m}}_{bc} \\ K^{\textcircled{m}}_{ca} & K^{\textcircled{m}}_{cb} & K^{\textcircled{m}}_{cc} \end{array} \right] \equiv sm(i,j)
\end{array}
\tag{4.79}
$$

Fig. 4.9 Algorithm for global coefficient matrix program is as follows

Using the coordinates of three edges, area of triangle and $b(i)$, $i = 1$ to 3, $c(i)$, $i = 1$ to 3 can be computed as in Eqs. (4.22) and (4.23). The array $sm(i, j)$, $i = 1$ to 3, $j = 1$ to 3, which is the local finite element matrix can be computed following Eq. (4.40).

Afterwards, the global finite element matrix is computed following the idea described in Sect. 4.12. In Table 4.4, edges a, b, and c are denoted by 1, 2, and 3. Therefore, the node number connected with edge a of element im is stored in $ie(1, im)$, edge b in $ie(2, im)$, and edge c in $ie(3, im)$. The algorithm explained in Sect. 4.12 can be implemented as in Fig. 4.9 and the following algorithm:

```
do i = 1, nx
  do j = 1, nx
    sk(i, j) = 0.d0
  end do j
end do i

do im = 1, mx
  do i = 1, 3
    ii = ie(i, im)
    do j = 1, 3
      jj = ie(j, im)
      sk(ii, jj) = sk(ii, jj) + sm(i, j)
    end do j
  end do i
end do im
```

It is important to clear $sk(i, j)$ before starting the superposition procedure. The index im shows the number of element. The index i corresponds to a, b, and c and the index j corresponds to a, b, and c. The indices ii and jj are the numbers of nodes to which the edges a, b, and c are connected. The local finite element matrix $sm(i, j)$, $i = 1$ to 3, $j = 1$ to 3 is superposed onto the global finite element matrix $sk(i, j)$, $i = 1$ to nx, $j = 1$ to nx.

4.13.5 Subroutine Potent

The purpose of this subroutine is to compute potentials on the whole flow field and to print out the computed potentials. The first step is to compute the known term $ff(i)$, $i = 1$ to nx, which is the right side of Eq. (4.50) or Eq. (4.63). The index i shows the number of nodes on which the potential values are given. The index ii is the actual number of nodes on which the potential values are given. The right side of Eq. (4.50) or Eq. (4.63) can be computed by the following algorithm:

```
do i = 1, nx
   ff(i) = 0.d0
end do

do i = 1, ix
   ii = ifx(i)
   do j = 1, nx
      ff(j) = ff(j) - sk(j, ii) * tfx(i)
   end do j
end do i
```

The next step is to treat the left side of Eq. (4.50) or Eq. (4.63). To obtain the finite element matrix in Eq. (4.50) or Eq. (4.63) from the matrix in Eq. (4.45) or Eq. (4.60), which is $sk(i, j)$ in the subroutine matrix, the following simple algorithm suffices. In this algorithm, the rows and columns of the corresponding nodes are eliminated and put the diagonal element 1. The actual algorithm is as follows:

$$
\begin{aligned}
&do\ i = 1, nx\\
&\quad ii = ifx(i)\\
&\quad do\ j = 1, nx\\
&\quad\quad sk(ii, j) = 0.d0\\
&\quad\quad sk(j, ii) = 0.d0\\
&\quad end\ do\ j\\
&\quad sk(ii, ii) = 1.d0\\
&end\ do\ i
\end{aligned}
$$

Then, the subroutine sweep can calculate the solution of the simultaneous equation using the finite element matrix and the known term. The potentials computed are stored in $ff(i),\ i = 1$ to nx. Again, the boundary conditions are treated in $ff(i)$. Finally, potentials are printed out.

4.13.6 Subroutine Velocity

This subroutine computes the velocities from the potentials; the velocities can be computed for each element. Consecutively, the potentials are in $ff(i),\ i = 1$ to nx. Thus at each element, the potentials at edges a, b, and c can be stored $p(i),\ i = 1$ to 3. Using Eqs. (4.52) and (4.53), the velocity at each element can be computed. The velocities x and y are constant at each element.

4.13.7 Program List of potential_flow

The computer program list is shown in Fig. 4.10. For the input data, the simple channel shown in Fig. 4.7 is used. The main identifiers that are used in the program are summarized in Table 4.6.

4.13.8 Output List

The output results are shown in Fig. 4.11. Those include input data, potential values, and x and y velocities. We observe that all x velocities are 1.0 and all y velocities are zero.

Computed potential and velocity are plotted in Fig. 4.12.

```
!*********************************************!
!*** finite element method for potential flow ***!
!***                   coded by Akihiro Aizawa ***!
!*********************************************!
!
!       nx-----total number of nodes
!       mx-----total nnumber of elements
!       xy-----node coodinates
!       ie-----conection index for each elements
!       ix-----total number of potentials
!              specified on the boundaries
!       ifx----node number of specified boundary condition
!       itx----specified values for potentials on the boundaries
!       sm-----stiffness matrix for each element
!       sk-----stiffness matrix for all domain


        program potential_flow

        implicit real(8) (a-h,o-z)
        implicit integer(4) (i-n)

        dimension  xy(2,40),ie(3,40),sk(40,40)
        dimension  ifx(10),tfx(10),pq(40)

!*********************************************!
!*******          Main Program          ******!
!*********************************************!

        open(10,file='input_pot.txt')
        open(20,file='check_input_pot.txt')
        open(30,file='output_pot.txt')
        open(40,file='check_matrix_pot.txt')

        call input(nx,mx,ix,xy,ie,ifx,tfx)
        call matrix(nx,mx,xy,ie,sk)
        call potent(nx,ix,ifx,tfx,pq,sk)
        call velcty(mx,xy,ie,pq)

        close(10)
        close(20)
        close(30)
        close(40)
        end program potential_flow

!*********************************************!
!*******           END Program          ******!
!*********************************************!
```

Fig. 4.10 Program list

```
!*********************************************!
!*********          Subprogram          ******!
!*********************************************!

!========== begin subroutine input ==========!
      subroutine input(nx,mx,ix,xy,ie,ifx,tfx)

      implicit real(8) (a-h,o-z)
      implicit integer(4) (i-n)

      dimension  xy(2,40),ie(3,40)
      dimension  ifx(10),tfx(10)
!     File input
      read(10,*) nx,mx,ix
      read(10,*) (i,xy(1,i),xy(2,i),j=1,nx)
      read(10,*) (i,(ie(j,i),j=1,3),k=1,mx)
      read(10,*) (i,ifx(i),tfx(i),j=1,ix)
!     Input data confirmation
      write(30,*)'*******  potential flow program  *******'
      write(30,*)
      write(30,*)'*******  input data  ******************'
      write(30,*)
      write(20,*)'*******  input data check  *************'
      write(20,*)
      write(30,*)'  num.ele     ia     ib     ic'
      write(20,*) nx,mx,ix
      write(20,*) (i,xy(1,i),xy(2,i),i=1,nx)
      write(20,*) (i,(ie(j,i),j=1,3),i=1,mx)

      do i=1,mx
        write(30,299) i,ie(1,i),ie(2,i),ie(3,i)
      end do
      write(30,*)
      write(20,*) (i,ifx(i),tfx(i),i=1,ix)

  299 format(5x,i5,3i7)

      return
      end subroutine input
!============ end subroutine input ===========!

!========== begin subroutine matrix ==========!
      subroutine matrix(nx,mx,xy,ie,sk)

      implicit real(8) (a-h,o-z)
      implicit integer(4) (i-n)
      dimension xy(2,40),ie(3,40),sk(40,40)
      dimension sm(3,3),b(3),c(3)
```

Fig. 4.10 (continued)

```
      do i=1,nx
        do j=1,nx
          sk(i,j)=0.0d0
        end do
      end do
      do i=1,3
        do j=1,3
          sm(i,j)=0.0d0
        end do
      end do

      do im=1,mx
        ia=ie(1,im)
        ib=ie(2,im)
        ic=ie(3,im)

        xa=xy(1,ia)
        xb=xy(1,ib)
        xc=xy(1,ic)

        ya=xy(2,ia)
        yb=xy(2,ib)
        yc=xy(2,ic)

        area=((yb-yc)*(xa-xc)-(xc-xb)*(yc-ya))*0.5d0
        area2=1.0d0/(area*2.0d0)

        b(1)=(yb-yc)*area2
        b(2)=(yc-ya)*area2
        b(3)=(ya-yb)*area2

        c(1)=(xc-xb)*area2
        c(2)=(xa-xc)*area2
        c(3)=(xb-xa)*area2

        do i=1,3
          do j=1,3
            sm(i,j)=(b(i)*b(j)+c(i)*c(j))*area
          end do
        end do

        do i=1,3
          ii=ie(i,im)
          do j=1,3
            jj=ie(j,im)
            sk(ii,jj)=sk(ii,jj)+sm(i,j)
          end do
        end do
      end do

      write(40,*)'***  check matrix of potential flow  ***'
```

Fig. 4.10 (continued)

```
      write(40,*)
      write(40,*)'check coeff.b coeff.c'

      do i=1,3
        write(40,*)'b',i,b(i)
      end do

      write(40,*)

      do i=1,3
        write(40,*)'c',i,c(i)
      end do

      write(40,*)
      write(40,*)'check area,area2'
      write(40,*)'area ',area
      write(40,*)'area2 ',area2
      write(40,*)
      write(40,*)'check matrix sm'

      do i=1,3
        write(40,*)i,sm(i,1),sm(i,2),sm(i,3)
      end do

      write(40,*)

      return
      end subroutine matrix
!===========  end subroutine matrix  ===========!

!==========  begin subroutine potent  ==========!
      subroutine potent(nx,ix,ifx,tfx,pq,sk)

      implicit real(8) (a-h,o-z)
      implicit integer(4) (i-n)

      dimension  ifx(10),tfx(10),pq(40),sk(40,40)

      do i=1,nx
        pq(i)=0.0d0
      end do

      do i=1,ix
        ii=ifx(i)
        do j=1,nx
          pq(j)=pq(j)-sk(j,ii)*tfx(i)
        end do
      end do

      do i=1,ix
        ii=ifx(i)
        do j=1,nx
```

Fig. 4.10 (continued)

```
         sk(ii,j)=0.0d0
         sk(j,ii)=0.0d0
         sk(ii,ii)=1.0d0
       end do
     end do

     call sweep(nx,sk,pq)

     do i=1,ix
       ii=ifx(i)
       pq(ii)=tfx(i)
     end do
     write(30,*)'*******  output results  ***************'
     write(30,*)
     write(30,*)'num.nodes   potential'
     write(40,*)'num.nodes   potential'

     do i=1,nx
       write(40,300) i,pq(i)
       write(30,300) i,pq(i)
     end do

     write(30,*)
 300 format(5x,i5,1f12.3)

     return
     end subroutine potent
!============  end subroutine potent  ===========!

!===========  begin subroutine velcty  ==========!
     subroutine velcty(mx,xy,ie,pq)

     implicit real(8) (a-h,o-z)
     implicit integer(4) (i-n)

     dimension  xy(2,40),ie(3,40),pq(40)
     dimension  b(3),c(3),p(3)

     write(40,*)'  num.ele  velocity-u  velocity-v'
     write(30,*)'  num.ele  velocity-u  velocity-v'

     do im=1,mx
       ia=ie(1,im)
       ib=ie(2,im)
       ic=ie(3,im)
       xa=xy(1,ia)
       xb=xy(1,ib)
       xc=xy(1,ic)
       ya=xy(2,ia)
       yb=xy(2,ib)
       yc=xy(2,ic)
```

Fig. 4.10 (continued)

```
          area=((yb-yc)*(xa-xc)-(xc-xb)*(yc-ya))*0.5d0
          area2=1.0d0/(area*2.0d0)
          b(1)=(yb-yc)*area2
          b(2)=(yc-ya)*area2
          b(3)=(ya-yb)*area2
          c(1)=(xc-xb)*area2
          c(2)=(xa-xc)*area2
          c(3)=(xb-xa)*area2
          do i=1,3
            ii=ie(i,im)
            p(i)=pq(ii)
          end do

          vx=0.0d0
          vy=0.0d0

          do i=1,3
            vx=vx-b(i)*p(i)
            vy=vy-c(i)*p(i)
          end do
          write(30,301) im,vx,vy
          write(40,301) im,vx,vy
        end do

  301   format(5x,i5,3f12.3)

        return
        end subroutine velcty
!============  end subroutine velcty  ===========!

!===========  begin subroutine sweep   ==========!
        subroutine sweep(nx,sk,pq)

        implicit real(8) (a-h,o-z)
        implicit integer(4) (i-n)

        dimension :: sk(40,40),pq(40)

        do i=1,nx
          p=1.0d0/sk(i,i)
          do j=i,nx
            sk(i,j)=sk(i,j)*p
          end do
          pq(i)    =pq(i)   *p
          do k=i+1,nx
            q=sk(k,i)
            do j=1,nx
              sk(k,j)=sk(k,j)-q*sk(i,j)
            end do
            pq(k)=pq(k)-pq(i)*q
          end do
        end do
```

Fig. 4.10 (continued)

```
      do i=nx-1,1,-1
        do j=i+1,nx
          pq(i)=pq(i)-pq(j)*sk(i,j)
        end do
      end do

      return
      end subroutine sweep
!============   end subroutine sweep   ===========!
```

Fig. 4.10 (continued)

Table 4.6 Main identifiers

Identifier	Explanation
ia	Node number to which edge a is connected
ib	Node number to which edge b is connected
ic	Node number to which edge c is connected
xa	x coordinate of edge a
xb	x coordinate of edge b
xc	x coordinate of edge c
ya	y coordinate of edge a
yb	y coordinate of edge b
yc	y coordinate of edge c
$area$	Area of triangular element
$b(i)$	b_a, b_b, b_c
$c(i)$	c_a, c_b, c_c
$sm(3,3)$	Local finite element matrix
$sk(nx,nx)$	Global finite element matrix
$ff(nx)$	Known terms of the global finite element equation and potential values computed
$p(3)$	Potential values of edges a, b, c
vx	Computed velocity u
vy	Computed velocity v

```
! *******  potential flow program  *******

! *******  input data  ******************

   num.ele       ia       ib       ic
         1        4        5        1
         2        2        1        5
         3        2        5        3
         4        6        3        5
         5        7        8        4
         6        5        4        8
         7        5        8        6
         8        9        6        8
         9       10       11        7
        10        8        7       11
        11        8       11        9
        12       12        9       11
        13       13       14       10
        14       11       10       14
        15       11       14       12
        16       15       12       14
! *******  output results  ***************

 num.nodes    potential
         1       10.000
         2       10.000
         3       10.000
         4        7.500
         5        7.500
         6        7.500
         7        5.000
         8        5.000
         9        5.000
        10        2.500
        11        2.500
        12        2.500
        13        0.000
        14        0.000
        15        0.000

   num.ele  velocity-u  velocity-v
         1       1.000       0.000
         2       1.000       0.000
         3       1.000       0.000
         4       1.000       0.000
         5       1.000       0.000
         6       1.000       0.000
         7       1.000       0.000
         8       1.000       0.000
         9       1.000       0.000
        10       1.000       0.000
        11       1.000       0.000
        12       1.000       0.000
        13       1.000       0.000
        14       1.000       0.000
        15       1.000       0.000
        16       1.000       0.000
```

Fig. 4.11 Output list

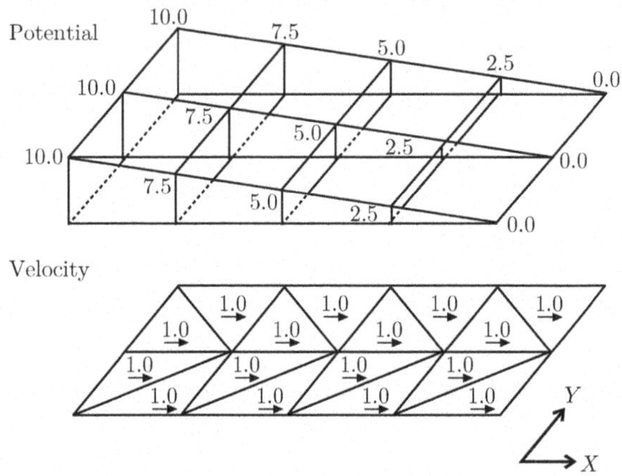

Fig. 4.12 Computed potential and velocity

4.14 Summary and Conclusion

This chapter has examined the finite element method of the potential flow. The procedures involve almost all of the basic concepts of the general finite element method. The idea developed in this chapter, in short, is the simple extension of the one-dimensional pipe element with two edges to the two-dimensional triangular element with three edges. Using the Galerkin method, the symmetric local finite element equation is derived. In addition, the coefficient matrices of the equilibrium and compatibility equations are mutually transpose. Therefore, the symmetric global finite element equation is obtained. The linear interpolation and weighting functions are used in order to simplify the integration operation. The systematic superposition procedure is also useful, which is the natural extension of those in spring and pipeline systems.

4.15 Exercises

1. Formulate the finite element equation using the quadrilateral element shown in Fig. 4.13 and the following interpolation function:

$$\phi = \Phi_a \phi_a + \Phi_b \phi_b + \Phi_c \phi_c + \Phi_d \phi_d$$

$$\Phi_a = \frac{1}{4}\left(1 + \frac{x}{l}\right)\left(1 + \frac{y}{h}\right)$$

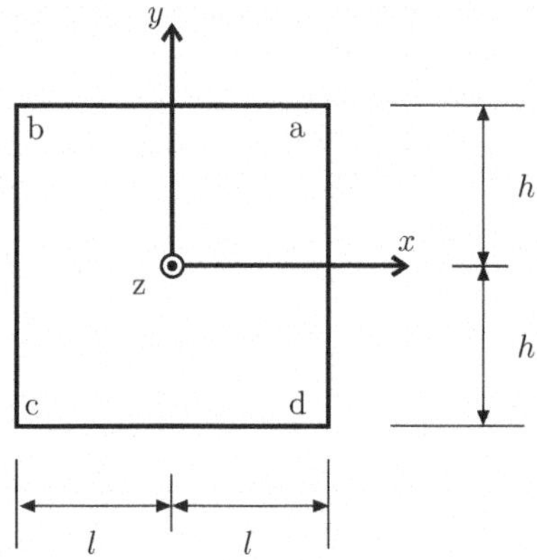

Fig. 4.13 Quadrilateral element

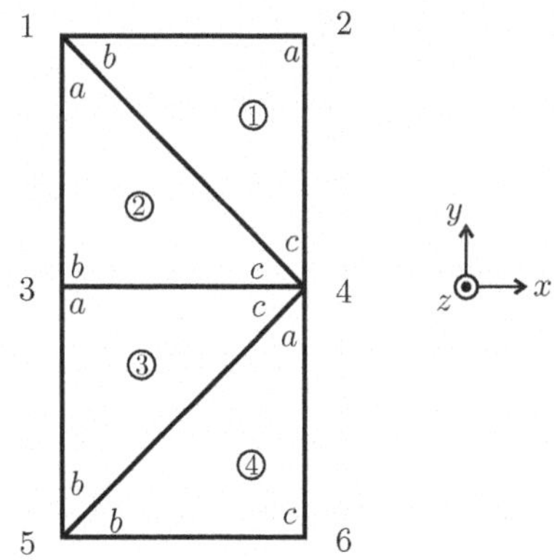

Fig. 4.14 Finite element mesh

$$\Phi_b = \frac{1}{4}\left(1 - \frac{x}{l}\right)\left(1 + \frac{y}{h}\right)$$

$$\Phi_c = \frac{1}{4}\left(1 - \frac{x}{l}\right)\left(1 - \frac{y}{h}\right)$$

$$\Phi_d = \frac{1}{4}\left(1 + \frac{x}{l}\right)\left(1 - \frac{y}{h}\right)$$

Caveat: in this problem, the integrand in the element is not constant.

2. Using the finite element mesh shown in Fig. 4.14, formulate the finite element equation. Examine the relationships of the local finite element matrices between elements ① and ②, between ② and ③, and between ① and ④.

3. Develop an algorithm of the automatic mesh generation for the mesh pattern shown in Fig. 4.14.
4. Write a computer algorithm based on the finite element matrix in Exercise 1, by modifying the program potential_flow in Sect. 4.13.

Chapter 5
Advection–Diffusion

5.1 Introduction

There are a plethora of important and practical phenomena in science and engineering concerning with advection and diffusion, such as thermal convection in a furnace, chemical substance diffusion in a solution, contaminant dispersion in the environmental waters, salt water diffusion in the sea, CO_2 dispersion in the atmosphere, etc. We must solve these problems not only in space but also in time, because they are time-dependent. The phenomena normally include two different and sometimes contradictory phenomena, i.e., advection and diffusion. The advection is a phenomenon in which a substance is conveyed by the flow, and the diffusion occurs when a substance spreads out by its own power over a large area and diluted. In general, those are not separable and occur at the same time, and are collectively called transport phenomena. In this chapter, to emphasize the fact that two phenomena are dealt with together, we use the terminology, advection–diffusion.

The computational methods to solve advection or diffusion are entirely different. Therefore, to solve the advection-dominated phenomena or diffusion-dominated ones, the solution methods are rather different. In this chapter, mainly advection-dominated phenomena are treated. This chapter starts with time-dependent processes. For the notation, indicial notation and summation convention are employed unless otherwise noted. In the case of time-dependent process analysis, important procedure for computation is stability. A considerable part of this session is devoted to describing stability. There are a plenty of stabilization schemes. Among those, an implicit scheme, an explicit scheme, a two-step explicit scheme, a SUPG scheme, a BTD scheme, a bubble function scheme, and a characteristic scheme are discussed in this section. The Courant–Fridrich–Levy (CFL) condition is one of the most widely used indicators of the stability. A number of examples are discussed. A simple computer program is attached. In this section, the advection–diffusion of substance is treated. By replacing substance concentration with other variables, the current analysis can easily be extended.

M. Kawahara, *Finite Element Methods in Incompressible, Adiabatic, and Compressible Flows*, Mathematics for Industry 19,
DOI 10.1007/978-4-431-55450-9_5

5.2 Advection–Diffusion of Substance

Denoting a concentration of substance by $\mathfrak{c}$ and diffusion coefficient by k, the conservation equation can be described as follows:

$$\dot{\mathfrak{c}} + u_i \mathfrak{c}_{,i} - k\mathfrak{c}_{,ii} = 0 \quad \text{in} \quad V. \tag{5.1}$$

Here and henceforth, the indicial notation and summation convention are used. Indices $i, j, \ldots$ are employed to express coordinates $x, y, \ldots$. Superposed $\dot{}$ means partial differentiation with respect to time and subscripted, followed by the coordinate is the partial differentiation with respect to coordinate. Subscripted indices $\alpha, \beta, \ldots$ are used to express edges of a finite element and number of nodes in the global finite element fields.

In Eq. (5.1), V means a considering volume of fluid and the boundary of V is denoted by S. The velocity of fluid is expressed by u_i and is assumed to be known. The boundary S is assumed to be divided into two parts, S_1 and S_2. The essential boundary condition is given on S_1:

$$\mathfrak{c} = \hat{\mathfrak{c}} \quad \text{on} \quad S_1 \tag{5.2}$$

where $\hat{\mathfrak{c}}$ is the specified value. The natural boundary condition in Eq. (5.3) is given on S_2:

$$\mathfrak{b} = k\mathfrak{c}_{,i} n_i = \hat{\mathfrak{b}} \quad \text{on} \quad S_2 \tag{5.3}$$

where $\mathfrak{b}$ is called flux, and n_i is the unit normal of the boundary. The specified value on S_2 is denoted by $\hat{\mathfrak{b}}$. The initial condition is as follows:

$$\mathfrak{c} = \mathfrak{c}^0 \quad \text{at} \quad t = t_0 \tag{5.4}$$

We assume that

$$\begin{aligned} S_1 \cup S_2 &= S \\ S_1 \cap S_2 &= \mathbf{0} \end{aligned} \tag{5.5}$$

where $\mathbf{0}$ is a null set.

Multiplying the weighting function $\mathfrak{c}^*$ to both sides of Eq. (5.1) and integrating over V, it is obtained that

$$\int_V \mathfrak{c}^*(\dot{\mathfrak{c}} + u_i \mathfrak{c}_{,i} - k\mathfrak{c}_{,ii})dV = 0 \tag{5.6}$$

Using the equality

$$(k\mathfrak{c}^*\mathfrak{c}_{,i})_{,i} = k\mathfrak{c}^*_{,i}\mathfrak{c}_{,i} + k\mathfrak{c}^*\mathfrak{c}_{,ii} \tag{5.7}$$

and integration by parts, Eq. (5.6) can be transformed into the following form:

$$\int_V (\mathfrak{c}^* \dot{\mathfrak{c}}) dV + \int_V (\mathfrak{c}^* u_i \mathfrak{c}_{,i}) dV + \int_V (k \mathfrak{c}^*_{,i} \mathfrak{c}_{,i}) dV = \int_S (\mathfrak{c}^* \mathfrak{b}) ds \tag{5.8}$$

The weighting function $\mathfrak{c}^*$ is an arbitrary function, and we assume

$$\mathfrak{c}^*_i = 0 \quad \text{on} \quad S_1 \tag{5.9}$$

Using Eqs. (5.5) and (5.9), the weighted residual equation over the whole flow field can be described in the following form:

$$\int_V (\mathfrak{c}^* \dot{\mathfrak{c}}) dV + \int_V (\mathfrak{c}^* u_i \mathfrak{c}_{,i}) dV + \int_V (k \mathfrak{c}^*_{,i} \mathfrak{c}_{,i}) dV = \int_{S_2} (\mathfrak{c}^* \hat{\mathfrak{b}}) dS \tag{5.10}$$

The local finite element equation can be formulated based on Eq. (5.8) considering V as an finite element and S as the boundary of the element. The global finite element equation can also be formulated based on Eq. (5.10) considering V as a whole flow field and S as the boundary of the field.

5.3 Finite Element Formulation

As we have previously stated in Chaps. 2, 3, and 4, we can formulate in a systematic manner the global finite element equation based on the local finite element equation. Therefore, in this section we concentrate on describing the local finite element equation of the substance advection–diffusion problem.

For the interpolation function, the following function is introduced:

$$\mathfrak{c} = \Phi_\alpha \mathfrak{c}_\alpha \tag{5.11}$$

where $\mathfrak{c}_\alpha$ represents the values at edge α of a finite element and Φ_α the shape function corresponding to the edge α of the finite element. When we use triangular elements, $\alpha = a$, b, and c, or 1, 2, and 3. Here and henceforth, indices α, β, γ, ... are used to express edges and nodes. If we would employ a higher order interpolation, the interpolation would be written in the same form as in Eq. (5.11).

In the case of the Galerkin method, the weighting function is

$$\mathfrak{c}^* = \Phi_\alpha \mathfrak{c}^*_\alpha \tag{5.12}$$

where $\mathfrak{c}^*_\alpha$ are arbitrary constants at edge α of the element. The velocity field is also interpolated in the following form:

$$u_i = \Phi_\alpha u_{\alpha i} \tag{5.13}$$

where $u_{\alpha i}$ are the velocities at edge α and in the i-direction.

Introducing Eqs. (5.11), (5.12), and (5.13) into Eq. (5.8) and arranging the terms, the following equation can be derived:

$$\begin{aligned}&\mathfrak{c}^*_\alpha \int_V (\Phi_\alpha \Phi_\beta) dV \dot{\mathfrak{c}}_\beta + \mathfrak{c}^*_\alpha \int_V (\Phi_\alpha \Phi_{\beta,i} \Phi_\gamma) dV\, u_{\gamma i} \mathfrak{c}_\beta \\ &+ \mathfrak{c}^*_\alpha k \int_V (\Phi_{\alpha,i} \Phi_{\beta,i}) dV\, \mathfrak{c}_\beta = \mathfrak{c}^*_\alpha \int_S (\Phi_\alpha \mathfrak{b}) dS \end{aligned} \tag{5.14}$$

Using the fact that $\mathfrak{c}^*_\alpha$ are arbitrary constants, the local finite element equation can be obtained as follows:

$$M_{\alpha\beta} \dot{\mathfrak{c}}_\beta + K_{\alpha\beta\gamma i} u_{\gamma i} \mathfrak{c}_\beta + S_{\alpha\beta} \mathfrak{c}_\beta = \Omega_\alpha \tag{5.15}$$

where

$$\begin{aligned} M_{\alpha\beta} &= \int_V (\Phi_\alpha \Phi_\beta) dV \\ K_{\alpha\beta\gamma i} &= \int_V (\Phi_\alpha \Phi_{\beta,i} \Phi_\gamma) dV \\ S_{\alpha\beta} &= k \int_V (\Phi_{\alpha,i} \Phi_{\beta,i}) dV \\ \Omega_\alpha &= \int_S (\Phi_\alpha \mathfrak{b}) dS \end{aligned} \tag{5.16}$$

In the case of the linear interpolation function based on the triangular element, such as Eq. (4.24), the coefficient matrices are as follows:

$$\begin{aligned} M_{\alpha\beta} &= \frac{\boldsymbol{\Delta}}{12} \begin{bmatrix} 2 & 1 & 1 \\ 1 & 2 & 1 \\ 1 & 1 & 2 \end{bmatrix} \\ K_{\alpha\beta\gamma 1} &= M_{\alpha\gamma} b_\beta \\ K_{\alpha\beta\gamma 2} &= M_{\alpha\gamma} c_\beta \\ S_{\alpha\beta} &= k(b_\alpha b_\beta + c_\alpha c_\beta) \boldsymbol{\Delta} \end{aligned} \tag{5.17}$$

where b_α and c_α are defined in Eq. (4.22) in Sect. 4.4 and $\boldsymbol{\Delta}$ is in Eq. (4.23).

On the boundary *a*–*b*, length of which is L, and if $\mathfrak{b}$ is a constant,

$$\Omega_\alpha = \int_{S_{a-b}} (\Phi_\alpha \mathfrak{b}) dS = \begin{bmatrix} \frac{L\mathfrak{b}}{2} \\ \frac{L\mathfrak{b}}{2} \\ 0 \end{bmatrix} \tag{5.18}$$

where $S_{a\text{–}b}$ indicates a part of boundary along a–b of a triangular element.

Equation (5.15) is formulated at each finite element. Using the same procedure stated previously in Sect. 4.7, the global finite element equation can also be expressed as in the similar form of Eq. (5.15). The global finite element equation can be written in the following form:

$$M_{\alpha\beta}\dot{C}_{\beta} + A_{\alpha\beta}C_{\beta} = \hat{\Omega}_{\alpha} \tag{5.19}$$

where indices α and β are $\alpha, \beta = 1, 2, \ldots$ (total number of nodes minus total number of nodes on which the concentration is specified). In Eq. (5.19),

$$A_{\alpha\beta} = K_{\alpha\beta\gamma i}U_{\gamma i} + S_{\alpha\beta} \tag{5.20}$$

postulating the velocity $U_{\gamma i}$ are the velocities at node γ and in the i-direction and known in the whole flow field, $M_{\alpha\beta}$ and $S_{\alpha\beta}$ are coefficient matrices on the global nodes in the whole flow field, and the known term $\hat{\Omega}_{\alpha}$ on the right side is formulated by superposing Ω_{α} on the boundary S. The last term is obtained from Eq. (5.5). Remember that the coefficient matrix $A_{\alpha\beta}$ is nonsymmetric, which means that the advection phenomenon is one way, i.e., the substance advected by the flow does not go back to the starting position. As we have stated in Chaps. 2 and 3, we can obtain the global finite element equation Eq. (5.19) from the local finite element equation Eq. (5.15). As explained in Sect. 4.7, the known term by the treatment of the boundary condition on S_1 is also included in $\hat{\Omega}_{\alpha}$. Thus, Eq. (5.19) can be considered as the final form of the global finite element equation. Because Eq. (5.19) is the time-dependent equation, the initial condition must be specified:

$$C_{\alpha} = C_{\alpha}^{0} \quad \text{at} \quad t = t_0 \quad \text{in } V \tag{5.21}$$

5.4 Integration in Time

In this section, Eq. (5.19) is solved starting from the initial condition Eq. (5.21). Total time [0, T] is divided into short time increment as shown in Fig. 5.1

We number time points consecutively as $0, 1, 2, \ldots, n, \ldots, \mathcal{N}$, where $\mathcal{N}$ is the total number of time points. All time increment is the same, Δt. Concentration at time point n is denoted by $C_{\alpha}{}^{n}$. In almost all cases, we can assume that the concentration C_{α}^{0}–$C_{\alpha}{}^{n}$ are all known. Namely, the concentration $C_{\alpha}{}^{n+1}$ at time point $n+1$ can be computed giving C_{α}^{0}–$C_{\alpha}{}^{n}$.

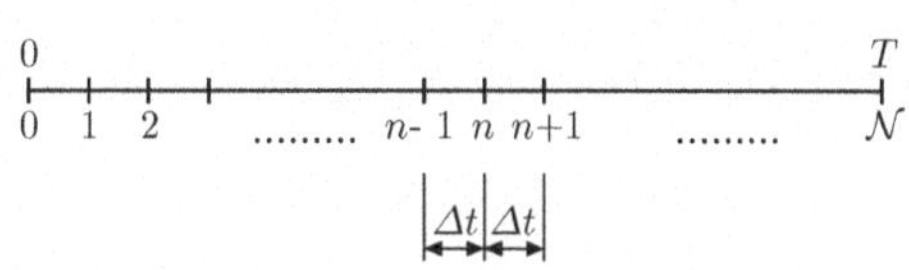

Fig. 5.1 Division in time

Most integration methods, which we will use the term "scheme," can be classified into two groups, an implicit scheme and an explicit scheme. The implicit scheme employs methods including procedures to solve simultaneous equation, whereas the explicit scheme excluding the procedures. Obviously, the implicit scheme is superior in computational stability, because a relatively long time increment can be used, but a large size of computational memory is required. On the other hand, the explicit scheme has a virtue too: fast computation for one-time cycle can be achieved and computational memories can be saved, but a relatively short time increment must be employed.

Consider the time integration of Eq. (5.19). The time derivative is approximated by the time difference as

$$M_{\alpha\beta}\dot{C}_\beta \fallingdotseq \frac{1}{\Delta t}M_{\alpha\beta}({C_\beta}^{n+1} - {C_\beta}^n) \tag{5.22}$$

The second term is approximated as

$$A_{\alpha\beta}C_\beta \fallingdotseq \theta A_{\alpha\beta}{C_\beta}^{n+1} + (1-\theta)A_{\alpha\beta}{C_\beta}^n \tag{5.23}$$

where θ is a parameter as

$$0 \leqq \theta \leqq 1 \tag{5.24}$$

Namely, the second term is approximated by the combination of the concentrations at time points n and $n+1$. As was stated, the concentration at time point n is already computed. For the known term,

$$\hat{\Omega}_\alpha \fallingdotseq {\hat{\Omega}_\alpha}^{n+1} \tag{5.25}$$

Introducing Eqs. (5.22), (5.23), and (5.25) into Eq. (5.19) and rearranging the terms, the following equation can be obtained:

$$(M_{\alpha\beta} + \theta\Delta t A_{\alpha\beta}){C_\beta}^{n+1} = M_{\alpha\beta}{C_\beta}^n - (1-\theta)\Delta t A_{\alpha\beta}{C_\beta}^n + \Delta t{\hat{\Omega}_\alpha}^{n+1} \tag{5.26}$$

In the case of $\theta = 1$, the scheme is called the full implicit scheme and can be written as

$$(M_{\alpha\beta} + \Delta t A_{\alpha\beta}){C_\beta}^{n+1} = M_{\alpha\beta}{C_\beta}^n + \Delta t{\hat{\Omega}_\alpha}^{n+1} \tag{5.27}$$

In the case of $\theta = \frac{1}{2}$, the scheme is the Crank–Nicolson scheme:

$$\left(M_{\alpha\beta} + \frac{1}{2}\Delta t A_{\alpha\beta}\right){C_\beta}^{n+1} = M_{\alpha\beta}{C_\beta}^n - \frac{1}{2}\Delta t A_{\alpha\beta}{C_\beta}^n + \Delta t{\hat{\Omega}_\alpha}^{n+1} \tag{5.28}$$

As we will discuss in further detail in Sect. 5.5.3, this scheme is neutral if the diffusion term is excluded. If $\theta = 0$, the scheme is the quasi-explicit scheme:

$$M_{\alpha\beta}C_\beta{}^{n+1} = M_{\alpha\beta}C_\beta{}^n - \Delta t A_{\alpha\beta}C_\beta{}^n + \Delta t \hat{\Omega}_\alpha{}^{n+1} \tag{5.29}$$

However, as we will show in Sect. 5.5.2, this scheme is unconditionally unstable, which means that we will never obtain the stable computation in the range of finite time increment.

To stabilize the scheme shown in Eq. (5.29), the "lumping technique" is employed. For coefficient matrix $M_{\alpha\beta}$, the non-diagonal elements are lumped together into the diagonal elements, and we obtain the lumped matrix as follows:

$$\overline{M}_{\alpha\beta} = \frac{\Delta}{3}\begin{bmatrix} 1 & & \\ & 1 & \\ & & 1 \end{bmatrix} \tag{5.30}$$

Using Eq. (5.30) on the left side of Eq. (5.29), the explicit scheme can be obtained:

$$\overline{M}_{\alpha\beta}C_\beta{}^{n+1} = M_{\alpha\beta}C_\beta{}^n - \Delta t A_{\alpha\beta}C_\beta{}^n + \Delta t \hat{\Omega}_\alpha{}^{n+1} \tag{5.31}$$

To save computational time, the two-step explicit scheme is effective, which is described as follows:

$$\overline{M}_{\alpha\beta}C_\beta{}^{n+1/2} = M_{\alpha\beta}C_\beta{}^n - \frac{\Delta t}{2} A_{\alpha\beta}C_\beta{}^n + \frac{\Delta t}{2} \hat{\Omega}_\alpha{}^{n+1} \tag{5.32}$$

$$\overline{M}_{\alpha\beta}C_\beta{}^{n+1} = M_{\alpha\beta}C_\beta{}^n - \Delta t A_{\alpha\beta}C_\beta{}^{n+1/2} + \Delta t \hat{\Omega}_\alpha{}^{n+1} \tag{5.33}$$

In this two-step model once the concentration at time point $n + 1/2$ is computed, then the concentration at time point $n + 1$ is obtained using the computed results. As we will show in Sect. 5.5.2, the longer time increment can be chosen than that in Eq. (5.31).

5.5 Stability Condition

5.5.1 CFL Condition

When computing time-dependent problems, how we get the stable computation is one of the main challenges, i.e., the longer the better. For this measure the Courant–Fridrich–Levy condition (the CFL condition) is one of the most convenient and widely used indicators. In this section, we consider the CFL condition using a simple one-dimensional linear equation. Let $u(x, t)$ be a function to be determined and u satisfy the following:

$$\frac{\partial u}{\partial t} + \frac{\partial u}{\partial x} = 0, \qquad 0 \leqq t \leqq T, \quad 0 \leqq x \leqq L \tag{5.34}$$

Fig. 5.2 One-dimensional finite element mesh and one-dimensional linear element

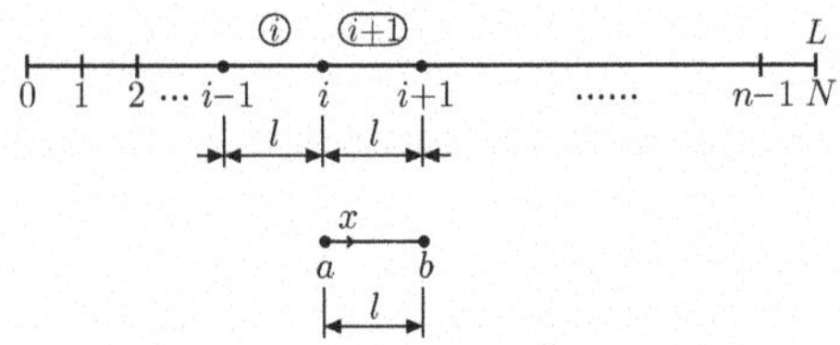

where T and L are the maxima of time t and coordinate x. Simple boundary and initial conditions are introduced as

$$u(0,t) = u(L,t) = 0 \tag{5.35}$$

$$u(x,0) = u^o(x) \tag{5.36}$$

where $u^o(x)$ is a function for the initial condition.

As in Fig. 5.2, total domain L is divided into small elements by node i ($i = 0, 1, \ldots, N$). One-dimensional element, which is the interval between node $i-1$ and node i, is denoted by ⓜ(ⓜ = ①, ②, . . . , Ⓝ).

The weighted residual equation can be obtained:

$$\int_o^l \left(u^* \frac{\partial u}{\partial t}\right) dx + \int_o^l \left(u^* \frac{\partial u}{\partial x}\right) dx = 0 \tag{5.37}$$

where u^* is the weighting function, and l is the length of the interval as shown in Fig. 5.2. Assume that the interpolation and weighting functions are as

$$u = \left(1 - \frac{x}{l}\right) u_a + \left(\frac{x}{l}\right) u_b \tag{5.38}$$

$$u^* = \left(1 - \frac{x}{l}\right) u_a^* + \left(\frac{x}{l}\right) u_b^* \tag{5.39}$$

where u_a and u_b are values of the function u at edges a and b, and u_a^* and u_b^* are values of the function u^* at edges a and b, the local finite element equation can be obtained:

$$\begin{bmatrix} \frac{l}{3} & \frac{l}{6} \\ \frac{l}{6} & \frac{l}{3} \end{bmatrix} \begin{bmatrix} \dot{u}_a \\ \dot{u}_b \end{bmatrix} + \begin{bmatrix} -\frac{1}{2} & \frac{1}{2} \\ -\frac{1}{2} & \frac{1}{2} \end{bmatrix} \begin{bmatrix} u_a \\ u_b \end{bmatrix} = \begin{bmatrix} 0 \\ 0 \end{bmatrix} \tag{5.40}$$

Using the finite element mesh shown in Fig. 5.2, the lower equation of Eq. (5.40) for element ⓘ is

$$\left(\frac{l}{6}\right) \dot{u}_{i-1} + \left(\frac{l}{3}\right) \dot{u}_i + \left(-\frac{1}{2}\right) u_{i-1} + \left(\frac{1}{2}\right) u_i = 0 \tag{5.41}$$

and the upper equation of Eq. (5.40) for element $\textcircled{\scriptsize i+1}$ is

$$\left(\frac{l}{3}\right)\dot{u}_i + \left(\frac{l}{6}\right)\dot{u}_{i+1} + \left(-\frac{1}{2}\right)u_i + \left(\frac{1}{2}\right)u_{i+1} = 0 \tag{5.42}$$

Adding Eqs. (5.41) and (5.42), the global finite element equation for node i can be obtained as in the following form:

$$\left(\frac{l}{6}\right)\dot{u}_{i-1} + \left(\frac{2l}{3}\right)\dot{u}_i + \left(\frac{l}{6}\right)\dot{u}_{i+1} + \left(-\frac{1}{2}\right)u_{i-1} + \left(\frac{1}{2}\right)u_{i+1} = 0 \tag{5.43}$$

Equation (5.43) corresponds to Eq. (5.19). Applying the integration schemes in time described in Sect. 5.3, Eq. (5.43) can be transformed into the following equation:

$$F_1\left({u_{i-1}}^{n+1}, {u_i}^{n+1}, {u_{i+1}}^{n+1}\right) = F_2\left({u_{i-1}}^{n}, {u_i}^{n}, {u_{i+1}}^{n}\right) \tag{5.44}$$

Generally, the function at time point $n+1$ can be obtained from the function at time point n. In Eq. (5.44), F_1 and F_2 are functions which depend on the schemes described in Sect. 5.4, and in some sense, artificial viscosity is included to stabilize the scheme.

Considering that Eq. (5.44) is a linear equation, there exists a solution in the following form:

$${u_i}^n = R^n e^{\hat{j}i\omega}, \quad \hat{j} = \sqrt{-1} \tag{5.45}$$

where ω is an arbitrary constant and expresses a frequency of a componential wave and R^n is an amplitude at time point n. Practical solution can be expressed by the summation of Eq. (5.45) with respect to variable ω.

Introducing Eq. (5.45) into Eq. (5.44) and rearranging the terms, the amplitude can be derived in the following form:

$$R^{n+1} = Q(\omega)R^n \tag{5.46}$$

If R^n includes an error ΔR^n at time point n, from Eq. (5.46), an error ΔR^{n+1} at time point $n+1$ satisfies

$$R^{n+1} + \Delta R^{n+1} = Q(\omega)(R^n + \Delta R^n) \tag{5.47}$$

thus

$$\Delta R^{n+1} = Q(\omega)\Delta R^n \tag{5.48}$$

Therefore, $Q(\omega)$ is an indicator expressing increase or non-increase of the errors. In this sense, $|Q(\omega)|$ is called as an amplification factor. According to the amplification factor, the computation is

$$\text{stable if} \quad |Q(\omega)| < 1, \tag{5.49}$$

$$\text{unstable if} \quad |Q(\omega)| > 1, \quad \text{and} \tag{5.50}$$

$$\text{neutral if} \quad |Q(\omega)| = 1 \tag{5.51}$$

This is because the error ΔR^{n+1} does not increase if $|Q(\omega)| < 1$ from Eq. (5.48). Being neutral is the state in which the computation is neither stable nor unstable. In this case, the artificial viscosity, which is introduced to stabilize the schemes, is minimum. Therefore, neutral scheme is preferable. However, in almost all cases, the neutral scheme turns out in fact to be unstable. For the practical computation, the scheme that is stable and, in addition, that has minimum artificial viscosity and artificial phase lag must be employed. However, the optimal scheme depends entirely on the problem to be solved.

5.5.2 Explicit Scheme

Applying the quasi-explicit scheme expressed in Eq. (5.29) to Eq. (5.43), Eq. (5.44) for node i turns out to be as follows:

$$\begin{aligned}&\left(\frac{1}{6}\right) u_{i-1}{}^{n+1} + \left(\frac{2}{3}\right) u_i{}^{n+1} + \left(\frac{1}{6}\right) u_{i+1}{}^{n+1} \\ &= \left(\frac{1}{6}\right) u_{i-1}{}^{n} + \left(\frac{2}{3}\right) u_i{}^{n} + \left(\frac{1}{6}\right) u_{i+1}{}^{n} - \mu \left\{ \left(-\frac{1}{2}\right) u_{i-1}{}^{n} + \left(\frac{1}{2}\right) u_{i+1}{}^{n} \right\}\end{aligned} \tag{5.52}$$

where

$$\mu = \frac{\Delta t}{l} \tag{5.53}$$

Introducing Eq. (5.45) into Eq. (5.52), we get

$$\left(\frac{1}{6}e^{-\hat{j}\omega} + \frac{2}{3} + \frac{1}{6}e^{\hat{j}\omega}\right) R^{n+1} = \left\{ \left(\frac{1}{6}e^{-\hat{j}\omega} + \frac{2}{3} + \frac{1}{6}e^{\hat{j}\omega}\right) - \frac{\mu}{2}(e^{-\hat{j}\omega} + e^{\hat{j}\omega}) \right\} R^{n} \tag{5.54}$$

From Eq. (5.54), the amplification factor is derived as

$$Q(\omega) = \frac{\left(\frac{2}{3} + \frac{1}{3}\cos\omega\right) - \hat{j}\mu \sin\omega}{\left(\frac{2}{3} + \frac{1}{3}\cos\omega\right)} \tag{5.55}$$

The absolute value of $Q(\omega)$ is

$$|Q(\omega)|^2 = 1 + \left(\frac{\mu \sin\ \omega}{\frac{2}{3} + \frac{1}{3} \cos\ \omega} \right)^2 \geqq 1 \tag{5.56}$$

Even if we put $\mu = 0$, we obtain $|Q(\omega)| = 1$. This type of scheme is referred to as "unconditionally unstable scheme." Because μ is dependent on Δt, it is impossible to obtain the computational results no matter how short the time increment Δt is.

Applying the explicit scheme expressed in Eq. (5.31) to Eq. (5.43), Eq. (5.44) for node i is as follows:

$$u_i^{n+1} = \left(\frac{1}{6}\right) u_{i-1}{}^n + \left(\frac{2}{3}\right) u_i{}^n + \left(\frac{1}{6}\right) u_{i+1}{}^n - \mu \left\{ \left(-\frac{1}{2}\right) u_{i-1}{}^n + \left(\frac{1}{2}\right) u_{i+1}{}^n \right\} \tag{5.57}$$

Equation (5.57) leads to

$$R^{n+1} = \left[\left(\frac{1}{6}\right) e^{-\hat{j}\omega} + \left(\frac{2}{3}\right) + \left(\frac{1}{6}\right) e^{\hat{j}\omega} - \mu \left\{ \left(-\frac{1}{2}\right) e^{-\hat{j}\omega} + \left(\frac{1}{2}\right) e^{\hat{j}\omega} \right\} \right] R^n \tag{5.58}$$

From Eq. (5.58), the amplification factor is

$$Q(\omega) = \frac{2}{3} + \frac{1}{3} \cos\ \omega - \hat{j} \mu \sin\ \omega \tag{5.59}$$

The absolute value of $Q(\omega)$ should be less than 1, as shown below:

$$\begin{aligned} |Q(\omega)|^2 &= \left(\frac{2}{3} + \frac{1}{3} \cos\ \omega \right)^2 + \mu^2 \sin^2 \omega \\ &= \left(\frac{1}{9} - \mu^2 \right) \cos^2 \omega + \frac{4}{9} \cos\ \omega + \left(\frac{4}{9} + \mu^2 \right) \leqq 1 \end{aligned} \tag{5.60}$$

for arbitrary wave number ω. The following stability condition is derived:

$$\mu \leqq \frac{1}{3} \tag{5.61}$$

If we use the time increment which satisfies Eq. (5.61), the stable computation can be obtained by the explicit scheme.

Applying the two-step explicit scheme expressed in Eqs. (5.32) and (5.33) to Eqs. (5.43) and (5.44) for node i becomes as follows:

$$
\begin{aligned}
{u_i}^{n+1/2} &= \left(\frac{1}{6}\right) {u_{i-1}}^n + \left(\frac{2}{3}\right) {u_i}^n + \left(\frac{1}{6}\right) {u_{i+1}}^n \\
&\quad - \frac{\mu}{2}\left\{\left(-\frac{1}{2}\right) {u_{i-1}}^n + \left(\frac{1}{2}\right) {u_{i+1}}^n\right\}
\end{aligned}
\tag{5.62}
$$

$$
\begin{aligned}
{u_i}^{n+1} &= \left(\frac{1}{6}\right) {u_{i-1}}^n + \left(\frac{2}{3}\right) {u_i}^n + \left(\frac{1}{6}\right) {u_{i+1}}^n \\
&\quad - \mu\left\{\left(-\frac{1}{2}\right) {u_{i-1}}^{n+1/2} + \left(\frac{1}{2}\right) {u_{i+1}}^{n+1/2}\right\}
\end{aligned}
\tag{5.63}
$$

Equations (5.62) and (5.63) leads to

$$
\begin{aligned}
R^{n+1/2} &= \left[\left\{\left(\frac{1}{6}\right) e^{-\hat{j}\omega} + \left(\frac{2}{3}\right) + \left(\frac{1}{6}\right) e^{\hat{j}\omega}\right\}\right. \\
&\quad \left. -\frac{\mu}{2}\left\{\left(-\frac{1}{2}\right) e^{-\hat{j}\omega} + \left(\frac{1}{2}\right) e^{\hat{j}\omega}\right\}\right] R^n
\end{aligned}
\tag{5.64}
$$

$$
\begin{aligned}
R^{n+1} &= \left[\left(\frac{1}{6}\right) e^{-\hat{j}\omega} + \left(\frac{2}{3}\right) + \left(\frac{1}{6}\right) e^{\hat{j}\omega}\right] R^n \\
&\quad - \mu\left[\left(-\frac{1}{2}\right) e^{-\hat{j}\omega} + \left(\frac{1}{2}\right) e^{\hat{j}\omega}\right] R^{n+1/2}
\end{aligned}
\tag{5.65}
$$

Equations (5.64) and (5.65) can be rewritten as

$$
\begin{bmatrix} R^{n+1/2} \\ R^{n+1} \end{bmatrix} = \begin{bmatrix} \dfrac{2}{3} + \dfrac{1}{3}\cos\,\omega - \hat{j}\dfrac{\mu}{2}\sin\,\omega & 0 \\ \dfrac{2}{3} + \dfrac{1}{3}\cos\,\omega & -\hat{j}\mu\sin\,\theta \end{bmatrix} \begin{bmatrix} R^n \\ R^{n+1/2} \end{bmatrix}
\tag{5.66}
$$

The absolute values of the amplification factor $|Q(\omega)|$ are derived as the eigenvalue of Eq. (5.66). In fact, eigenvalue λ can be computed as

$$
\begin{vmatrix} (a - \hat{j}\mu b/2) - \lambda & 0 \\ a & (-\hat{j}\mu b) - \lambda \end{vmatrix} = 0
\tag{5.67}
$$

where

$$
a = \frac{2}{3} + \frac{1}{3}\cos\,\omega
\tag{5.68}
$$

$$
b = \sin\,\omega
\tag{5.69}
$$

The eigenvalue satisfies

$$\left\{\lambda - \left(a - \hat{j}\mu\frac{b}{2}\right)\right\}\left\{\lambda - (-\hat{j}\mu b)\right\} = 0 \tag{5.70}$$

Thus, the stability condition can be derived by the fact that inequalities

$$\begin{aligned} |\lambda_1|^2 &= \left(\frac{2}{3} + \frac{1}{3}\cos\omega\right)^2 + \frac{\mu^2}{4}\sin^2\omega \leqq 1 \\ |\lambda_2|^2 &= \mu^2 \sin^2\omega \leqq 1 \end{aligned} \tag{5.71}$$

should hold for any arbitrary ω. The stability condition is

$$|\mu| \leqq \frac{2}{\sqrt{3}} \tag{5.72}$$

As is seen in Eq. (5.56), it is impossible to obtain the stable computation by the quasi-explicit scheme in Eq. (5.29). As also revealed in Eqs. (5.61) and (5.72), the explicit scheme can be stabilized using the lumping coefficient technique. This is one of the simplest techniques, although there are many other stabilization techniques, some of which we will discuss later.

Comparing Eqs. (5.72) and (5.61), the time increment for the two-step explicit scheme is $2\sqrt{3} \fallingdotseq 3.4$ times longer than that for the one-step explicit scheme. This is a drastical improvement. However, effects of the artificial viscosity to stabilize the computation cannot be disregarded.

5.5.3 *Implicit Scheme*

In Sect. 5.5.2, we have shown that there are some schemes, which could not be adopted for the computation. In this section, the stability of the implicit scheme is discussed. Applying the implicit scheme expressed in Eq. (5.26) to Eq. (5.43), Eq. (5.44) for node i becomes as

$$\begin{aligned} &\left(\frac{1}{6}\right) {u_{i-1}}^{n+1} + \left(\frac{2}{3}\right) {u_i}^{n+1} + \left(\frac{1}{6}\right) {u_{i+1}}^{n+1} \\ &\quad + \theta\mu\left\{\left(-\frac{1}{2}\right) {u_{i-1}}^{n+1} + \left(\frac{1}{2}\right) {u_{i+1}}^{n+1}\right\} \\ &= \left(\frac{1}{6}\right) {u_{i-1}}^{n} + \left(\frac{2}{3}\right) {u_i}^{n} + \left(\frac{1}{6}\right) {u_{i+1}}^{n} \\ &\quad - (1-\theta)\mu\left\{\left(-\frac{1}{2}\right) {u_{i-1}}^{n} + \left(\frac{1}{2}\right) {u_{i+1}}^{n}\right\} \end{aligned} \tag{5.73}$$

Substituting Eq. (5.45) into Eq. (5.73), we get

$$\left[\left(\frac{1}{6}\right)e^{-\hat{j}\omega}+\left(\frac{2}{3}\right)+\left(\frac{1}{6}\right)e^{\hat{j}\omega}+\theta\mu\left\{\left(-\frac{1}{2}\right)e^{-\hat{j}\omega}+\left(\frac{1}{2}\right)e^{\hat{j}\omega}\right\}\right]R^{n+1}$$
$$=\left[\left(\frac{1}{6}\right)e^{-\hat{j}\omega}+\left(\frac{2}{3}\right)+\left(\frac{1}{6}\right)e^{\hat{j}\omega}-(1-\theta)\mu\left\{\left(-\frac{1}{2}\right)e^{-\hat{j}\omega}+\left(\frac{1}{2}\right)e^{\hat{j}\omega}\right\}\right]R^{n} \tag{5.74}$$

From Eq. (5.74), the amplification factor is derived as

$$Q(\omega)=\frac{a-\hat{j}(1-\theta)\mu b}{a+\hat{j}\theta\mu b} \tag{5.75}$$

where a and b are in Eqs. (5.68) and (5.69). The absolute value of $Q(\omega)$ is

$$|Q(\omega)|^2=\left(\frac{a^2-(1-\theta)\theta\mu^2b^2}{a^2+\theta^2\mu^2b^2}\right)^2+\left(\frac{(1-2\theta)\mu ab}{a^2+\theta^2\mu^2b^2}\right)^2 \tag{5.76}$$

By precise calculation, it is shown that

$$|Q(\omega)|^2 \leqq 1 \quad \text{if} \quad \theta \geqq \frac{1}{2} \tag{5.77}$$

Therefore, as long as $\theta \geqq \frac{1}{2}$, the computation is stable. In principle, we can obtain numerical results using any long time increment. This type of scheme is referred to as "unconditionally stable scheme." However, in practice, the time increment is limited by some other factors.

In general, stability of the implicit scheme is superior to that of the explicit scheme. Although large computational memories are required, the implicit scheme is often used in practical computation. However, we should bear in mind that considerable quantities of artificial viscosity are included to stabilize the computation. It is often the case that much smaller values of computational results than those to be attained may be computed. To reduce the artificial viscosity, the scheme as close as possible to the neutral scheme could be used. In this sense, the Crank–Nicolson scheme ($\theta = 0.5$) is preferable. However, this scheme sometimes results in the numerical instability. Thus, for instance, $\theta = 0.55$ or any other number close to 0.5 is used.

The time increment is longer at the implicit scheme than at the explicit scheme. Thus, it is sometimes possible to reduce computational time. On the other hand, in the case of computation of suddenly changing phenomena in a short time interval, there is a possibility that shorter time increment should be chosen because the suddenly changing phenomena are sometimes missed and dulled results are computed. We should pay close attention to the computation by the implicit scheme of the phenomenon that changes rapidly with time.

5.5.4 Stability with Viscosity

In the former sections, we have stated the problems of stability without considering physical viscosity, for the sake of simplicity, i.e., the governing equation is Eq. (5.34). In this section, we will treat the equation including both advection and diffusion:

$$\frac{\partial u}{\partial t} + \frac{\partial u}{\partial x} - \nu \frac{\partial^2 u}{\partial x^2} = 0 \qquad 0 < t \leqq T, \quad 0 \leqq x \leqq L \tag{5.78}$$

$$u(0, t) = u(L, t) = 0, \quad \frac{\partial}{\partial x} u(0, t) = \frac{\partial}{\partial x} u(L, t) = 0 \tag{5.79}$$

$$u(x, 0) = u^0(x) \tag{5.80}$$

where ν is the normalized viscosity. Formulating the weighted residual equation from Eq. (5.78), and employing the interpolation and weighting functions as Eqs. (5.38) and (5.39), the local finite element equation can be derived as

$$\begin{bmatrix} \frac{l}{3} & \frac{l}{6} \\ \frac{l}{6} & \frac{l}{3} \end{bmatrix} \begin{bmatrix} \dot{u}_a \\ \dot{u}_b \end{bmatrix} + \begin{bmatrix} -\frac{1}{2} & \frac{1}{2} \\ -\frac{1}{2} & \frac{1}{2} \end{bmatrix} \begin{bmatrix} u_a \\ u_b \end{bmatrix} + \begin{bmatrix} \frac{\nu}{l} & -\frac{\nu}{l} \\ -\frac{\nu}{l} & \frac{\nu}{l} \end{bmatrix} \begin{bmatrix} u_a \\ u_b \end{bmatrix} = \begin{bmatrix} 0 \\ 0 \end{bmatrix} \tag{5.81}$$

The global finite element equation for node i can be described based on the mesh shown in Fig. 5.2 as follows:

$$\begin{aligned} &\left(\frac{l}{6}\right) \dot{u}_{i-1} + \left(\frac{2l}{3}\right) \dot{u}_i + \left(\frac{l}{6}\right) \dot{u}_{i+1} + \left(-\frac{1}{2}\right) u_{i-1} + \left(\frac{1}{2}\right) u_{i+1} \\ &+ \left(-\frac{\nu}{l}\right) u_{i-1} + \left(\frac{2\nu}{l}\right) u_i + \left(-\frac{\nu}{l}\right) u_{i+1} = 0 \end{aligned} \tag{5.82}$$

Introducing the lumping coefficient technique, the explicit scheme can be denoted as

$$u_i^{n+1} = u_i^n - \mu \left\{ \left(-\frac{1}{2}\right) u_{i-1}{}^n + \left(\frac{1}{2}\right) u_{i+1}{}^n \right\} - \mu \left(\frac{\nu}{l}\right) (-u_{i-1}{}^n + 2u_i{}^n - u_{i+1}{}^n) \tag{5.83}$$

Assuming the solution as Eq. (5.45) and substituting it into Eq. (5.83), we obtain

$$R^{n+1} = R^n - \mu \left\{ \left(-\frac{1}{2}\right) e^{-\hat{j}\omega} + \left(\frac{1}{2}\right) e^{\hat{j}\omega} \right\} R^n - \mu \left(\frac{\nu}{l}\right) \left(-e^{-\hat{j}\omega} + 2 - e^{\hat{j}\omega}\right) R^n \tag{5.84}$$

The absolute value of the amplification factor can be expressed as in the following form:

$$|Q(\omega)|^2 = \left\{1 + 2\mu\left(\frac{\nu}{l}\right)(\cos\,\omega - 1)\right\}^2 + \mu^2 \sin^2 \omega \tag{5.85}$$

In case that $\nu = 0$, Eq. (5.85) becomes

$$|Q(\omega)|^2 = 1 + \mu^2 \sin^2 \omega \geqq 1 \tag{5.86}$$

which is unconditionally unstable, i.e., if the scheme shown in Eq. (5.83) has no viscosity, we are unable to compute it using the scheme. If we can formulate the condition for μ, which satisfies $|Q(\omega)|^2 \leqq 1$, we obtain the stability condition. In principle, we must consider any arbitrary ω. However, to avoid complex calculations, we, for the moment, consider a specific ω.

If $\omega = 0$,

$$|Q(\omega)|^2 = 1 \tag{5.87}$$

The wave whose wave number $\omega = 0$ is the flat wave, i.e., it is hard to compute the flat wave by the numerical analysis in general. In case of $\omega = \dfrac{\pi}{2}$, from the condition

$$|Q(\omega)|^2 = \left\{1 - 2\mu\left(\frac{\nu}{l}\right)\right\}^2 + \mu^2 \leqq 1 \tag{5.88}$$

we get

$$\mu \leqq \frac{\left(\frac{\nu}{l}\right)}{\left(\frac{\nu}{l}\right)^2 + 1} \tag{5.89}$$

and in case of $\omega = \pi$, from the condition

$$|Q(\omega)|^2 = \left\{1 - 4\mu\left(\frac{\nu}{l}\right)\right\}^2 \leqq 1 \tag{5.90}$$

we also get

$$\mu \leqq \frac{1}{2}\frac{l}{\nu} \tag{5.91}$$

Equations (5.89) and (5.91) are necessary conditions, but not sufficient conditions. We must use the time increment shorter than the limits in Eqs. (5.89) and (5.91). However, we do not know how short the value of the increment needs to be. On the contrary, the explicit scheme, which is unconditionally unstable, becomes conditionally stable with the viscosity term. Therefore, the limits shown in Eqs. (5.89) and (5.91) are certain kinds of indicators for the selection of time increment. In general, the viscosity term has a tendency to stabilize the computation. However, because it may be the case that the viscosity term destroys the stability condition, we must pay careful attention to the selection of the time increment in the practical computation.

5.5.5 Stability with Second-Order Interpolation Function

It is one of the most interesting research questions what the stability with the second-order interpolation function is. We will investigate the stability of the one-dimensional scheme using the second-order interpolation based on the mesh in Fig. 5.3.

For the basic equation, Eqs. (5.34)–(5.36) are used. The interpolation function is

$$u = \left(1 - \frac{3}{l}x + \frac{2}{l^2}x^2\right) u_a + \left(-\frac{x}{l} + \frac{2}{l^2}x^2\right) u_b + \left(\frac{4}{l}x - \frac{4}{l^2}x^2\right) u_c \tag{5.92}$$

where edges a, b, and c are shown in Fig. 5.3, i.e., edges a and b are connected to the global nodes and edge c is the middle node. For the weighting function, the same type of Eq. (5.92) is utilized. The local finite element equation is derived as follows:

$$\frac{l}{60}\begin{bmatrix} 8 & -2 & 4 \\ -2 & 8 & 4 \\ 4 & 4 & 32 \end{bmatrix}\begin{bmatrix} \dot{u}_a \\ \dot{u}_b \\ \dot{u}_c \end{bmatrix} + \frac{1}{12}\begin{bmatrix} -6 & -2 & 8 \\ 2 & 6 & -8 \\ -8 & 8 & 0 \end{bmatrix}\begin{bmatrix} u_a \\ u_b \\ u_c \end{bmatrix} = \begin{bmatrix} 0 \\ 0 \\ 0 \end{bmatrix} \tag{5.93}$$

Superimposing over the entire nodes as shown in Fig. 5.3, the global finite element equation become for node i as

$$\begin{aligned} &\left(-\frac{1}{30}\right)\dot{u}_{i-2} + \left(\frac{1}{15}\right)\dot{u}_{i-1} + \left(\frac{4}{15}\right)\dot{u}_i + \left(\frac{1}{15}\right)\dot{u}_{i+1} + \left(-\frac{1}{30}\right)\dot{u}_{i+2} \\ &- 2\mu\left\{\left(\frac{1}{6}\right)u_{i-2} + \left(-\frac{2}{3}\right)u_{i-1} + \left(\frac{2}{3}\right)u_{i+1} + \left(-\frac{1}{6}\right)u_{i+2}\right\} = 0 \end{aligned} \tag{5.94}$$

and for middle node i as

$$\left(\frac{4}{60}\right)\dot{u}_{i-1} + \left(\frac{32}{60}\right)\dot{u}_i + \left(\frac{4}{60}\right)\dot{u}_{i+1} - 2\mu\left\{\left(-\frac{8}{12}\right)u_{i-1} + \left(\frac{8}{12}\right)u_{i+1}\right\} = 0 \tag{5.95}$$

i.e., the global finite element equations are dependent on nodes and on middle nodes. Making use of the explicit scheme for the integration in time, the following equations can be derived in for node i

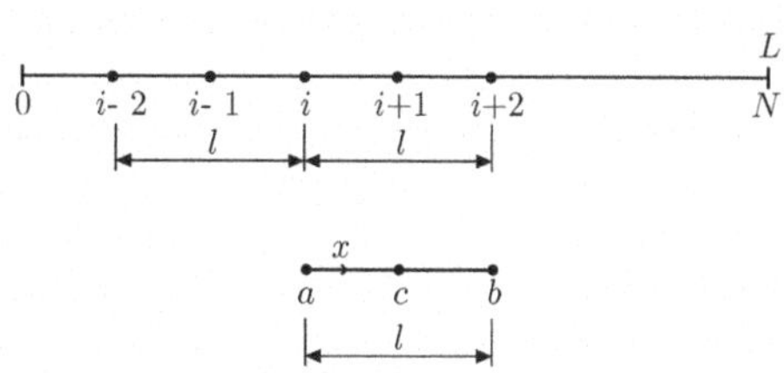

Fig. 5.3 One-dimensional finite element mesh and one-dimensional quadratic element

$$
\begin{aligned}
u_i^{n+1} &= \left\{\left(-\frac{1}{10}\right)u_{i-2}{}^n + \left(\frac{1}{5}\right)u_{i-1}{}^n + \left(\frac{4}{5}\right)u_i{}^n + \left(\frac{1}{5}\right)u_{i+1}{}^n + \left(-\frac{1}{10}\right)u_{i+2}{}^n\right\} \\
&\quad - 2\mu\left\{\left(\frac{1}{2}\right)u_{i-2}{}^n + (-2)u_{i-1}{}^n + (2)u_{i+1}{}^n + \left(-\frac{1}{2}\right)u_{i+2}{}^n\right\}
\end{aligned} \tag{5.96}
$$

and for middle node i

$$
u_i^{n+1} = \left\{\left(\frac{1}{10}\right)u_{i-1}{}^n + \left(\frac{4}{5}\right)u_i{}^n + \left(\frac{1}{10}\right)u_{i+1}{}^n\right\} - 2\mu\left(u_{i+1}{}^n - u_{i-1}{}^n\right) \tag{5.97}
$$

The solution Eq. (5.45) leads to amplification factors, for node i

$$
R^{n+1} = \left[\left\{\left(\frac{4}{5}\right) - \left(\frac{1}{5}\right)\cos 2\omega + \left(\frac{2}{5}\right)\cos \omega\right\} - 2\mu\hat{j}(4\sin\omega - \sin 2\omega)\right]R^n \tag{5.98}
$$

and for middle node i

$$
R^{n+1} = \left[\left(\frac{4}{5} + \frac{1}{5}\cos\omega\right) - 4\mu\hat{j}\sin\omega\right]R^n \tag{5.99}
$$

The absolute values of the amplification factors are for node i

$$
|Q(\omega)|^2 = \left\{\left(\frac{4}{5}\right) - \left(\frac{1}{5}\right)\cos 2\omega + \left(\frac{2}{5}\right)\cos\omega\right\}^2 + \mu^2(8\sin\omega - 2\sin 2\omega)^2 \leqq 1 \tag{5.100}
$$

and for middle node i

$$
|Q(\omega)|^2 = \left(\frac{4}{5} + \frac{1}{5}\cos\omega\right)^2 + \mu^2(4\sin\omega)^2 \leqq 1 \tag{5.101}
$$

If we calculate the condition for μ from Eqs. (5.100) and (5.101), we get the stability condition. However, the calculation is very complex, and useful conditions are not yet easily obtained. If we chose a specified μ, i.e., the time increment, before computation and replace it in Eqs. (5.100) and (5.101), then we can verify the availability of μ changing ω for a certain range. If we put $\omega = \frac{\pi}{2}$, then we get $\mu = 0$, which means unconditionally unstable. Therefore, it is safe to say that the stability condition at least includes the unconditionally unstable part. In general, the computational schemes based on the higher order interpolation function tend to destabilize the computation. The stability sometimes sacrifices to the accuracy.

5.6 SUPG Scheme

To stabilize the computational integration in time, the SUPG scheme is one of the most widely used schemes. In this section, we will investigate the SUPG scheme. SUPG stands for Streamline Upstream Petrov Galerkin. The Petrov–Galerkin method is the approximation method using different types of the interpolation and weighting functions. Because the upstream effect has a significant impact on the stability, an upstream effect in the streamline direction is considered. We will show the SUPG formulation based on the governing equation:

$$\dot{\mathfrak{c}} + u_i \mathfrak{c}_{,i} = 0 \qquad \text{in} \quad V \tag{5.102}$$

i.e., the viscosity term is subtracted from Eq. (5.1). The initial condition is

$$\mathfrak{c} = \mathfrak{c}^o \qquad \text{at} \quad t = t_0 \tag{5.103}$$

Multiplying weighting function $\tilde{\mathfrak{c}}$ to both sides of Eq. (5.102), the weighted residual equation is obtained:

$$\int_V (\tilde{\mathfrak{c}}\dot{\mathfrak{c}})dV + \int_V (\tilde{\mathfrak{c}} u_i \mathfrak{c}_{,i})dV = 0 \tag{5.104}$$

For the weighting function, the following type of function is used:

$$\tilde{\mathfrak{c}} = \mathfrak{c}^* + \tau_s u_i \mathfrak{c}^*_{,i} \tag{5.105}$$

where $\mathfrak{c}^*$ is the Galerkin-type normal weighting function, i.e., the same type of the interpolation function, and τ_s is a small parameter which adjusts the artificial viscosity. The coefficient τ_s is referred to as the stabilization parameter. Replacing $\tilde{\mathfrak{c}}$ in Eq. (5.104) by Eq. (5.105), we get

$$\int_V (\mathfrak{c}^*\dot{\mathfrak{c}})dV + \int_V (\mathfrak{c}^* u_i \mathfrak{c}_{,i})dV + \tau_s \int_V (\mathfrak{c}^*_{,i} u_i \dot{\mathfrak{c}})dV + \tau_s \int_V (\mathfrak{c}^*_{,i} u_i u_j \mathfrak{c}_{,j})dV = 0 \tag{5.106}$$

In Eq. (5.106), the first two terms are those of the normal Galerkin formulation, the effect of the third term disappears shortly because $\dot{c}$ is included, and the last term expresses the artificial viscosity. To distinguish that the last two terms are the SUPG terms, we can rewrite Eq. (5.106) as in the following form:

$$\int_V (\mathfrak{c}^*\dot{\mathfrak{c}})dV + \int_V (\mathfrak{c}^* u_i \mathfrak{c}_{,i})dV + \mathscr{H} = 0 \tag{5.107}$$

where

$$\mathscr{H} = \tau_s \int_V (\mathfrak{c}^*_{,i} u_i \dot{\mathfrak{c}})dV + \tau_s \int_V (\mathfrak{c}^*_{,i} u_i u_j \mathfrak{c}_{,j})dV \tag{5.108}$$

Equations (5.107) and (5.108) are the weighted residual equation for the SUPG method.

Substituting Eqs. (5.11), (5.12), and (5.13) into Eqs. (5.107) and (5.108), and arranging the terms, the following equations can be derived:

$$\mathrm{c}_\alpha^* \int_V (\Phi_\alpha \Phi_\beta) dV \dot{\mathrm{c}}_\beta + \mathrm{c}_\alpha^* \int_V (\Phi_\alpha \Phi_\gamma \Phi_{\beta,i}) dV u_{\gamma i} \mathrm{c}_\beta + \mathrm{c}_\alpha^* \mathscr{H}_\alpha = 0 \tag{5.109}$$

where

$$\mathscr{H}_\alpha = \tau_s \int_V (\Phi_{\alpha,i} \Phi_\beta \Phi_\gamma) dV u_{\beta i} \dot{\mathrm{c}}_\gamma + \tau_s \int_V (\Phi_{\alpha,i} \Phi_\beta \Phi_\gamma \Phi_{\delta,j}) dV u_{\beta i} u_{\gamma j} \mathrm{c}_\delta \tag{5.110}$$

Based on the fact that c_α^* are arbitrary constants, the local finite element equation can be described as follows:

$$M_{\alpha\beta} \dot{\mathrm{c}}_\beta + K_{\alpha\beta\gamma i} u_{\gamma i} \mathrm{c}_\beta + \mathscr{H}_\alpha = 0 \tag{5.111}$$

$$\mathscr{H}_\alpha = \tau_s K_{\beta\alpha\gamma i} u_{\beta i} \dot{\mathrm{c}}_\gamma + \tau_s L_{\alpha i \beta\gamma\delta j} u_{\beta i} u_{\gamma j} \mathrm{c}_\delta \tag{5.112}$$

where

$$K_{\beta\alpha\gamma i} = \int_V (\Phi_{\alpha,i} \Phi_\beta \Phi_\gamma) dV$$

$$L_{\alpha i \beta\gamma\delta j} = \int_V (\Phi_{\alpha,i} \Phi_\beta \Phi_\gamma \Phi_{\delta,j}) dV$$

Equation (5.112) is normally referred to as the SUPG term. The first term in SUPG term, i.e., Eq. (5.112), decays quickly and does not have significant effects. The second term in Eq. (5.112) is the artificial viscosity, which can stabilize the computation.

In the case of the linear interpolation function based on the triangular element, the coefficient matrices in Eq. (5.111) are the same as those in Eq. (5.15) except the SUPG term. Comparing $K_{\alpha\beta\gamma i}$ and $K_{\beta\alpha\gamma i}$, the elements of the coefficient are the same, but indices are exchanged. The second term in Eq. (5.112) can easily be integrated as follows:

$$L_{\alpha 1 \beta\gamma\delta 1} = b_\alpha M_{\beta\gamma} b_\delta$$

$$L_{\alpha 1 \beta\gamma\delta 2} = b_\alpha M_{\beta\gamma} c_\delta$$

$$L_{\alpha 2 \beta\gamma\delta 1} = c_\alpha M_{\beta\gamma} b_\delta$$

$$L_{\alpha 2 \beta\gamma\delta 2} = c_\alpha M_{\beta\gamma} c_\delta$$

where b_α and c_α are in Eq. (4.22) in Sect. 4.4 with $(\alpha, \beta, \gamma, \delta) = (a, b, c)$, and $M_{\alpha\beta}$ are in Eq. (5.17).

Rearranging the coefficients of Eqs. (5.111) and (5.112), we get

$$M_{\alpha\beta}\dot{C}_{\beta} + A_{\alpha\beta}C_{\beta} + \mathscr{H}_{\alpha} = 0 \tag{5.113}$$

$$\mathscr{H}_{\alpha} = \tau_s D_{\alpha\beta}\dot{C}_{\beta} + \tau_s E_{\alpha\beta}C_{\beta} \tag{5.114}$$

where

$$\begin{aligned} A_{\alpha\beta} &= K_{\alpha\beta\gamma i}u_{\gamma i} \\ D_{\alpha\beta} &= K_{\gamma\alpha\beta i}u_{\gamma i} \\ E_{\alpha\beta} &= L_{\alpha i\delta\gamma\beta j}u_{\delta i}u_{\gamma j} \end{aligned}$$

Equations (5.111) and (5.112) are local finite element equations. As stated in the previous chapters, we also use Eqs. (5.113) and (5.114) as the global finite element equation.

For the numerical integration in time, the explicit scheme can be adapted. Applying Eq. (5.22) to Eq. (5.113) and the lumping technique, the numerical integration formula can be obtained as

$$\overline{M}_{\alpha\beta}C_{\beta}{}^{n+1} = \overline{M}_{\alpha\beta}C_{\beta}{}^{n} - \Delta t A_{\alpha\beta}C_{\beta}{}^{n} + \Delta t \mathscr{H}_{\alpha}{}^{n} \tag{5.115}$$

Starting from Eq. (5.21), the advection computation can be carried out. The parameter τ_s can be given as in Eq. (5.167) or Eq. (5.168).

The stability of the SUPG scheme is almost obvious. To simplify the problem, one-dimensional equation is discussed first here:

$$\frac{\partial \mathfrak{c}}{\partial t} + v\frac{\partial \mathfrak{c}}{\partial x} = 0 \tag{5.116}$$

Let us normalize Eq. (5.116) using

$$u = \frac{\mathfrak{c}}{\mathfrak{C}}, \quad y = \frac{x}{\mathcal{X}}, \quad \tau = \frac{v}{\mathcal{X}}t \tag{5.117}$$

where $\mathfrak{C}$ and $\mathcal{X}$ are the fundamental values, then we get

$$\frac{\partial u}{\partial \tau} + \frac{\partial u}{\partial y} = 0 \tag{5.118}$$

Equation (5.118) is obtained. Applying the SUPG scheme to Eq. (5.118), we get

$$\int_0^l \left(u^* \frac{\partial u}{\partial \tau}\right) dy + \int_0^l \left(u^* \frac{\partial u}{\partial y}\right) dy + \tau_s \int_0^l \left(\frac{\partial u^*}{\partial y}\frac{\partial u}{\partial \tau}\right) dy + \tau_s \int_0^l \left(\frac{\partial u^*}{\partial y}\frac{\partial u}{\partial y}\right) dy = 0 \tag{5.119}$$

Deferring the discussion of the third term, the discussion in Sect. 5.5.4, i.e., stability with viscosity, is adaptable to Eq. (5.119). Thus, SUPG scheme is suitable for practical computations.

5.7 BTD Scheme

In this section, we will describe the balancing tensor diffusivity (BTD) scheme. Consider the advection equation Eq. (5.102):

$$\dot{\mathfrak{c}} + u_j \mathfrak{c}_{,j} = 0 \tag{5.102}$$

where u_j is assumed to be constant for the sake of simplicity. From Eq. (5.102), we can derive the first- and second-order derivatives without time derivative as follows:

$$\dot{\mathfrak{c}} = -u_j \mathfrak{c}_{,j} \tag{5.120}$$

$$\begin{aligned} \ddot{\mathfrak{c}} &= \frac{\partial}{\partial t}(-u_j \mathfrak{c}_{,j}) = -u_j \dot{\mathfrak{c}}_{,j} \\ &= u_j u_i \mathfrak{c}_{,ij} \end{aligned} \tag{5.121}$$

Taylor expansion with short time increment Δt leads to

$$\mathfrak{c}^{n+1} = \mathfrak{c}^n + \Delta t \dot{\mathfrak{c}} + \frac{\Delta t^2}{2} \ddot{\mathfrak{c}} \tag{5.122}$$

abstracting away from the higher and equal to the terms than order of Δt^3. In Eq. (5.122), $\mathfrak{c}^n$ means $\mathfrak{c}$ at time point n. Introducing Eqs. (5.120) and (5.121) into Eq. (5.122) and rearranging the terms, the following equation can be derived:

$$\mathfrak{c}^{n+1} = \mathfrak{c}^n + \Delta t(-u_j \mathfrak{c}^n_{,j}) + \frac{\Delta t^2}{2}(u_j u_i \mathfrak{c}^n_{,ij}) = 0 \tag{5.123}$$

From Eq. (5.123), we get

$$\frac{\mathfrak{c}^{n+1} - \mathfrak{c}^n}{\Delta t} + u_j \mathfrak{c}^n_{,j} - \frac{\Delta t}{2}(u_i u_j \mathfrak{c}^n_{,ij}) = 0 \tag{5.124}$$

Using the weighting function $\mathfrak{c}^*$, the weighted residual equation can be written as follows:

$$\int_{V_e} \mathfrak{c}^* \left(\frac{\mathfrak{c}^{n+1} - \mathfrak{c}^n}{\Delta t} \right) dV + \int_{V_e} (\mathfrak{c}^* u_j \mathfrak{c}^n_{,j}) dV - \frac{\Delta t}{2} \int_{V_e} (\mathfrak{c}^* u_i u_j \mathfrak{c}^n_{,ij}) dV = 0 \tag{5.125}$$

Integration by parts is applied to the last term of Eq. (5.125), and we get

$$\int_{V_e} c^* \left(\frac{c^{n+1} - c^n}{\Delta t} \right) dV + \int_{V_e} (c^* u_j c^n_{,j}) dV + \frac{\Delta t}{2} \int_{V_e} (c^*_{,i} u_i u_j c^n_{,j}) dV$$
$$= \frac{\Delta t}{2} \int_{S_e} (c^* u_i u_j c^n_{,j}) n_i dS \tag{5.126}$$

The right side term is zero with $u_i n_i = 0$.

The finite element equation can be described as follows:

$$M_{\alpha\beta} c^{n+1}_\beta = M_{\alpha\beta} c^n_\beta - \Delta t K_{\alpha\beta\gamma i} u_{\gamma i} c^n_\beta - \frac{\Delta t^2}{2} L_{\alpha i \beta\gamma\delta j} u_{\beta i} u_{\gamma j} c^n_\delta \tag{5.127}$$

where $M_{\alpha\beta}$, $K_{\alpha\beta\gamma i}$, $L_{\alpha i\beta\gamma\delta j}$ are written in Eqs. (5.111) and (5.112). The global form of the finite element equation can be derived as in the following form:

$$M_{\alpha\beta} C^{n+1}_\beta = M_{\alpha\beta} C^n_\beta - \Delta t A_{\alpha\beta} C^n_\beta - \frac{\Delta t^2}{2} E_{\alpha\beta} C^n_\beta \tag{5.128}$$

where $A_{\alpha\beta}$ and $E_{\alpha\beta}$ are written in Eqs. (5.113) and (5.114).

Equation (5.128) is the computational algorithm, which is referred to as the BTD scheme. We can see that the last term in Eq. (5.128) is the same as the last term in Eq. (5.114) replacing the parameter τ_s with $\Delta t/2$. Therefore, this term has the stabilizing effect, which is almost the same as that of the SUPG scheme. Actually, the stable computation can be obtained by Eq. (5.128). This also corresponds to the one-step algorithm of the two-step explicit scheme expressed by Eqs. (5.32) and (5.33).

The BTD scheme expressed by Eq. (5.128) is an explicit scheme. We can also formulate an implicit BTD scheme. From Eq. (5.120),

$$\ddot{c} = -u_j \dot{c}_{,j} \fallingdotseq -u_j \frac{c^{n+1}_{,j} - c^n_{,j}}{\Delta t} \tag{5.129}$$

Substituting Eqs. (5.120) and (5.129) into Eq. (5.122) and rearranging the terms, we get

$$\frac{c^{n+1} - c^n}{\Delta t} = -u_j c^n - \frac{1}{2} u_j c^{n+1}_{,j} + \frac{1}{2} u_j c^n_{,j} \tag{5.130}$$

Thus

$$\left(\frac{1}{\Delta t} c^{n+1} + \frac{1}{2} u_j c^{n+1}_{,j} \right) = \left(\frac{1}{\Delta t} c^n - \frac{1}{2} u_j c^n_{,j} \right) \tag{5.131}$$

From Eq. (5.131), the following finite element scheme can be obtained:

$$\left(\frac{1}{\Delta t}M_{\alpha\beta} + \frac{1}{2}A_{\alpha\beta}\right) C_{\alpha\beta}^{n+1} = \left(\frac{1}{\Delta t}M_{\alpha\beta} - \frac{1}{2}A_{\alpha\beta}\right) C_{\beta}^{n} \tag{5.132}$$

which is the same scheme expressed in Eq. (5.28). This means that the BTD scheme and the Crank–Nicolson scheme have the same sort of stability.

5.8 Bubble Function Scheme

The computation of advection–diffusion equation can be stabilized based on the bubble function interpolation. We will show in this section the derivation of the stabilization scheme based on the bubble function interpolation. First, we treat the steady equation to clarify the characteristics of the scheme.

Consider the following Eqs. (5.133), (5.134) and (5.135):

$$u_i\mathfrak{c}_{,i} - \nu\mathfrak{c}_{,ii} = 0 \quad \text{in} \quad V \tag{5.133}$$

$$\mathfrak{c} = 0 \quad \text{on} \quad S_1 \tag{5.134}$$

$$\mathfrak{b} = \nu\mathfrak{c}_{,i}n_i = \hat{\mathfrak{b}} \quad \text{on} \quad S_2 \tag{5.135}$$

where u_i is advection velocity and assumed to be constant. Multiplying the weighting function $\mathfrak{c}^*$ to both sides of Eq. (5.133) and rearranging the terms, we get

$$\int_{V_e} (\mathfrak{c}^* u_i \mathfrak{c}_{,i})dV + \nu \int_{V_e} (\mathfrak{c}^*_{,i}\mathfrak{c}_{,i})dV = \int_{S_e} (\mathfrak{c}^*\mathfrak{b})ds \tag{5.136}$$

where V_e and S_e represent volume and boundary of an element, respectively. For the interpolation function, we will use the bubble function interpolation function, which can be expressed as follows:

$$\mathfrak{c} = \tilde{\Phi}_\alpha \mathfrak{c}_\alpha + \phi\tilde{\mathfrak{c}}_4 \tag{5.137}$$

where $\tilde{\Phi}_\alpha$ represents the following set of functions, which is based on the bubble function element as shown in Fig. 5.4, and ϕ is the bubble function based on the barycentric node shown also in Fig. 5.4:

$$\begin{aligned}
\tilde{\Phi}_1 &= \eta_1 - \frac{27}{3}\eta_1\eta_2\eta_3 \\
\tilde{\Phi}_2 &= \eta_2 - \frac{27}{3}\eta_1\eta_2\eta_3 \\
\tilde{\Phi}_3 &= \eta_3 - \frac{27}{3}\eta_1\eta_2\eta_3 \\
\phi &= \Phi_4 = 27\eta_1\eta_2\eta_3
\end{aligned} \tag{5.138}$$

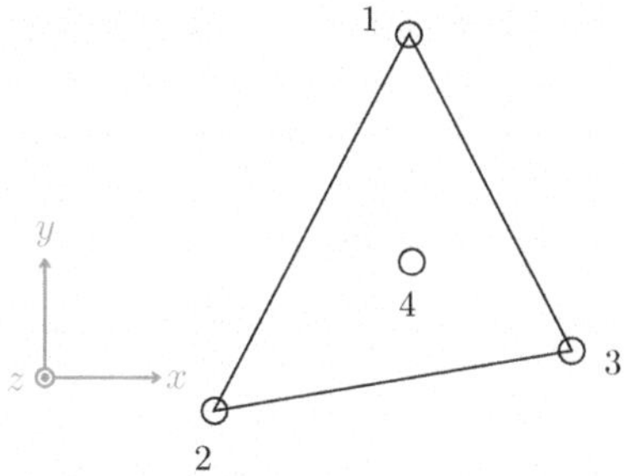

Fig. 5.4 Bubble function element

where η_1, η_2, and η_3 are the area coordinates, which will be defined in Eq. (6.28) in Sect. 6.5. Those are the same functions already used in Eq. (4.25) in Sect. 4.4. The functions in Eq. (5.138) are shown in Fig. 5.5. Note that the nodal values of $\tilde{\Phi}_\alpha$ in Eq. (5.138) are 1 on the considering node and 0 on the other nodes.

The most distinctive feature of the bubble function ϕ is that the value on the boundary of element is zero, i.e.,

$$\phi = 0 \quad \text{on} \quad S_e \tag{5.139}$$

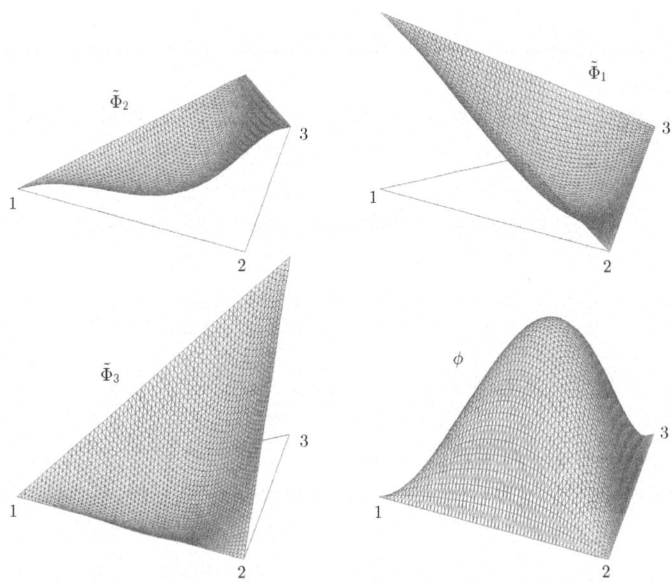

Fig. 5.5 Interpolation functions

The nodal values of the bubble function in Eq. (5.139) are denoted by $\mathfrak{c}_\alpha$ and $\tilde{\mathfrak{c}}_4$, where $\mathfrak{c}_\alpha$ ($\alpha = 1$ to 3) are the values on the three vertices, and $\tilde{\mathfrak{c}}_4$ is the value on the barycentric node. Equation (5.137) can be redefined as follows:

$$\mathfrak{c} = \Phi_\alpha \mathfrak{c}_\alpha + \phi \mathfrak{c}_4 \tag{5.140}$$

where

$$\mathfrak{c}_4 = \tilde{\mathfrak{c}}_4 - \frac{1}{3}(\mathfrak{c}_1 + \mathfrak{c}_2 + \mathfrak{c}_3) \tag{5.141}$$

we can denote Eq. (5.140) as

$$\mathfrak{c} = \Phi_\alpha \mathfrak{c}_\alpha + \Phi_4 \mathfrak{c}_4 = \Phi_\alpha \mathfrak{c}_\alpha \quad (\alpha = 1 \text{ to } 4) \tag{5.142}$$

Equation (5.142) can also be rewritten as follows:

$$\mathfrak{c} = \bar{\mathfrak{c}} + \phi \mathfrak{c}_4 \tag{5.143}$$

where $\bar{\mathfrak{c}}$ means the standard linear interpolation function and $\mathfrak{c}_4$ is the value defined in Eq. (5.141).

For the weighting function, the same type of Eq. (5.143) is used, which is

$$\mathfrak{c}^* = \bar{\mathfrak{c}}^* + \phi \mathfrak{c}_4^* = \Phi_\alpha \mathfrak{c}_\alpha^* + \phi \mathfrak{c}_4^* \tag{5.144}$$

where $\bar{\mathfrak{c}}^*$ is the standard linear weighting function, and $\mathfrak{c}_4^*$ is the constant on the barycentric node.

Consider the case in which the weighting function is the bubble function itself. Replacing $\mathfrak{c}^*$ in Eq. (5.136) by the bubble function ϕ, we get

$$\int_{V_e} (\phi u_i \mathfrak{c}_{,i}) dV + \nu \int_{V_e} (\phi_{,i} \mathfrak{c}_{,i}) dV = \int_{S_e} (\phi \mathfrak{b}) dS \tag{5.145}$$

Using Eq. (5.143), Eq. (5.145) can be transformed as follows:

$$\int_{V_e} (\phi u_i \bar{\mathfrak{c}}_{,i}) dV + \int_{V_e} \phi u_i (\phi \mathfrak{c}_4)_{,i} dV + \nu \int_{V_e} (\phi_{,i} \bar{\mathfrak{c}}_{,i}) dV + \nu \int_{V_e} (\phi_{,i} \phi_{,i}) dV \, \mathfrak{c}_4 = 0 \tag{5.146}$$

where $Q_\alpha = 0$ with $\hat{\mathfrak{b}} = 0$.

Let us reformulate the second and third terms in Eq. (5.146). The second term is

$$\int_{V_e} \phi u_i (\phi \mathfrak{c}_4)_{,i} dV = u_i \int_{V_e} (\phi \phi_{,i}) dV \, \mathfrak{c}_4 \tag{5.147}$$

because we assume u_i is constant in the element. Equation (5.147) is

$$\int_{V_e} \phi u_i (\phi \mathfrak{c}_i)_{,i} dV = 0 \tag{5.148}$$

because

$$\int_{V_e} (\phi\phi_{,i})dV = \int_{S_e} (\phi\phi n_i)dS - \int_{V_e} (\phi\phi_{,i})dV = -\int_{V_e} (\phi\phi_{,i})dV \tag{5.149}$$

where we use Eq. (5.139). The third term in Eq. (5.146) is

$$\nu \int_{V_e} (\bar{\mathfrak{c}}_{,i}\phi_{,i})dV = \nu \int_{S_e} (\bar{\mathfrak{c}}_{,i}\phi)n_i dS - \nu \int_{V_e} (\bar{\mathfrak{c}}_{,ii}\phi)dV = 0 \tag{5.150}$$

because we assume $\bar{\mathfrak{c}}$ is the linear interpolation function. Thus, Eq. (5.146) can be rewritten as

$$\int_{V_e} (u_i\bar{\mathfrak{c}}_{,i}\phi)dV + \nu \int_{V_e} (\phi_{,i}\phi_{,i})dV\,\mathfrak{c}_4 = 0 \tag{5.151}$$

Finally, we obtain $\mathfrak{c}_4$ as in the following form:

$$\mathfrak{c}_4 = -\frac{\displaystyle\int_{V_e} (u_i\bar{\mathfrak{c}}_{,i}\phi)dV}{\displaystyle\nu \int_{V_e} (\phi_{,i}\phi_{,i})dV} \tag{5.152}$$

Then, we can eliminate $\mathfrak{c}_4$ from the computational scheme.

Replacing the weighting function $\mathfrak{c}^*$ by the linear weighting function $\bar{\mathfrak{c}}^*$ in Eq. (5.136), introducing Eq. (5.143) into the resulted equation, and rearranging the terms, we get

$$\int_{V_e} (\bar{\mathfrak{c}}^* u_i\bar{\mathfrak{c}}_{,i})dV + \int_{V_e} (\bar{\mathfrak{c}}^* u_i\phi_{,i})dV\,\mathfrak{c}_4 + \nu \int_{V_e} (\bar{\mathfrak{c}}^*_{,i}\bar{\mathfrak{c}}_{,i})dV + \nu \int_{V_e} (\bar{\mathfrak{c}}^*_{,i}\phi_{,i})dV\,\mathfrak{c}_4$$
$$= \int_{S_e} (\bar{\mathfrak{c}}^*\mathfrak{b})dS \tag{5.153}$$

The fourth term in Eq. (5.153) is

$$\int_{V_e} (\bar{\mathfrak{c}}^*_{,i}\phi_{,i})dV = \int_{S_e} \bar{\mathfrak{c}}^*_{,i}\phi n_i dS - \int_{V_e} (\bar{\mathfrak{c}}^*_{,ii}\phi)dV = 0 \tag{5.154}$$

because $\bar{\mathfrak{c}}^*$ is the linear weighting function. Substituting Eq. (5.152) into the second term of Eq. (5.153), and using Eq. (5.154), the result is

$$\int_{V_e} (\bar{\mathfrak{c}}^* u_i\bar{\mathfrak{c}}_{,i})dV - \frac{\int_{V_e} (\bar{\mathfrak{c}}^* u_i\phi_{,i})dV \cdot \int_{V_e} (u_j\bar{\mathfrak{c}}_{,j}\phi)dV}{\nu \int_{V_e} (\phi_{,k}\phi_{,k})dV} + \nu \int_{V_e} (\bar{\mathfrak{c}}^*_{,i}\bar{\mathfrak{c}}_{,i})dV = \int_{S_e} (\bar{\mathfrak{c}}^*\mathfrak{b})dS \tag{5.155}$$

Equation (5.155) can be transformed into the following form:

$$\int_{V_e}(\bar{c}^* u_i \bar{c}_{,i})dV + \nu\int_{V_e}(\bar{c}^*_{,i}\bar{c}_{,i})dV + \frac{(\int_{V_e}\phi dV)^2}{\nu\int_{V_e}(\phi_{,k}\phi_{,k})dV\cdot\boldsymbol{\Delta}}\cdot\int_{V_e}(\bar{c}^*_{,i}u_i u_j \bar{c}_{,j})dV$$
$$= \int_{S_e}(\bar{c}^*\bar{\mathfrak{b}})dS \tag{5.156}$$

where $\boldsymbol{\Delta}$ is the area of a triangular element. In the derivation of Eq. (5.156), we use the equality

$$\int_{V_e}(\bar{c}^* u_i \phi_{,i})dV = \int_{S_e}(\bar{c}^* u_i \phi n_i)dS - \int_{V_e}(\bar{c}^*_{,i}u_i\phi)dV = -\int_{V_e}(\bar{c}^*_{,i}u_i\phi)dV \tag{5.157}$$

and the fact that u_i, $\bar{c}^*_{,i}$, $\bar{c}_{,j}$ are constants. Equation (5.156) can be reformulated as in the following form:

$$\int_{V_e}(\bar{c}^* u_i \bar{c}_{,i})dV + \nu\int_{V_e}(\bar{c}^*_{,i}\bar{c}_{,i})dV + \tau_{eB}\int_{V_e}(\bar{c}^*_{,i}u_i u_j \bar{c}_{,j})dV = \int_{S_e}(\bar{c}^* b)dS \tag{5.158}$$

where

$$\tau_{eB} = \frac{(\int_{V_e}\phi dV)^2}{\nu\int_{V_e}(\phi_{,k}\phi_{,k})dV\cdot\boldsymbol{\Delta}} \tag{5.159}$$

Equation (5.158) is similar to Eq. (5.106), i.e., the SUPG scheme. Therefore, the computational scheme based on the bubble function is the stable scheme similar to the SUPG scheme.

For practical computation, the stability of the bubble function scheme is not satisfactory. Therefore, we must add the artificial viscosity to the scheme. Equation (5.136) can be reformulated as follows:

$$\int_{V_e}(c^* u_i c_{,i})dV + \nu\int_{V_e}(\bar{c}^*_{,i}\bar{c}_{,i})dV + \nu\int_{V_e}(\bar{c}^*_{,i}\phi_{,i})dV\, c_4 + \nu\int_{V_e}(\phi_{,i}\bar{c}_{,i})dV\, c_4^*$$
$$+\nu\int_{V_e}(\phi_{,i}\phi_{,i})dV\, c_4^* c_4$$
$$= \int_{S_e}(c^*\mathfrak{b})dS \tag{5.160}$$

using Eqs. (5.143) and (5.144). The third and fourth terms in Eq. (5.160) are zero. We can superimpose Eq. (5.160) onto the whole finite element field and we obtain the equation system on the whole flow field:

$$\int_V (C^* u_i C_{,i}) dV + \nu \int_V (\bar{C}^*_{,i} \bar{C}_{,i}) dV + \nu \sum_{e=1}^{Ne} \int_{V_e} (\phi_{,i}\phi_{,i}) dV \, \mathfrak{c}_4^* \mathfrak{c}_4 = \int_{S_2} (C^* \hat{\mathfrak{b}}) dS \tag{5.161}$$

where C and C^* are interpolation and weighting functions on the whole flow field V, $\bar{C}$ and $\bar{C}^*$ are linear interpolation and weighting functions, and the third term means that viscosity term of the bubble function exists only on the barycentric node of each element and it consists of the sum of the terms over the total elements of which total number is Ne. To secure the stability, the artificial viscosity ν' is added to Eq. (5.161) as follows:

$$\int_V (C^* u_i C_{,i}) dV + \nu \int_V (\bar{C}^*_{,i} \bar{C}_{,i}) dV + (\nu + \nu') \sum_{e=1}^{Ne} \int_{V_e} (\phi_{,i}\phi_{,i}) dV \, \mathfrak{c}_4^* \mathfrak{c}_4$$
$$= \int_{S_2} (C^* \hat{\mathfrak{b}}) dS \tag{5.162}$$

To determine the artificial viscosity ν', several strategies are available, one of which is to use Eqs. (5.166) and (5.167) abstracting away from time increment Δt. The steady flow computation can be carried out by Eq. (5.162) with the artificial viscosity. The practical form of the global finite element method can easily be obtained by Eq. (5.162). The global finite element equation consists of the normal finite element equation based on the three node linear finite element and the term based on the artificial viscosity added on the barycentric node. Computational code can be easily obtained.

In the case of the unsteady flow problems, we use Eq. (5.1) instead of Eq. (5.133) as the governing equation. Following similar procedures written in the first part of this section, we get

$$\mathfrak{c}_4 = -\frac{\int_{V_e} (\phi \dot{\bar{\mathfrak{c}}}) dV}{\nu \int_{V_e} (\phi_{,k}\phi_{,k}) dV} - \frac{\int_{V_e} (\phi u_i \bar{\mathfrak{c}}_{,i}) dV}{\nu \int_{V_e} (\phi_{,k}\phi_{,k}) dV} \tag{5.163}$$

which is equivalent to Eq. (5.152). For the purpose of the stabilization of computation, the first term on the right side of Eq. (5.163) can be disregarded. Then, the discussions of the last part of this section are also valid in the unsteady flow field. The final form of the weighted residual equation corresponding to Eq. (5.162) can be described as in the following form:

$$\int_V (C^* \dot{C}) dV + \int_V (C^* u_i C_{,i}) dV + \nu \int_V (\bar{C}^*_{,i} \bar{C}_{,i}) dV + (\nu + \nu') \sum_{e=1}^{Ne} \int_{V_e} (\phi_{,i}\phi_{,i}) dV \, \mathfrak{c}_4^* \mathfrak{c}_4$$
$$= \int_{S_2} (C^* \hat{\mathfrak{b}}) dS \tag{5.164}$$

where ν' is the artificial viscosity, which can be the same as that of the SUPG scheme, i.e.,

$$\tau_{eB} = \tau_s \tag{5.165}$$

where τ_s is the stabilization parameter in the SUPG scheme expressed in Eq. (5.105). In general, it is known that the artificial viscosity of the SUPG scheme is sufficient to perform practical computations. The artificial viscosity expressed by the relation Eq. (5.165) can be determined as follows:

$$(\nu + \nu') \int_{V_e} (\phi_{,i}\phi_{,i}) dV = \frac{\left(\int_{V_e} \phi dV\right)^2}{\Delta} \cdot \frac{1}{\tau_s} \tag{5.166}$$

where

$$\tau_s = \left[\left(\frac{2}{\Delta t}\right)^2 + \left(\frac{2|u_i|}{he}\right)^2 + \left(\frac{4\nu}{he^2}\right)^2 \right]^{-\frac{1}{2}} \tag{5.167}$$

In Eq. (5.167), Δt is the time increment, $|u_i|$ is the magnitude of velocity, and he is the characteristic length of the element. When we use only the first term of Eq. (5.167), i.e.,

$$\tau_s = \frac{\Delta t}{2} \tag{5.168}$$

the artificial viscosity corresponds to that of the BTD scheme described in Sect. 5.7. Using the finite element method based on Eq. (5.164), unsteady advection–diffusion problem can be computed. Equation (5.164) means that the artificial viscosity is added on only the coefficients of the degrees of freedom $\mathfrak{c}_4$ on the barycentric node and the stable computation can be obtained. For the time discretization, the methods described in the first part of this chapter can be adopted.

5.9 Characteristic Scheme

When considering the stabilization of the computation, the characteristic scheme is one of the most important and optimal methods. The computation by the scheme is based on the idea of characteristic line.

Consider the advection equation:

$$\frac{\partial \mathfrak{c}}{\partial t} + u_i \mathfrak{c}_{,i} = 0 \tag{5.169}$$

where $\mathfrak{c}$ is a function of coordinate x_i and time t, and u_i is a known function independent of t. Transform variables (x_i, t) to (ξ_i, t) where

$$\xi_i = x_i - u_i t \tag{5.170}$$

then

$$\frac{\partial \mathfrak{c}}{\partial t} = \frac{\partial \mathfrak{c}}{\partial \xi_i}\frac{\partial \xi_i}{\partial t} + \frac{\partial \mathfrak{c}}{\partial t}\frac{\partial t}{\partial t} = \frac{\partial \mathfrak{c}}{\partial \xi_i}(-u_i) + \frac{\partial \mathfrak{c}}{\partial t}1 = -u_i\frac{\partial \mathfrak{c}}{\partial \xi_i} + \frac{\partial \mathfrak{c}}{\partial t} \tag{5.171}$$

$$\frac{\partial \mathfrak{c}}{\partial x_i} = \frac{\partial \mathfrak{c}}{\partial \xi_k}\frac{\partial \xi_k}{\partial x_i} + \frac{\partial \mathfrak{c}}{\partial t}\frac{\partial t}{\partial x_i} = \frac{\partial \mathfrak{c}}{\partial \xi_k}\delta_{ki} + \frac{\partial \mathfrak{c}}{\partial t}0 = \frac{\partial \mathfrak{c}}{\partial \xi_i} \tag{5.172}$$

Substituting Eqs. (5.171) and (5.172) into Eq. (5.169), and rearranging the terms, the result is that

$$\frac{\partial \mathfrak{c}}{\partial t} + u_i \mathfrak{c}_{,i} = -u_i\frac{\partial \mathfrak{c}}{\partial \xi_i} + \frac{\partial \mathfrak{c}}{\partial t} + u_i\frac{\partial \mathfrak{c}}{\partial \xi_i} = 0 \tag{5.173}$$

Equation (5.173) can be rearranged as follows:

$$\frac{\partial}{\partial t}\mathfrak{c}(\xi_i, t) = 0 \tag{5.174}$$

This is why Eqs. (5.169) and (5.174) are referred to as "transport equations." Namely, $\mathfrak{c}(\xi_i, t)$ is transported with respect to time keeping its shape as it is.

From Eq. (5.170),

$$\frac{dx_i}{dt} = u_i \tag{5.175}$$

Equation (5.175) is referred to as the characteristic equation. Integrating Eq. (5.175), we get

$$x_i = x_i{}^0 + u_i t \tag{5.176}$$

where $x_i{}^0$ is the position of x_i at time $t = t_0$, in which t_0 is the starting time. In order to solve Eq. (5.169), we solve Eq. (5.174) and the position can be computed by Eq. (5.176). The solution is

$$\mathfrak{c}(x_i, t) = \mathfrak{c}(x_i^0 + u_i t, t) \tag{5.177}$$

which means that the solution of Eq. (5.169) is the solution of Eq. (5.174) at the position $x_i{}^0 + u_i t$. Therefore, if we know $x_i{}^0$, we obtain the solution.

To clarify the idea of characteristic line, we consider the following one-dimensional equation:

$$\frac{\partial}{\partial t}\mathfrak{c}(x, t) + u\frac{\partial}{\partial x}\mathfrak{c}(x, t) = 0 \tag{5.178}$$

The solution of Eq. (5.178) is expressed as

$$\mathfrak{c}(x, t) = \mathfrak{c}(x - ut) \tag{5.179}$$

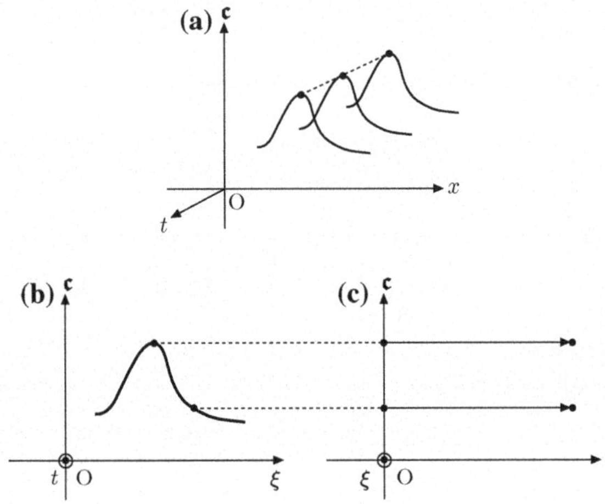

Fig. 5.6 Solution of transport equation. **a** Transport in c–x plane. **b** Transport in c–ξ plane. **c** Transport in c(ξ, t)–t plane

which is shown in Fig. 5.6a. Equation (5.178) can be transformed into the following equation:

$$\frac{\partial}{\partial t} c(\xi, t) = 0 \tag{5.180}$$

where

$$\xi = x - ut \tag{5.181}$$

The idea is illustrated in Figs. 5.6b, c. In c-ξ plane, the solution looks like in non-motion, and in c-t plane, the solution is constant.

Consider the advection–diffusion equation:

$$\frac{\partial}{\partial t} c(x_k, t) + u_i \frac{\partial}{\partial x_i} c(x_k, t) - \nu \frac{\partial^2}{\partial x_i \partial x_i} c(x_k, t) = 0 \tag{5.182}$$

Applying the characteristic scheme to the first two terms, Eq. (5.182) can be transformed into the following equation:

$$\frac{\partial}{\partial t} c(\xi_k, t) - \nu \frac{\partial^2}{\partial \xi_i \partial \xi_i} c(\xi_k, t) = 0 \tag{5.183}$$

where

$$\xi_k = x_k - u_k t \tag{5.184}$$

Discretizing total time into the sum of short time increments as shown in Fig. 5.1, the discretized equation in time can be expressed as follows:

$$c^{n+1} = \bar{\bar{c}}^{\,n} - \Delta t \nu \frac{\partial^2 c^n}{\partial \xi_i \partial \xi_i} \tag{5.185}$$

where

$$\mathfrak{c}^{n+1} = \mathfrak{c}(x_i{}^{n+1}, t^{n+1}) \tag{5.186}$$

$$\bar{\bar{\mathfrak{c}}}^{n} = \mathfrak{c}(\xi_i{}^{n}, t^{n}) \tag{5.187}$$

$$\xi_i{}^{n} = x_i{}^{n+1} - \Delta t u_i{}^{n+1} \tag{5.188}$$

assuming that the velocity $u_i{}^{n+1}$ is known in all $x_i{}^{n+1}$ and $\xi_i{}^{n}$. We also assume

$$\nu \frac{\partial^2 \mathfrak{c}}{\partial \xi_i \partial \xi_i} = \nu \frac{\partial^2 \mathfrak{c}}{\partial x_i \partial x_i} \tag{5.189}$$

Equation (5.185) is rewritten as

$$\mathfrak{c}^{n+1} = \bar{\bar{\mathfrak{c}}}^{n} - \Delta t \nu \mathfrak{c}_{,ii}{}^{n} \tag{5.190}$$

Multiplying both sides of Eq. (5.190) by the weighting function $\mathfrak{c}^*$, integrating over the volume V, and using the integration by parts, we get

$$\int_V (\mathfrak{c}^* \mathfrak{c}^{n+1}) dV = \int_V (\mathfrak{c}^* \bar{\bar{\mathfrak{c}}}^{n}) dV + \Delta t \int_V \nu (\mathfrak{c}^*_{,i} \mathfrak{c}_{,i}) dV = \int_S (\mathfrak{c}^* \mathfrak{b}) dS \tag{5.191}$$

Using the interpolation and weighting functions in Eqs. (5.11) and (5.12), and assuming $\hat{\mathfrak{b}} = 0$, the global finite element equation can be derived as follows:

$$M_{\alpha\beta} C_\beta{}^{n+1} = M_{\alpha\beta} \bar{\bar{C}}_\beta{}^{n} - \Delta t S_{\alpha\beta} C_\beta{}^{n} \tag{5.192}$$

where

$$C_\beta{}^{n+1} = C_\beta(x_{\gamma i}{}^{n+1}, t^{n+1}) \tag{5.193}$$

$$\bar{\bar{C}}_\beta{}^{n} = C_\beta(\xi_{\gamma i}{}^{n}, t^{n}) \tag{5.194}$$

$$\xi_{\gamma i}{}^{n} = x_{\gamma i}{}^{n+1} - \Delta t u_{\gamma i}{}^{n+1} \tag{5.195}$$

In fact, we need long computational time for the computation of Eq. (5.194). Therefore, we can employ the explicit scheme as follows:

$$\overline{M}_{\alpha\beta} C_\beta{}^{n+1} = \overline{M}_{\alpha\beta} \bar{\bar{C}}_\beta{}^{n} - \Delta t S_{\alpha\beta} C_\beta{}^{n} \tag{5.196}$$

where $\overline{M}_{\alpha\beta}$ is the lumped coefficient. The position computed by Eq. (5.195) is not necessarily be coincident with node $x_{\gamma i}{}^{n}$. The algorithm to compute Eq. (5.194) is as follows.

(1) Compute $\xi_{\gamma i}{}^{n} = x_{\gamma i}{}^{n+1} - \Delta t u_{\gamma i}{}^{n+1}$
(2) Search a finite element in which $\xi_{\gamma i}{}^{n}$ is included.

Fig. 5.7 Interpolation of C

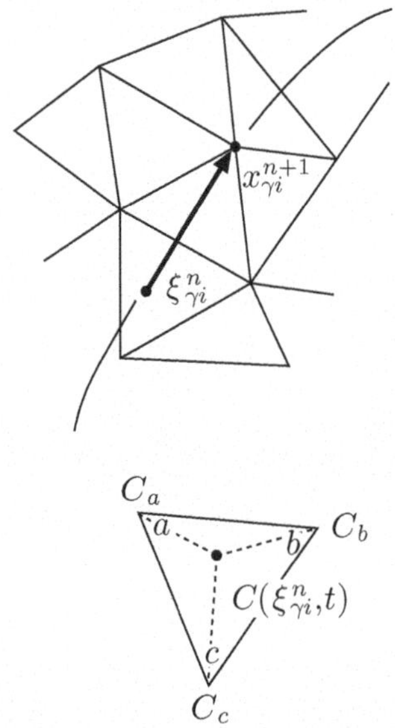

(3) Interpolate

$$C(\xi^n_{\gamma i}, t^n) = \Phi_\alpha C_\alpha({\xi_{\gamma i}}^n, t^n) \tag{5.197}$$

where $C_\alpha({\xi_{\gamma i}}^n, t^n)$ represents C at edge α ($\alpha = a, b, c$) of the finite element including ${\xi_{\gamma i}}^n$ as shown in Fig. 5.7.

4) Use $C({\xi_{\gamma i}}^n, t^n)$ obtained in (3) for $\overline{\overline{C}}_\beta{}^n({\xi_{\gamma i}}^n, t^n)$.
5) Iterate (1)–(4) for all node γ in the whole flow field.

Because the computational stability of the present scheme is not satisfactory, Maury (1996) used the combination of the bubble function scheme to solve the Navier–Stokes equation.

5.10 Computer Program Advection_Diffusion

To get a comprehensive picture of the finite element method of advection–diffusion problem, we present the computer program advection_diffusion programed by FORTRAN 90 in this section. For the numerical integration in time, the time-marching scheme, i.e., the scheme in which the concentration at time point $(n + 1)$ is obtained giving the concentration at time point n, is used. The basic concept is θ scheme which is shown in Eq. (5.26). For data of θ, $\theta = 1/2$ is used. To solve the problem shown in Fig. 5.10 or Fig. 5.14, Eq. (5.26) is simplified. The working equation is Eq. (5.203). Looking at Eq. (5.203), coefficient matrices of both sides are time independent if the conveying velocity in the field is time independent. At

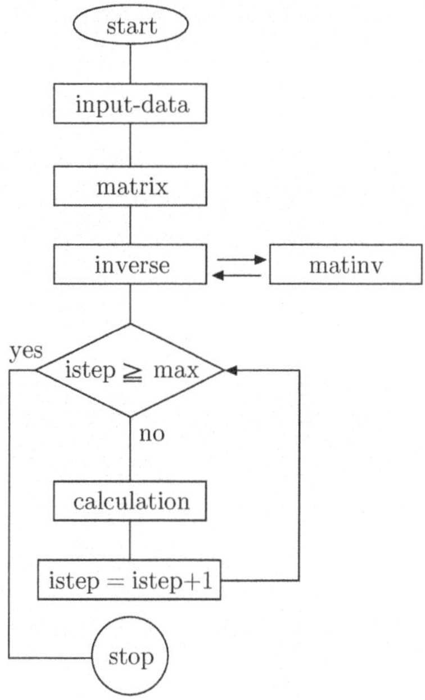

Fig. 5.8 Flow chart

that situation, those coefficient matrices are not necessary to compute at each time point. The known terms on the right side of Eq. (5.203) are constant if the boundary specified values are also time independent.

The flow chart of this program is shown in Fig. 5.8. After data is fed into subroutine input data, the coefficient matrices of both sides of Eq. (5.203) are programed in subroutine matrix. The inverse matrix of the coefficient matrix of the left side of Eq. (5.203) is computed in subroutine inverse. The main index is *istep* which ranges from 1 to *max*, i.e., maximum number of time points. The index *istep* expresses the time point as shown in Fig. 5.1. At each *istep*, the subroutine calculation is performed to compute the concentration and the results are output.

In this program, for simplicity, the velocities u and v in the field are given on all nodes, notwithstanding that they are time independent. Namely, this computer program can solve the problem of the time-dependent advection–diffusion on the steady conveying flow.

5.10.1 Governing Equation of the Program Advection_Diffusion

To solve the problem shown in Fig. 5.10 or Fig. 5.14, we suppose that the boundary concentration in Eq. (5.2) $\mathfrak{c}$ on S_1 and flux in Eq. (5.3) $\mathfrak{b}$ on S_2 are given as zero.

Therefore, the basic equation for the global concentration C_β can be simplified as follows:

$$M_{\alpha\beta}\dot{C}_\beta + A_{\alpha\beta}C_\beta = 0 \tag{5.198}$$

where $M_{\alpha\beta}$ and $A_{\alpha\beta}$ are in Eq. (5.17) and Eq. (5.20). The matrix $A_{\alpha\beta}$ includes the conveying velocity and we assume that this is also time independent. Discretizing

$$\dot{C}_\beta \fallingdotseq \frac{1}{\Delta t}(C_\beta^{n+1} - C_\beta^n) \tag{5.199}$$

and

$$C_\beta = \theta C_\beta^{n+1} + (1-\theta)C_\beta^n \tag{5.200}$$

where C_β^n is the concentration at time point n, Δt is the time increment, and θ is the parameter. Equation (5.198) can be described as follows:

$$S_{\alpha\beta}C_\beta^{n+1} = B_{\alpha\beta}C_\beta^n \tag{5.201}$$

where

$$\begin{aligned} S_{\alpha\beta} &= M_{\alpha\beta} + \Delta t\theta A_{\alpha\beta} \\ B_{\alpha\beta} &= M_{\alpha\beta} - \Delta t(1-\theta)A_{\alpha\beta} \end{aligned} \tag{5.202}$$

Let the specified values on S_1 be $\hat{C}_\beta$ and unknown values on all nodes be $\overline{C}_\beta$, Eq. (5.201) can be written as follows:

$$\overline{S}_{\alpha\beta}\overline{C}_\beta^{n+1} = B_{\alpha\beta}C_\beta^n - \hat{w}_\alpha \tag{5.203}$$

In Eq. (5.203), $\overline{S}_{\alpha\beta}$ is the coefficient matrix, to which the boundary treatment is introduced. $\overline{C}_\beta^{n+1}$ is C_β^{n+1}, from which the boundary values are subtracted. The procedure for the boundary treatment is the same as described in Sect. 4.7, and the practical algorithm is written in Sect. 4.13.5. The known terms are

$$\hat{w}_\alpha = \hat{S}_{\alpha\beta}\hat{C}_\beta \tag{5.204}$$

where $\hat{S}_{\alpha\beta}$ is the part of matrix $S_{\alpha\beta}$ corresponding to the boundary points on which concentration values are specified. The procedure of $\hat{w}_\alpha$ is expressed in Sect. 5.10.4 as subroutine inverse. The first term on the right side of Eq. (5.203) can be computed element-wise, which is explained in Sect. 5.10.5 subroutine calculation. Starting with the initial concentration, the concentration at time point $(n+1)$ can be obtained from Eq. (5.203).

5.10.2 Subroutine Input Data

Input data are listed in Table 5.1. The geometrical configuration is input. Then, consequently, data of total number of time iteration, those of initial condition, those of boundary condition, those of velocity, diffusion coefficient, time increment, and numerical integration parameter are given. Finally, data for initial conditions, boundary conditions, and velocity components are input in.

It is necessary to specify the initial conditions on all nodal points. However, only nonzero initial conditions are prespecified because zero values can be assigned by the 0 assign statement. The total number of nonzero initial condition is $ic0$. Sometimes, the values are zero. The nonzero initial conditions are in $cc0(i)$, $i = 1$ to $ic0$, which are given on node $nc0(i)$, $i = 1$ to $ic0$. In the subroutine calculation, $cc1(i)$, $i = 1$ to nx is used for the array of values at time point n. For the use of the starting values, the initial conditions $cc0(i)$, $i = 1$ to nx are used. Thus, $cc1(i)$, $i = 1$ to nx at the

Table 5.1 Input data

Identifier	Maximum number	Explanation
nx	1	Total number of nodes
mx	1	Total number of elements
$xy(i,j)$	2, nx	x, y coordinates of node j
$ie(i,j)$	3, mx	Connectivity index of edge i of element j
max	1	Maximum number of time points
$iout$	1	For every $iout$, the computed results are output
$ic0$	1	Total number of initial concentration
icb	1	Total number of nodes on which concentration are given
iub	1	Total number of nodes on which velocities u and v are given except 0.0
dd	1	Diffusion coefficient
dt	1	Time increment Δt
$theta$	1	Parameter θ
$nc0(i)$	$ic0$	Number of nodes on which initial concentrations are given
$cc0(i)$	$ic0$	Initial concentration on node $nc0(i)$
$ncb(i)$	icb	Number of nodes on which concentrations are given
$fcb(i)$	icb	Specified concentration on node $ncb(i)$
$nub(i)$	iub	Number of nodes on which velocities u and v are given
$uv(i)$	iub	x velocity on which node $nub(i)$
$vu(i)$	iub	y velocity on which node $nub(i)$

initial time point must be replaced by $cc0(i)$, $i = 1$ to nx. Usually, the boundary conditions should be specified. The total number of the boundary condition is *icb*. The boundary conditions are in *fcb*(i), $i = 1$ to *icb*, which are given on node $ncb(i)$, $i = 1$ to *icb*. The arrays $uu(i)$ and $vv(i)$, $i = 1$ to nx mean the velocity on the whole nodes and assumed to be time independent. Those input data are written out as the print forms of input data.

5.10.3 Subroutine Matrix

The main identifiers in the subroutine matrix are listed in Table 5.2. In this subroutine, the coefficient matrices of both sides of Eq. (5.203) are programed. Essentially, those matrices should be programed at each time point during the time-marching

Table 5.2 Main identifier

Identifier	Maximum number	Explanation
$uu(i)$	nx	x velocity on node i
$vv(i)$	nx	y velocity on node i
im	1	Number of element
ia	1	Number of node to which edge 1 of element *im* is connected
ib	1	Number of node to which edge 2 is connected
ic	1	Number of node to which edge 3 is connected
xa	1	x coordinate of edge 1
xb	1	x coordinate of edge 2
xc	1	x coordinate of edge 3
ya	1	y coordinate of edge 1
yb	1	y coordinate of edge 2
yc	1	y coordinate of edge 3
ua	1	x velocity on edge 1
ub	1	x velocity on edge 2
uc	1	x velocity on edge 3
va	1	y velocity on edge 1
vb	1	y velocity on edge 2
vc	1	y velocity on edge 3

(continued)

Table 5.2 (continued)

Identifier	Maximum number	Explanation
area	1	Area of element
*area*06	1	area/6.0
*area*12	1	area/12.0
rarea	1	1/(2*area)
*b*1	1	b_a
*b*2	1	b_b
*b*3	1	b_c
*c*1	1	c_a
*c*2	1	c_b
*c*3	1	c_c
*e*1	1	$2u_a + u_b + u_c$
*e*2	1	$2v_a + v_b + v_c$
*f*1	1	$u_a + 2u_b + u_c$
*f*2	1	$v_a + 2v_b + v_c$
*g*1	1	$u_a + u_b + 2u_c$
*g*2	1	$v_a + v_b + 2v_c$
da	1	Diffusion coefficient times area
$am(i, j)$	3,3	Coefficient matrix $M_{\alpha\beta}$
$sk(i, j)$	3,3	Coefficient matrix $A_{\alpha\beta}$
$bm(i, j, im)$	3, 3, *mx*	Right side term in Eq. (5.201)
$ss(i, j)$	*nx*, *nx*	Coefficient matrix in the left side of Eq. (5.201)
$cc1(i)$	*nx*	Concentration on node *i* at *n*th time point
$cc2(i)$	*nx*	Concentration on node *i* at $(n + 1)$th time point
$cc3(i)$	*nx*	Right side of Eq. (5.203)
$cc4(i)$	*nx*	Known term of Eq. (5.204)

computation. Those matrices can be time independent if the velocities in the field u and v are time independent. The computation is carried out according to *im*, $im = 1$ to *mx*, which corresponds to element number. The left side matrix $(M_{\alpha\beta} + \Delta t\theta A_{\alpha\beta})$ in Eq. (5.203) is stored in $ss(i, j)$, $i = 1$ to $nx, j = 1$ to nx. The matrix $ss(i, j)$ should be the global form because the inverse computation is performed in the next subroutine inverse. The right side matrix $(M_{\alpha\beta} - \Delta t(1-\theta)A_{\alpha\beta})$ in Eq. (5.203) is stored in $bm(i, j, im)$, $i = 1$ to 3, $j = 1$ to 3 and $im = 1$ to mx. As the index im corresponds to element number, the right side matrix of each element is stored. This stored array will be simply used in the computation in the subroutine calculation.

To program $ss(i, j)$ and $bm(i, j, im)$, the local coefficient matrices $M_{\alpha\beta}$ and $A_{\alpha\beta}$ in Eq. (5.19) are needed. These are stored in $am(i, j)$, $i = 1$ to 3, and $j = 1$ to 3 and $sk(i, j)$, $i = 1$ to 3 and $j = 1$ to 3. The precise forms of those matrices are given in Eq. (5.17). The matrix $am(i, j)$ can be simply programed following Eq. (5.17). The array $sk(i, j)$ is $A_{\alpha\beta}$ in Eq. (5.20) and programed as follows. The second part of $A_{\alpha\beta}$, i.e., $S_{\alpha\beta}$, is shown in Eq. (5.17), which is the same one as in Eq. (4.40), in Sect. 4.6. The first part of $A_{\alpha\beta}$ in Eq. (5.20) can be transformed in the following manner. For example, $s(1, 1)$ is expressed as

$$\begin{aligned} s(1,1) &= M_{aa}b_a u_a + M_{ab}b_a u_b + M_{ac}b_a u_c \\ &\quad + M_{aa}c_a v_a + M_{ab}c_a v_b + M_{ac}c_a v_c \\ &= \frac{\boldsymbol{\Delta}}{12}(2u_a + u_b + u_c)b_a \\ &\quad + \frac{\boldsymbol{\Delta}}{12}(2v_a + v_b + v_c)c_a \\ &= \frac{\boldsymbol{\Delta}}{12}e_1 b_a + \frac{\boldsymbol{\Delta}}{12}e_2 c_a \end{aligned}$$

where

$$\begin{aligned} e_1 &= 2u_a + u_b + u_c \\ e_2 &= 2v_a + v_b + v_c \end{aligned}$$

and $\boldsymbol{\Delta}$ is an area of a triangle. Almost identical expressions hold for $s(1, 2)$ to $s(3, 3)$, which can be seen in the subroutine matrix. In those expressions, u_a, u_b, u_c and v_a, v_b, v_c are u and v velocities at edges a, b, c and b_a, b_b, b_c, c_a, c_b, c_c are in Eq. (4.22). Those values can be calculated if we know the coordinates of three edges a, b, c of a triangular element.

5.10.4 *Subroutine Inverse*

In this subroutine, the inverse matrix of the coefficient matrix of the left side of Eq. (5.203) is computed. The coefficient matrix is made and stored in $ss(i, j)$, $i = 1$ to nx, $j = 1$ to nx in the subroutine matrix. The arrays $cc1(i)$, $cc2(i)$, $cc3(i)$, $cc4(i)$,

$i = 1$ to nx are used in the computation in the subroutine calculation. One of them: the array $cc4(i)$, $i = 1$ to nx, is used to store the known terms, which is based on the assumption that both conveying velocity and the boundary concentration specified are time independent. The procedure to compute the known terms is the same as the one we used in the subroutine potent in Sect. 4.13.5. The boundary treatment is imposed on $ss(i, j), i = 1$ to $nx, j = 1$ to nx. The inverse matrix of $ss(i, j)$ is computed by the subroutine matinv and obtained in $ss(i, j)$. Then, again the boundary treatment is imposed. The procedure is also the same as expressed in the subroutine potent in Sect. 4.13.5.

5.10.5 Subroutine Calculation

This procedure is the principal part of the program advection_diffusion. The main index is *istep* which means time point ranges from 1 to *max*, as shown in Fig. 5.1. The subroutine calculation is called out at every time point to compute the concentration. The time-marching scheme based on Eq. (5.203) computes the concentration at time point $(n + 1)$ from the concentration at time point n. The known concentration on a whole field at time point n is stored in $cc1(i)$, $i = 1, nx$. The array $cc3(i)$, $i = 1, nx$ is the first term of the right side of Eq. (5.203). The coefficient matrix $bm(i, j, im)$ was obtained beforehand in the subroutine matrix since the coefficient is time independent. We obtain the sum of $cc3(i)$ and $cc4(i)$ in the array $cc3(i)$, $i = 1$ to nx, which is the right side of Eq. (5.203), since $cc4(i)$ is the known term programed in the subroutine inverse. The product of matrix multiplication of $ss(i, j)$ and $cc3(j)$ is the concentration at time point $(n + 1)$, $cc2(i)$, $i = 1, nx$. Notice that $ss(i, j)$ is the inverse of the left side of Eq. (5.203), which was made in the subroutine inverse. After the boundary treatment, the concentration $cc2(i)$ is printed out. Replacing $cc1(i)$ with $cc2(i)$, the computation is iterated until the time point reaches to *max*. The main identifier is listed in Table 5.2.

5.10.6 Program List of Advection_Diffusion

The computer program list is shown in Fig. 5.9.

```
!***********************************************************************
!***********************************************************************
!************                               ***************************
!************  Advection - Diffusion        ***************************
!************  coded by M.Wakui             ***************************
!************                               ***************************
!***********************************************************************
!***********************************************************************
!
!      nx-----------total number of global nodes
!      mx-----------total number of elements
!      xx,yy--------node coordinates
!      ie-----------connectivity indices for elements
!      dd-----------dispersion coefficient
!      icb----------total number of boundary concentration
!      ncb----------node numbers on which boundary condition is specified
!      fcb----------specified values for boundary condition
!      iub----------total number of specified velocity
!      nub----------node numbers for specified velocity
!      uv,vu--------specified values for velocity
!      max----------maximum number of time iteration
!
!***********************************************************************
!***********************************************************************


!***********************************************************************
!****                ***************************************************
!****  main program  ***************************************************
!****                ***************************************************
!***********************************************************************

       program advection_diffusion

       implicit real(8) (a-h,o-z)
       implicit integer(4) (i-n)

       real(8) dd,dt,theta
       real(8) xy,uu,vv,cc1,cc2,cc3,cc4,fcb
       real(8) ss,bm
       integer(4) nx,mx,max,iout
       integer(4) ie,ncb,icb

       parameter(inod=20,iele=50)

       dimension xy(2,inod),ie(3,iele),ncb(inod),fcb(inod)
       dimension uu(inod),vv(inod)
       dimension cc1(inod),cc2(inod),cc3(inod),cc4(inod)
       dimension bm(3,3,iele),ss(inod,inod)

       open(5,file='input_data.txt')
       open(10,file='node_data.txt')
       open(20,file='check_input_data.txt')
       open(30,file='output_diffusion.txt')
       open(40,file='output_dat.dat')

       call input_data(nx,mx,max,iout,dd,dt,theta,xy,uu,vv,
     &                        ie,ncb,icb,fcb,cc1,inod,iele)
       call matrix(nx,mx,ie,dd,dt,theta,xy,uu,vv,ss,bm,inod,iele)

       call inverse(nx,icb,ncb,fcb,ss,inod,cc2,cc3,cc4)

       do istep = 1, max
         call calculation(nx,mx,istep,iout,ie,icb,ncb,fcb,dt,ss,bm,
     &                        cc1,cc2,cc3,cc4,inod,iele)
       end do
```

Fig. 5.9 Program list advection_diffusion

```
      close(5)
      close(10)
      close(20)
      close(30)
      close(40)

      stop
      end program advection_diffusion
!************************************************************************

!************************************************************************
!****                    ***********************************************
!****  data input        ***********************************************
!****                    ***********************************************
!************************************************************************

      subroutine input_data(nx,mx,max,iout,dd,dt,theta,xy,uu,vv,
     &                    ie,ncb,icb,fcb,cc1,inod,iele)

      implicit real(8) (a-h,o-z)
      implicit integer(4) (i-n)

      real(8) dd,dt,theta,xy,uu,vv,uv,vu,fcb,cc1,cc0
      integer(4) nx,mx,max,iout
      integer(4) ie,icb,ncb,iub,nub,ic0,nc0

      dimension xy(2,inod),ie(3,iele)
      dimension nub(inod),uu(inod),vv(inod),uv(inod),vu(inod)
      dimension ncb(inod),fcb(inod),nc0(inod),cc0(inod),cc1(inod)

      do i = 1, nx
        uu(i) = 0.0d0
        vv(i) = 0.0d0
        uv(i) = 0.0d0
        vu(i) = 0.0d0
        cc0(i) = 0.0d0
        cc1(i) = 0.0d0
        fcb(i) = 0.0d0
      end do

!****  coordinate xy(i,j) and connectivity ie(i,j)  ********************

      read(10,*) nx,mx
      read(10,*) (i,xy(1,i),xy(2,i),j=1,nx)
      read(10,*) (i,ie(1,i),ie(2,i),ie(3,i),j=1,mx)

!****  data for time iteration  ****************************************

      read(5,*) max,iout
      read(5,*) ic0,icb,iub
      read(5,*) dd,dt,theta

!****  initial conditions cc0(i) and starting conditions cc1(i)  *******

      if(ic0.ne.0)then
        read(5,*) (i,nc0(i),cc0(i),j=1,ic0)
        do i = 1, ic0
          ii = nc0(i)
          cc1(ii) = cc0(i)
        end do
      end if
```

Fig. 5.9 (continued)

```
!**** boundary conditions fcb(i)  **************************************

      if(icb.ne.0)then
        read(5,*) (i,ncb(i),fcb(i),j=1,icb)
      end if

!****  velocitities u and v  *******************************************

      if(iub.ne.0)then
        read(5,*) (i,nub(i),uv(i),vu(i),j=1,iub)
        do i = 1, iub
          ii = nub(i)
          uu(ii) = uv(i)
          vv(ii) = vu(i)
        end do
      end if

!****  check write  ****************************************************

      write(20,*)'*********  Advection - Diffusion program  **********'
      write(20,*)
      write(20,*)'*********  check input data  ***********************'
      write(20,*)
      write(20,'(4(a,i3,3x))')'nx=',nx,'mx=',mx,'max=',max,'iout=',iout
      write(20,'(3(a,i3,3x))') 'ic0=',ic0,'icb=',icb,'iub=',iub
      write(20,'(3(a,f7.3,3x))') 'dd=',dd,'dt=',dt,'theta=',theta
      write(20,*)
      write(20,*)'*********  X,Y cordinate  **************************'
      write(20,*)
      write(20,*)'No.         X(m/s)           Y(m/s)'
      write(20,'(i5,3x,2f16.10)') (i,xy(1,i),xy(2,i),i=1,nx)
      write(20,*)
      write(20,*)'*********  node connectivity  *********************'
      write(20,*)
      write(20,*)'No.'
      write(20,'(4i5)') (i,ie(1,i),ie(2,i),ie(3,i),i=1,mx)
      write(20,*)
      write(20,*)'*********  specified values  ***********************'
      if(ic0.ne.0)then
        write(20,*)
        write(20,*)'No.           PPM'
        write(20,'(2i5,f16.10)') (i,nc0(i),cc1(i),i=1,iub)
        write(20,*)
      end if
      if(iub.ne.0)then
        write(20,*)
        write(20,*)'No.           UU(m/s)          VV(m/s)'
        write(20,'(2i5,2f16.10)') (i,nub(i),uu(i),vv(i),i=1,iub)
        write(20,*)
      end if
      write(20,*)'*********  boundary condition  ********************'
      if(icb.ne.0)then
        write(20,*)
        write(20,*)'No.           PPM'
        write(20,'(2i5,f16.10)') (i,ncb(i),fcb(i),i=1,icb)
      end if

      end subroutine input_data
!***********************************************************************
```

Fig. 5.9 (continued)

```
!***********************************************************************
!****                                      *****************************
!****  calculation of coefficient matrix   *****************************
!****                                      *****************************
!***********************************************************************

      subroutine matrix(nx,mx,ie,dd,dt,theta,xy,uu,vv,ss,bm,inod,iele)

      implicit real(8) (a-h,o-z)
      implicit integer(4) (i-n)
      real(8) dd,dt,theta
      real(8) xy
      real(8) uu,vv
      real(8) sk,am,ss,bm
      integer(4) nx,mx,ie

      dimension xy(2,inod),ie(3,iele),uu(inod),vv(inod)
      dimension sk(3,3),am(3,3),bm(3,3,iele),ss(inod,inod)

      do i = 1, 3
        do j = 1, 3
          sk(i,j) = 0.0d0
          am(i,j) = 0.0d0
        end do
      end do

      do i = 1, 3
        do j = 1, 3
          do k = 1, mx
            bm(i,j,k) = 0.0d0
          end do
        end do
      end do

      do i = 1, nx
        do j = 1, nx
          ss(i,j) = 0.0d0
        end do
      end do

!****  matrices am and sk  *********************************************

      do im = 1, mx
        ia = ie(1,im)
        ib = ie(2,im)
        ic = ie(3,im)
        xa = xy(1,ia)
        xb = xy(1,ib)
        xc = xy(1,ic)
        ya = xy(2,ia)
        yb = xy(2,ib)
        yc = xy(2,ic)
        ua = uu(ia)
        ub = uu(ib)
        uc = uu(ic)
        va = vv(ia)
        vb = vv(ib)
        vc = vv(ic)
        area = ((xb-xa)*(yc-ya)-(xc-xa)*(yb-ya))*0.5d0
        rarea = 0.5d0/area
        area06 = area / 6.0d0
        area12 = area / 12.0d0
        b1 = (yb-yc)*rarea
        b2 = (yc-ya)*rarea
```

Fig. 5.9 (continued)

```
         b3 = (ya-yb)*rarea
         c1 = (xc-xb)*rarea
         c2 = (xa-xc)*rarea
         c3 = (xb-xa)*rarea
         e1 = 2.0d0*ua + ub + uc
         e2 = 2.0d0*va + vb + vc
         f1 = ua + 2.0d0*ub + uc
         f2 = va + 2.0d0*vb + vc
         g1 = ua + ub + 2.0d0*uc
         g2 = va + vb + 2.0d0*vc
         da = dd*area
         am(1,1) = area06
         am(2,2) = area06
         am(3,3) = area06
         am(1,2) = area12
         am(1,3) = area12
         am(2,1) = area12
         am(2,3) = area12
         am(3,1) = area12
         am(3,2) = area12
         sk(1,1) = (b1*e1+c1*e2)*area12 + (b1*b1+c1*c1)*da
         sk(1,2) = (b2*e1+c2*e2)*area12 + (b1*b2+c1*c2)*da
         sk(1,3) = (b3*e1+c3*e2)*area12 + (b1*b3+c1*c3)*da
         sk(2,1) = (b1*f1+c1*f2)*area12 + (b2*b1+c2*c1)*da
         sk(2,2) = (b2*f1+c2*f2)*area12 + (b2*b2+c2*c2)*da
         sk(2,3) = (b3*f1+c3*f2)*area12 + (b2*b3+c2*c3)*da
         sk(3,1) = (b1*g1+c1*g2)*area12 + (b3*b1+c3*c1)*da
         sk(3,2) = (b2*g1+c2*g2)*area12 + (b3*b2+c3*c2)*da
         sk(3,3) = (b3*g1+c3*g2)*area12 + (b3*b3+c3*c3)*da

!****   matrices ss and bm   *********************************************

         do i = 1,3
           ii = ie(i,im)
           do j = 1,3
             jj = ie(j,im)
             bm(i,j,im) = -dt*(1.0-theta)*sk(i,j) + am(i,j)
             ss(ii,jj) = ss(ii,jj) + dt*theta*sk(i,j) + am(i,j)
           end do
         end do
       end do

       end subroutine matrix
!************************************************************************


!************************************************************************
!****                                   *********************************
!****   inverse matrix of ss            *********************************
!****                                   *********************************
!************************************************************************

       subroutine inverse(nx,icb,ncb,fcb,ss,inod,cc2,cc3,cc4)

       implicit real(8) (a-h,o-z)
       implicit integer(4) (i-n)
       real(8) cc2,cc3,cc4
       real(8) ss,fcb
       integer(4) nx
       integer(4) icb,ncb

       dimension fcb(inod),ncb(inod)
       dimension cc2(inod),cc3(inod),cc4(inod)
       dimension ss(inod,inod)
```

Fig. 5.9 (continued)

```
!****  boundary treatment  *********************************************

       do i=1,nx
         cc2(i) = 0.0d0
         cc3(i) = 0.0d0
         cc4(i) = 0.0d0
       end do

!****  computation of known term  **************************************

       do i = 1, icb
         ii = ncb(i)
         do j = 1, nx
           cc4(j) = cc4(j) - ss(j,ii)*fcb(i)
         end do

!****  boundary treatment  *********************************************

         do j = 1,nx
           ss(ii,j) = 0.0d0
           ss(j,ii) = 0.0d0
         end do
         ss(ii,ii) = 1.0d0
       end do

!****  inverse of matrix ss(i,j)  **************************************

       call matinv(ss,nx,inod)

!****  boundary treatment  *********************************************

       do i = 1,icb
         ii = ncb(i)
         ss(ii,ii) = 0.0d0
       end do

       write(30,*) '*************  output results  ********************'
       write(30,*)

       return
       end subroutine inverse
!***********************************************************************

!***********************************************************************
!****                                  ********************************
!****  calculation of concentration    ********************************
!****                                  ********************************
!***********************************************************************

       subroutine calculation(nx,mx,istep,iout,ie,icb,ncb,fcb,dt,ss,bm,
     &                        cc1,cc2,cc3,cc4,inod,iele)

       implicit real(8) (a-h,o-z)
       implicit integer(4) (i-n)

       real(8) dt,cc1,cc2,cc3,cc4
       real(8) ss,bm,fcb
       integer(4) nx,mx,istep,iout
       integer(4) ie,icb,ncb

       dimension ie(3,iele),fcb(inod),ncb(inod)
       dimension cc1(inod),cc2(inod),cc3(inod),cc4(inod)
       dimension bm(3,3,iele),ss(inod,inod)

!***********************************************************************
```

Fig. 5.9 (continued)

```
      do i = 1, nx
        cc3(i) = 0.0d0
      end do

      do im = 1,mx
        do i = 1,3
          ii = ie(i,im)
          do j = 1,3
            jj = ie(j,im)
            cc3(ii) = cc3(ii) + bm(i,j,im)*cc1(jj)
          end do
        end do
      end do

!****  adtion of known terms  *******************************************

      do i = 1, nx
        cc3(i) = cc3(i) + cc4(i)
      end do

!****  multiplication of the inverse of the right side  ****************
!****  times concentration at n th time point           ****************

      do i = 1, nx
        wn = 0.0d0
        do j = 1, nx
          wn = wn + ss(i,j)*cc3(j)
        end do
        cc2(i) = wn
      end do

!****  treatment of boundary condition  ********************************

      do i = 1, icb
        ii = ncb(i)
        cc2(ii) = fcb(i)
      end do

!****  output calculated result  ***************************************

      it=mod(istep,iout)
      if(it.eq.0)then
        tt = float(istep)
        ttime = tt*dt
        write(30,*) 'istep ',istep,' time ',ttime
        write(30,*)
        write(30,300) (i*6-5,cc2(i*6-5),i*6-4,cc2(i*6-4),i*6-3,
     &  cc2(i*6-3),i*6-2,cc2(i*6-2),i*6-1,cc2(i*6-1),i*6,cc2(i*6),i=1,3)
        write(30,*)
      end if
      write(40,400) (cc2(i*6-5),cc2(i*6-4),cc2(i*6-3),
     &                   cc2(i*6-2),cc2(i*6-1),cc2(i*6),i=1,3)
      write(40,*)

!****  replacement of conection cc1(i) with concentration cc2(i)  ******

      do i = 1, nx
        sc = cc2(i)
        cc2(i) = 0.0d0
        cc1(i) = sc
      end do

  300 format(6(i5,f8.3))
  400 format(6(f11.3,3x))
```

Fig. 5.9 (continued)

```
      end subroutine calculation
!************************************************************************


!************************************************************************
!****          **********************************************************
!****  matinv  **********************************************************
!****          **********************************************************
!************************************************************************

      subroutine matinv(ss,n,ida)

      implicit real(8) (a-h,o-z)
      implicit integer(4) (i-n)

      dimension ss(ida,ida)
      iitt = 100

      do i = 1, n
        p = ss(i,i)
        if(p.eq.0.0)then
          write(*,*)i,i,'zero diagonal term'
          goto 500
        end if
        ss(i,i) = 1.0d0
        do j=1,n
          ss(i,j)=ss(i,j)/p
        end do
        ttii=mod(i,iitt)
        if(ttii.eq.0.0)then
          write(*,*)i
        end if
        do 300 k = 1, n
          if(k.eq.i) goto 300
          q = ss(k,i)
          ss(k,i) = 0.0
          do j = 1, n
            ss(k,j) = ss(k,j) - q*ss(i,j)
          end do
 300      continue
      end do

      goto 600

 500  stop 999

 600  return
      end
```

Fig. 5.9 (continued)

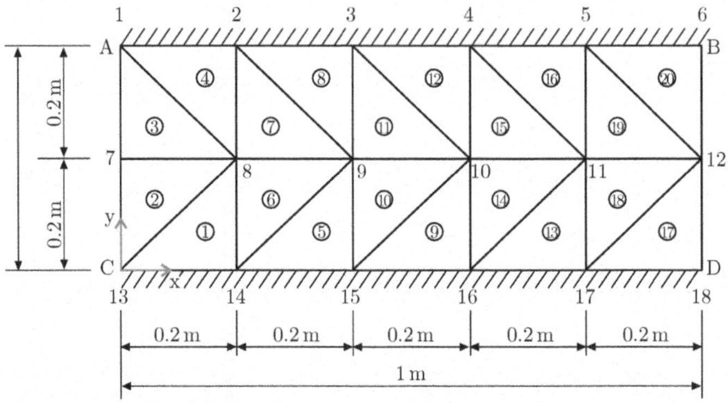

Fig. 5.10 Channel for trial computation

5.10.7 Trial Computation

Numerical results of the trial computation are shown in this section. The simple channel represented in Fig. 5.10 is employed. The output of the input data is shown in Fig. 5.11. An example of output data of computed results is shown in Fig. 5.12. Altogether, four cases of computations are carried out changing the conditions, i.e., with and without advection velocity and zero concentration is specified or not on the boundary B–D. The computational conditions are listed in Table 5.3. The computed results are plotted in Fig. 5.13a–d.

5.11 Numerical Study

To illustrate a numerical study of the unsteady advection of concentration problem, the rotating cone is calculated by five schemes, i.e., the Crank–Nicolson scheme, the SUPG scheme, the characteristic scheme, and the bubble function scheme. The schematic view of the problem is shown in Fig. 5.14. On the four side of the field A-B-C-D, the concentration is set to be

$$\mathfrak{c} = 0 \qquad \text{on A-B-C-D} \tag{5.205}$$

An advection velocity v_i is $(-y, x)$, where x and y are coordinates of the nodes. The origin of coordinate is the center of the field. The initial condition of the concentration $\mathfrak{c}$ is expressed as

```
*********  Advection - Diffusion program  **********

*********  check input data  ***********************

nx= 18   mx= 20   max=100   iout= 10
ic0=  0   icb=  6   iub= 18
dd=  0.500   dt=  0.020   theta=  0.500

*********  X,Y cordinate  *************************

 No.          X(m/s)          Y(m/s)
    1         0.0000000000    0.4000000000
    2         0.2000000000    0.4000000000
    3         0.4000000000    0.4000000000
    4         0.6000000000    0.4000000000
    5         0.8000000000    0.4000000000
    6         1.0000000000    0.4000000000
    7         0.0000000000    0.2000000000
    8         0.2000000000    0.2000000000
    9         0.4000000000    0.2000000000
   10         0.6000000000    0.2000000000
   11         0.8000000000    0.2000000000
   12         1.0000000000    0.2000000000
   13         0.0000000000    0.0000000000
   14         0.2000000000    0.0000000000
   15         0.4000000000    0.0000000000
   16         0.6000000000    0.0000000000
   17         0.8000000000    0.0000000000
   18         1.0000000000    0.0000000000

*********  node connectivity  *********************

 No.
    1   14    8   13
    2    7   13    8
    3    7    8    1
    4    2    1    8
    5    2    8    3
    6    9    3    8
    7    9    8   15
    8   14   15    8
    9   16   10   15
   10    9   15   10
   11    9   10    3
   12    4    3   10
   13    4   10    5
   14   11    5   10
   15   11   10   17
   16   16   17   10
   17   18   12   17
   18   11   17   12
   19   11   12    5
   20    6    5   12

*********  specified values  **********************

 No.             UU(m/s)          VV(m/s)
    1    1       1.0000000000     0.0000000000
    2    2       1.0000000000     0.0000000000
    3    3       1.0000000000     0.0000000000
    4    4       1.0000000000     0.0000000000
    5    5       1.0000000000     0.0000000000
    6    6       1.0000000000     0.0000000000
    7    7       1.0000000000     0.0000000000
```

Fig. 5.11 Input data

```
   8    8    1.0000000000    0.0000000000
   9    9    1.0000000000    0.0000000000
  10   10    1.0000000000    0.0000000000
  11   11    1.0000000000    0.0000000000
  12   12    1.0000000000    0.0000000000
  13   13    1.0000000000    0.0000000000
  14   14    1.0000000000    0.0000000000
  15   15    1.0000000000    0.0000000000
  16   16    1.0000000000    0.0000000000
  17   17    1.0000000000    0.0000000000
  18   18    1.0000000000    0.0000000000

*********  boundary condition  ********************

No.            PPM
   1    1    1.0000000000
   2    6    0.0000000000
   3    7    1.0000000000
   4   12    0.0000000000
   5   13    1.0000000000
   6   18    0.0000000000
```

Fig. 5.11 (continued)

$$c = \begin{cases} 0.5(\cos(2\pi r) + 1) & r \leqq 0.5 \\ 0 & r > 0.5 \end{cases} \tag{5.206}$$

where

$$r = \sqrt{x^2 + (y + 0.5)^2} \tag{5.207}$$

The initial condition is illustrated in Fig. 5.15. The cone is conveyed by the flow and the shape of the cone should be conserved. The finite element mesh is shown in Fig. 5.16. The total numbers of nodes and elements are 1681 and 3200, respectively. The time increment Δt is 0.001. The computed results obtained by the Crank–Nicolson scheme are represented in Fig. 5.17a–d. The rotating configurations of the concentration are shown in the figures. The comparison of the peak values computed by the different schemes is summarized in Table 5.4. The ratio of the peak value of one cycle later to the initial peak value is shown in Table 5.4. It is interesting that the peak value ratio by the characteristic scheme is 1, which means exact match. The ratio by the Crank–Nicolson scheme, which we found out in the Sect. 5.5.3 to be neutral, is slightly larger than 1. Contrary to this, the ratio of the peak value computed by the SUPG scheme is slightly smaller than 1, which means stable scheme. In the results of Fig. 5.17d, the small nonzero concentration can be seen in the place where essentially they should be zero.

```
*************  output results  ********************

istep 10  time 0.2000000000000000

   1   1.000    2   0.768    3   0.513    4   0.294    5   0.127    6   0.000
   7   1.000    8   0.771    9   0.516   10   0.289   11   0.130   12   0.000
  13   1.000   14   0.768   15   0.513   16   0.294   17   0.127   18   0.000

istep 20  time 0.4000000000000000

   1   1.000    2   0.872    3   0.714    4   0.521    5   0.300    6   0.000
   7   1.000    8   0.876    9   0.712   10   0.523   11   0.292   12   0.000
  13   1.000   14   0.872   15   0.714   16   0.521   17   0.300   18   0.000

istep 30  time 0.6000000000000000

   1   1.000    2   0.906    3   0.780    4   0.598    5   0.360    6   0.000
   7   1.000    8   0.909    9   0.776   10   0.603   11   0.349   12   0.000
  13   1.000   14   0.906   15   0.780   16   0.598   17   0.360   18   0.000

istep 40  time 0.8000000000000000

   1   1.000    2   0.917    3   0.802    4   0.623    5   0.380    6   0.000
   7   1.000    8   0.920    9   0.797   10   0.629   11   0.368   12   0.000
  13   1.000   14   0.917   15   0.802   16   0.623   17   0.380   18   0.000

istep 50  time 1.000000000000000

   1   1.000    2   0.920    3   0.809    4   0.632    5   0.387    6   0.000
   7   1.000    8   0.924    9   0.804   10   0.638   11   0.374   12   0.000
  13   1.000   14   0.920   15   0.809   16   0.632   17   0.387   18   0.000

istep 60  time 1.200000000000000

   1   1.000    2   0.922    3   0.811    4   0.635    5   0.389    6   0.000
   7   1.000    8   0.925    9   0.807   10   0.641   11   0.376   12   0.000
  13   1.000   14   0.922   15   0.811   16   0.635   17   0.389   18   0.000

istep 70  time 1.400000000000000

   1   1.000    2   0.922    3   0.812    4   0.636    5   0.390    6   0.000
   7   1.000    8   0.926    9   0.807   10   0.642   11   0.377   12   0.000
  13   1.000   14   0.922   15   0.812   16   0.636   17   0.390   18   0.000

istep 80  time 1.600000000000000

   1   1.000    2   0.922    3   0.812    4   0.636    5   0.390    6   0.000
   7   1.000    8   0.926    9   0.808   10   0.643   11   0.377   12   0.000
  13   1.000   14   0.922   15   0.812   16   0.636   17   0.390   18   0.000

istep 90  time 1.800000000000000

   1   1.000    2   0.922    3   0.812    4   0.636    5   0.390    6   0.000
   7   1.000    8   0.926    9   0.808   10   0.643   11   0.377   12   0.000
  13   1.000   14   0.922   15   0.812   16   0.636   17   0.390   18   0.000

istep 100  time 2.000000000000000

   1   1.000    2   0.922    3   0.812    4   0.636    5   0.390    6   0.000
   7   1.000    8   0.926    9   0.808   10   0.643   11   0.377   12   0.000
  13   1.000   14   0.922   15   0.812   16   0.636   17   0.390   18   0.000
```

Fig. 5.12 An example of output data

Table 5.3 Computation condition

Case	Dif. coef.	Boundary cond.		vel. u	vel. v	Δt	
		AC	BD				
1	$0.5\,m^2/s$	1.0 ppm	0.0 ppm	0.0 m/s	0.0 m/s	0.02 s	Fig. 5.13a
2	0.5	1.0	0.0	1.0	0.0	0.02	Fig. 5.13b
3	0.5	1.0	–	0.0	0.0	0.02	Fig. 5.13c
4	0.5	1.0	–	1.0	0.0	0.02	Fig. 5.13d

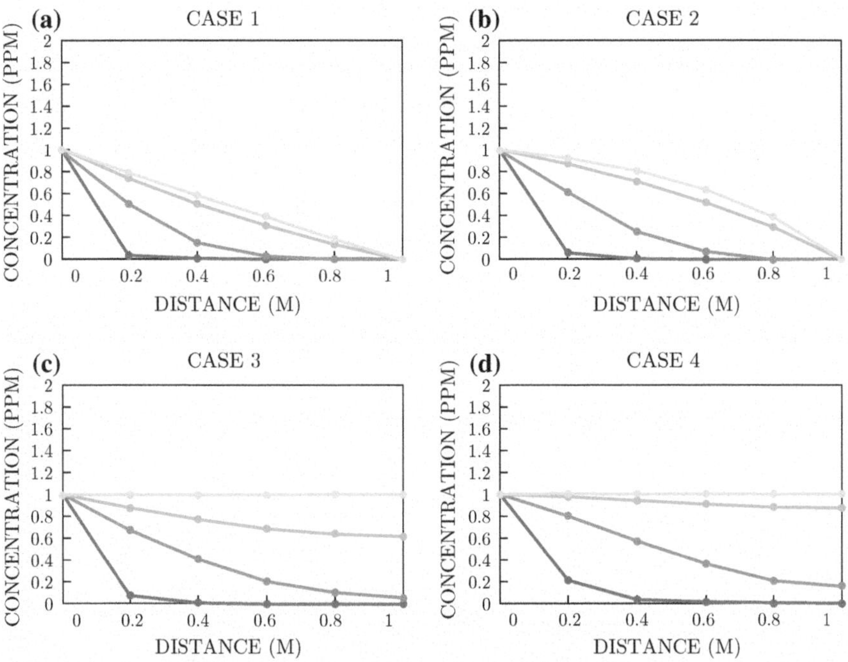

Fig. 5.13 Computed results of trial computation, **a** without advection velocity and zero boundary value, **b** with advection velocity and zero boundary value, **c** without advection velocity and non-boundary value, **d** with advection velocity and non-boundary value

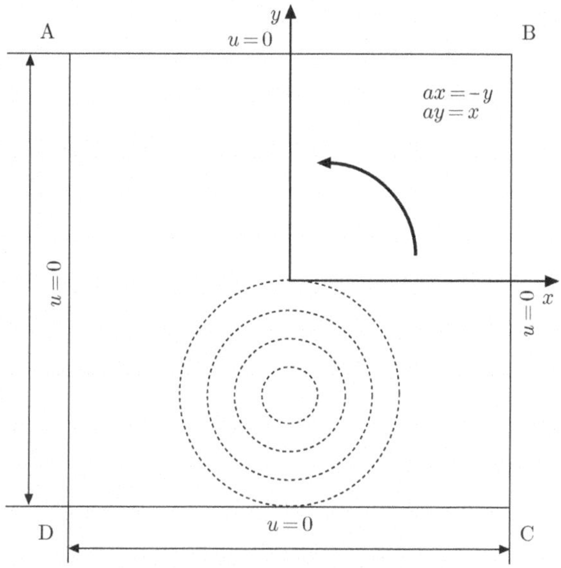

Fig. 5.14 Schematic view of rotating cone problem

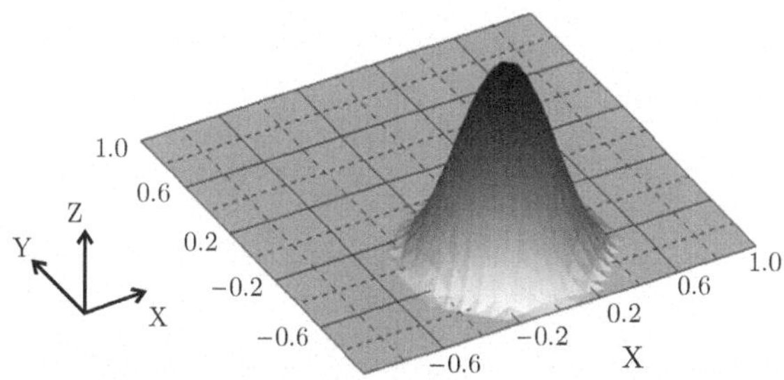

Fig. 5.15 Initial condition of rotating cone

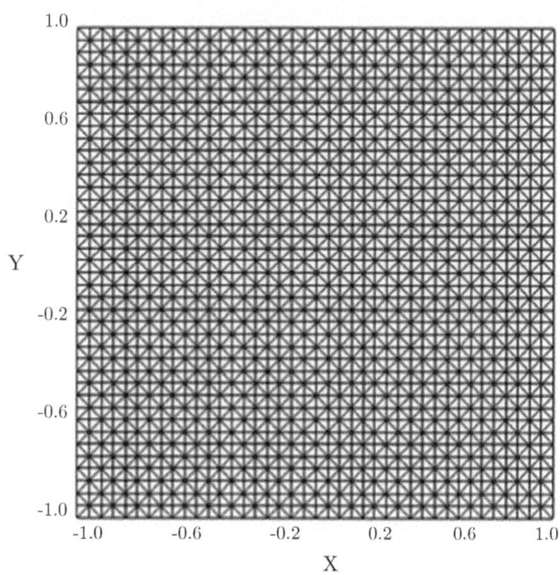

Fig. 5.16 Finite element mesh

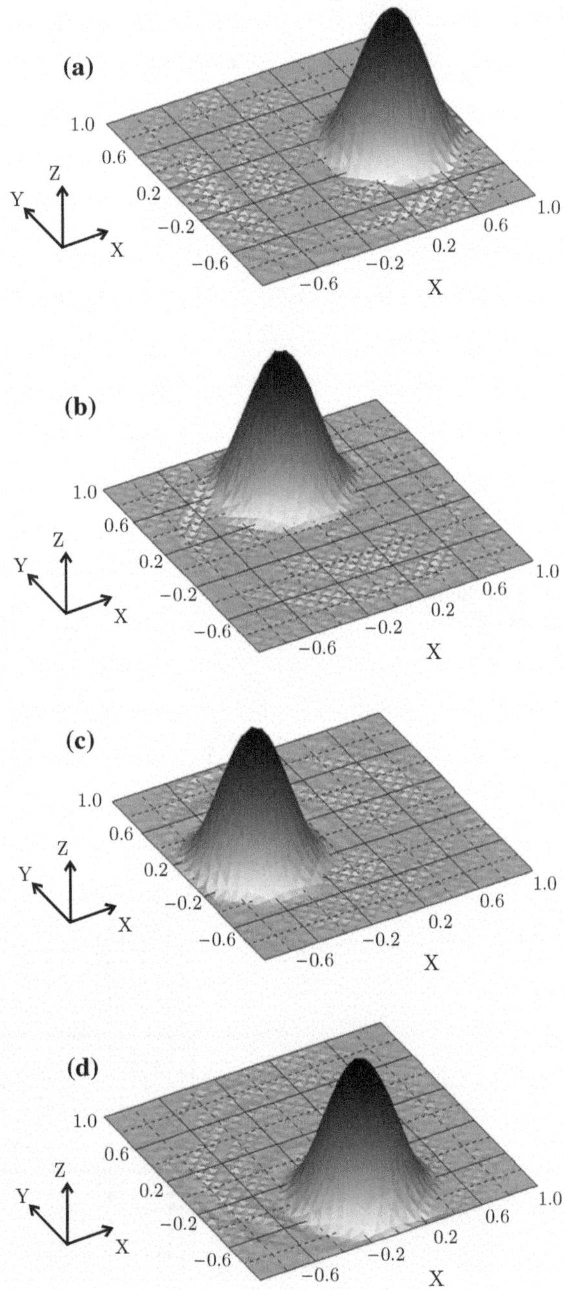

Fig. 5.17 Computed cone configuration. **a** Cone at step 16. **b** Cone at step 32. **c** Cone at step 48. **d** Cone at step 64

Table 5.4 Peak values of concentration

Scheme	Ratio of peak value
Crank–Nicolson	1.00307
SUPG	0.987375
Characteristic	1
Bubble f. + SUPG s.[a]	1.001065
Bubble f. + BTD s.[b]	1.001090

[a]Bubble function scheme with SUPG stabilization
[b]Bubble function scheme with BTD stabilization

5.12 Summary and Conclusion

In this chapter, we have discussed transport problem, which includes mutually contradictory phenomena, i.e., advection and diffusion. The main point is that the phenomena are time-dependent. To solve time-dependent problems, although it is possible to use the finite element method in time, the time-marching schemes are usually used considering stability condition. We discussed the explicit scheme, the Crank–Nicolson scheme, the implicit scheme, the SUPG scheme, the BTD scheme, the bubble function scheme, and the characteristic scheme. The explicit scheme does not employ the inverse operation of the coefficient matrix. To do this, a limit of time increment should be introduced. Thus, in general it is rather difficult to use long time increment. Contrary to this, the implicit scheme uses the inverse operation of the coefficient matrix and rather long time increment can be adaptable. To use unreasonably long time increment is not recommended because the phenomena with abrupt changes cannot be properly analyzed. The Crank–Nicolson scheme is neutral and we have shown the artificial viscosity is small by the numerical study. The SUPG scheme is stable; however, it is necessary to use long computational time for the computation of the SUPG term. The standard bubble function scheme has not sufficient artificial viscosity. We have shown that the artificial viscosity of the bubble function schemes can be compensated to be equal to those of the SUPG and BTD schemes using stabilization parameters. The characteristic scheme needs computational time to search for the upstream point on the characteristic line. A plenty of computational schemes have been presented in addition to the schemes discussed in this chapter. How to choose the best scheme depends on the problem to be solved. Computational mesh is also an important factor to obtain genuine numerical results.

Chapter 6
Creeping Flow

6.1 Introduction

As the last topic of the basic concepts of the finite element method in fluid flows, creeping flow problem of the incompressible fluid is discussed in this chapter. The flow is highly viscous and moves with slow velocity, whose acceleration effect is negligible. Thus, the problem turns out to be linear. This type of flow is not practical, i.e., we can only count melting candle flow and few other examples. However, the analysis of the flow includes important technique, namely, the mixed interpolation. The interpolation for velocity is one order or more higher than that for pressure. The incompressibility is assumption of some sort of limit state. Therefore, special treatment should be required. To simplify the problem, we confine our attention to the two-dimensional analysis. However, extension to the three-dimensional analysis is straightforward.

In Sects. 6.2–6.4, the indicial notation and summation convention with repeated indices are used. After Sect. 6.5, until the end of this chapter, coordinate is expressed by the pair (x, y) without indicial notation. The summation convention is used for subscripts α, β, ξ, η, μ, λ to express matrix multiplication.

6.2 Creeping Flow

Let a whole flow field be denoted by V, whose boundary S is shown in Fig. 6.1. Assuming the fluid is incompressible, the equation of continuity is expressed by the following equation:

$$v_{i,i} = 0 \quad \text{in} \quad V \tag{6.1}$$

M. Kawahara, *Finite Element Methods in Incompressible, Adiabatic, and Compressible Flows*, Mathematics for Industry 19,
DOI 10.1007/978-4-431-55450-9_6

Fig. 6.1 Flow field

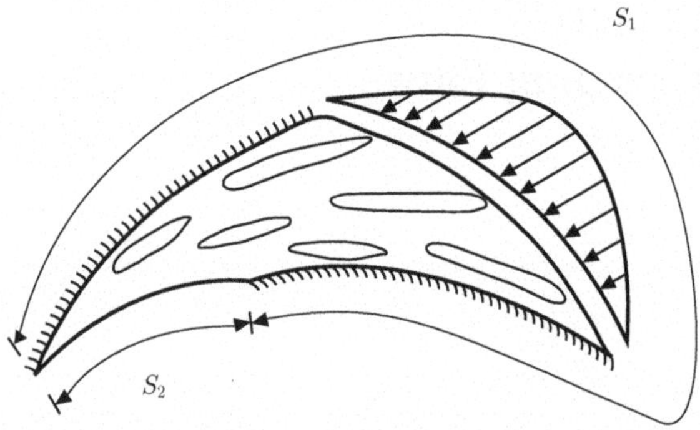

where v_i is velocity. In Sects. 6.2–6.4, we use indicial notation and summation convention. Equilibrium of total stress can be described as follows disregarding acceleration:

$$\sigma_{ji,j} = 0 \quad \text{in} \quad V \tag{6.2}$$

where σ_{ij} is total stress, which is given by the constitutive equation as

$$\sigma_{ij} = -p\delta_{ij} + \mu(v_{i,j} + v_{j,i}) \quad \text{in} \quad V \tag{6.3}$$

where p is pressure and μ is a viscosity coefficient. Introducing Eq. (6.3) into Eq. (6.2), and using Eq. (6.1), the Stokes equation is obtained as

$$-p,_i + \mu v_{i,jj} = 0 \,. \tag{6.4}$$

The flow field is surrounded by the boundary S, which is assumed to be divided into two nonoverlapped parts, S_1 and S_2. On the boundary S_1, the velocity is given as

$$v_i = \hat{v}_i \quad \text{on} \quad S_1 \tag{6.5}$$

where $\hat{v}_i$ is the given value on the boundary; for instance, $\hat{v}_i = 0$ if S_1 is a wall of channel. If the velocity is forced to flow in or out on S_1, $\hat{v}_i$ is the forced velocity. On the boundary S_2, surface force t_i is specified as

$$t_i = \sigma_{ij} n_j = \hat{t}_i \quad \text{on} \quad S_2 \tag{6.6}$$

where $\hat{t}_i$ represents surface force given on the boundary. If the channel is cut at the boundary and flow is forced to flow out into a vacuum, the surface force may be zero. If the flow is forced to flow in by a piston, for instance, the surface force may be given.

We confine our attention, in this section, to two-dimensional steady flow. Thus, the governing equations are equation of continuity Eq. (6.1), and equilibrium equation Eq. (6.2) with constitutive equation Eq. (6.3) under the boundary conditions Eqs. (6.5) and (6.6).

6.3 Weighted Residual Equation

Multiplying both sides of Eq. (6.1) by a weighting function p^* and integrating over a flow field V, the weighted residual equation for continuity is obtained as follows:

$$\int_V (p^* v_{i,i}) dV = 0 \tag{6.7}$$

where we can use any kind of continuous function for p^*. Multiplying both sides of Eq. (6.2) by a weighting function v_i^* and integrating over the flow field V, the following equation is obtained:

$$\int_V (v_i^* \sigma_{ji,j}) dV = 0 \tag{6.8}$$

where v_i^* is an arbitrary continuous function. Replacing the integrand of Eq. (6.8) by the equality

$$(v_i^* \sigma_{ji})_{,j} = v_i^* \sigma_{ji,j} + v_{i,j}^* \sigma_{ji} \tag{6.9}$$

Equation (6.8) can be transformed into the following form:

$$\int_V (v_{i,j}^* \sigma_{ji}) dV = \int_V (v_i^* \sigma_{ji})_{,j} dV \; . \tag{6.10}$$

Applying the Stokes theorem to the right side of Eq. (6.10), the volume integral can be converted to the surface integral. Then, it is obtained as

$$\int_V (v_{i,j}^* \sigma_{ij}) dV = \int_S (v_i^* \sigma_{ij} n_j) dS \tag{6.11}$$

Note that the indices are exchanged using the symmetry of σ_{ij}. Introducing the first relation in Eq. (6.6), the weighted residual equation for equilibrium can be derived as follows:

$$\int_V (v_{i,j}^* \sigma_{ij}) dV = \int_S (v_i^* t_i) dS \tag{6.12}$$

Recall that volume V is arbitrary. Therefore, it can be considered as an each finite element or as a whole flow field, which consists of total finite elements. As described in Sect. 4.3, by summing up Eq. (6.12) for each element, Eq. (6.12) for a whole flow field can be obtained.

Let V be the whole flow field, whose boundary is S. Recall that boundaries S_1 and S_2 satisfy the relation

$$\begin{aligned} S_1 \cup S_2 &= S \\ S_1 \cap S_2 &= \mathbf{0} \end{aligned} \tag{6.13}$$

and assume

$$v_i^* = 0 \quad \text{on} \quad S_1 \tag{6.14}$$

then, using Eq. (6.6), the weighted residual equation for the whole flow field is derived as

$$\int_V (v_{i,j}^* \sigma_{ij}) dV = \int_{S_2} (v_i^* \hat{t}_i) dS \tag{6.15}$$

We would emphasize again that the global form, Eq. (6.15), can be derived by summing up equation system of Eq. (6.12) established on each finite element, by which the whole flow field is divided piecewise.

Substituting Eq. (6.3) into Eq. (6.12) and rearranging the terms, we get

$$-\int_V (v_{i,i}^* p) dV + \mu \int_V (v_{i,j}^* v_{i,j}) dV + \mu \int_V (v_{i,j}^* v_{j,i}) dV = \int_S (v_i^* t_i) dS \tag{6.16}$$

This is the weighted residual equation in terms of velocity v_i, pressure p, and surface force t_i. Equation (6.16) is equivalent to Eqs. (6.2) and (6.3). Substituting Eq. (6.6) into Eq. (6.16) and rearranging the terms, we get

$$-\int_V (v_{i,i}^* p) dV + \mu \int_V (v_{i,j}^* v_{i,j}) dV + \mu \int_V (v_{i,j}^* v_{j,i}) dV = \int_{S_2} (v_i^* \hat{t}_i) dS \tag{6.17}$$

assuming that $v_i^* = 0$ on S_1. Equation (6.17) with boundary condition Eq. (6.5) is equivalent to Eqs. (6.2) and (6.3) with boundary conditions Eqs. (6.5) and (6.6). We will formulate the local finite element equation based on Eqs. (6.7) and (6.16), then, derive the global finite element equation by the superimposing technique, which is already precisely described in Chaps. 2–4.

6.4 Finite Element Equation

Based on Eqs. (6.7) and (6.16), the local finite element equation can be deduced in this section. The two-dimensional flow field is supposed to be divided into a sum of small finite elements as shown in Fig. 6.2. For the configuration of the finite element, triangular shape is used. The field variables are velocity v_i and pressure p. For the analysis of the incompressible flow, the mixed interpolation should be used. Namely, the interpolation employed for velocity is one order or more higher interpolation than that for pressure. In this section, for the interpolation of velocity, six-node triangular

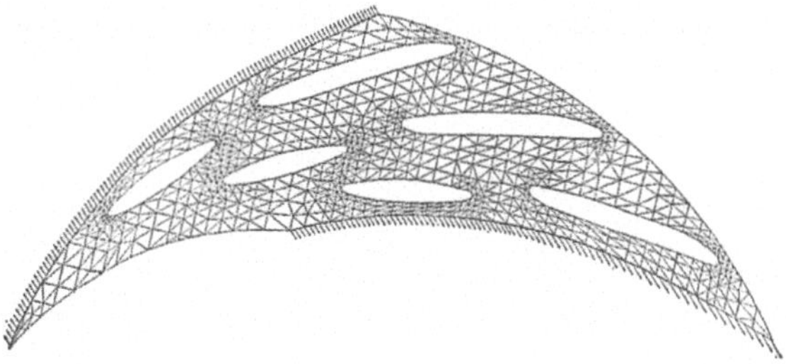

Fig. 6.2 Finite element mesh

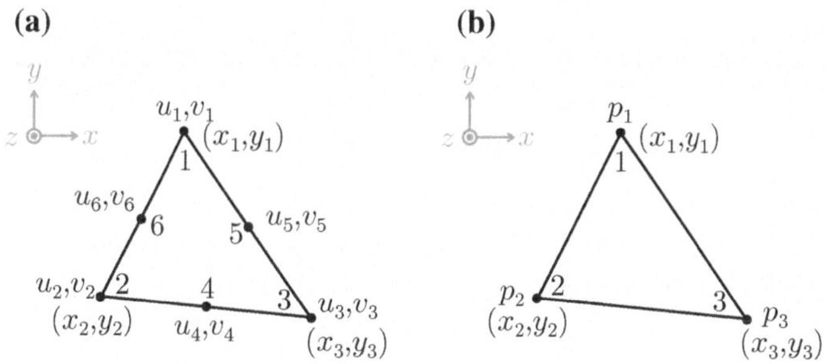

Fig. 6.3 Triangular finite element. **a** six-node triangular element for velocity. **b** three-node triangular element for pressure

element based on the quadratic polynomial and for the interpolation of pressure, three-node triangular element based on the linear polynomial are utilized as shown in Fig. 6.3. We have distinguished clearly between node and edge; however, in this chapter and later, we use only node, and we do not distinguish node and edge.

Denoting velocity at node α in the i-direction $v_{\alpha i}$, the interpolation relation of velocity can be expressed as follows:

$$v_i = \Phi_\beta v_{\beta i} \quad (\beta = 1 \text{ to } 6) \tag{6.18}$$

where Φ_β is the shape function, which consists of quadratic polynomials whose details are defined and explained in Sect. 6.6. Denoting pressure at node λ with p_λ, the interpolation relation of pressure can be expressed as follows:

$$p = \Psi_\lambda p_\lambda \quad (\lambda = 1 \text{ to } 3) \tag{6.19}$$

where Ψ_λ is the shape function, which consists of linear polynomials defined in Sect. 6.6. The following relations are used for the weighting function:

$$v_i^* = \Phi_\alpha v_{\alpha i}^* \tag{6.20}$$

$$p^* = \Psi_\mu p_\mu^* \tag{6.21}$$

Introducing Eqs. (6.18)–(6.21) into Eqs. (6.7) and (6.16), and rearranging the terms, we get

$$p_\mu^* \left[\int_V (\Psi_\mu \Phi_{\beta,i}) dV \right] v_{\beta i} = 0 \tag{6.22}$$

$$\begin{aligned} v_{\alpha i}^* \Bigg\{ & \left[-\int_V (\Phi_{\alpha,i} \Psi_\lambda) dV \right] p_\lambda + \left[\mu \int_V (\Phi_{\alpha,k} \Phi_{\beta,k}) \delta_{ij} dV \right] v_{\beta j} \\ & + \left[\mu \int_V (\Phi_{\alpha,j} \Phi_{\beta,i}) dV \right] v_{\beta j} - \int_S (\Phi_\alpha t_i) dS \Bigg\} = 0 \end{aligned} \tag{6.23}$$

From Eqs. (6.22) and (6.23), the local finite element equation is derived as in the following form taking advantage of the fact that p_μ^* and $v_{\alpha i}^*$ are arbitrary constants.

$$\begin{bmatrix} S_{\alpha i \beta j} & -H_{\lambda \alpha i} \\ H_{\mu \beta j} & 0 \end{bmatrix} \begin{bmatrix} v_{\beta j} \\ p_\lambda \end{bmatrix} = \begin{bmatrix} \Omega_{\alpha i} \\ 0 \end{bmatrix} \tag{6.24}$$

where

$$S_{\alpha i \beta j} = \mu \int_V (\Phi_{\alpha,k} \Phi_{\beta,k}) \delta_{ij} dV + \mu \int_V (\Phi_{\alpha,j} \Phi_{\beta,i}) dV \tag{6.25}$$

$$H_{\mu \alpha i} = \int_V (\Psi_\mu \Phi_{\alpha,i}) dV . \tag{6.26}$$

$$\Omega_{\alpha i} = \int_S (\Phi_\alpha t_i) dS \tag{6.27}$$

The global finite element equation can be obtained by the superposition procedure as described in Sects. 2.6 and 4.7 based on Eq. (6.24). In the case of the incompressible flows, the calculation becomes complicated, because Eq. (6.24) contains zero diagonal terms. To compute the concrete forms of coefficients of Eq. (6.24), the area coordinate is useful for differentiation and integration over triangular area, which is discussed in further detail in the next section.

6.5 Area Coordinate

To complete the calculation of coefficients of Eq. (6.24), the interpolation functions Eqs. (6.18) and (6.19) must be determined in concrete terms. For the purpose, it is useful to introduce the area coordinate. We should denote three nodes as a, b, c in Fig. 6.4 to avoid confusion in the former sections.

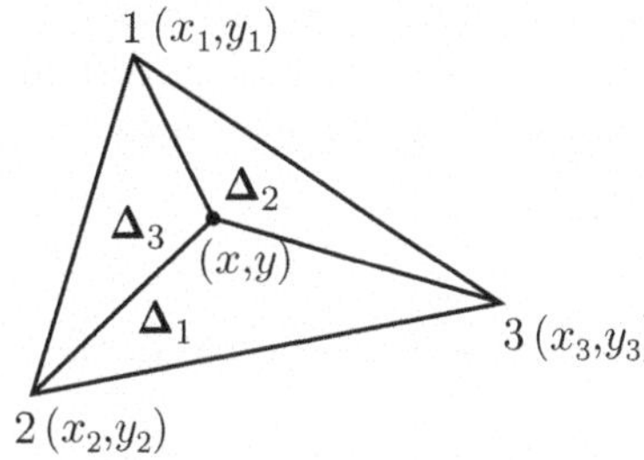

Fig. 6.4 Area coordinate

Therefore, from now on, we denote three nodes of triangle as 1, 2, 3. This is because not only it is traditional to use numbers in the area coordinate, but also numbers are much easier to handle in the computer program.

The coordinate of arbitrary point is shown by (x, y) in Fig. 6.4. The coordinates of three nodes are expressed by (x_i, y_i), $i = 1$ to 3. In this chapter, subscripts i, j, k do not show coordinates x, y, z as were in the former sections. They instead shows nodes 1, 2, 3. We will use middle points as shown in Fig. 6.3a, and those are referred to as nodes 4, 5, 6. The triangle whose three nodes (x, y), (x_2, y_2), (x_3, y_3) is denoted by $\boldsymbol{\Delta}_1$, whose three nodes (x, y), (x_3, y_3), (x_1, y_1) by $\boldsymbol{\Delta}_2$, and whose three nodes (x, y), (x_1, y_1), (x_2, y_2) by $\boldsymbol{\Delta}_3$ as represented in Fig. 6.4.

The area coordinates η_1, η_2, η_3 are defined as

$$\begin{aligned} \eta_1 &= \frac{\boldsymbol{\Delta}_1}{\boldsymbol{\Delta}} = a_1 + b_1 x + c_1 y \\ \eta_2 &= \frac{\boldsymbol{\Delta}_2}{\boldsymbol{\Delta}} = a_2 + b_2 x + c_2 y \\ \eta_3 &= \frac{\boldsymbol{\Delta}_3}{\boldsymbol{\Delta}} = a_3 + b_3 x + c_3 y \end{aligned} \tag{6.28}$$

where $\boldsymbol{\Delta}$ is the area of the triangle and a_i, b_i, c_i are as follows:

$$\begin{aligned} a_i &= \frac{1}{2\boldsymbol{\Delta}}(x_j y_k - x_k y_j) \\ b_i &= \frac{1}{2\boldsymbol{\Delta}}(y_j - y_k) \\ c_i &= \frac{1}{2\boldsymbol{\Delta}}(x_k - x_j) \end{aligned} \tag{4.23}$$

$$\boldsymbol{\Delta} = \frac{1}{2}\left\{(x_j - x_i)(y_k - y_i) - (x_k - x_i)(y_j - y_i)\right\} \tag{4.24}$$

where i, j, k are permutations of 1, 2, 3. Those are already discussed in Sect. 4.4. The following relations are derived from the area coordinate. By definition, three components are linearly dependent, i.e.,

$$\eta_1 + \eta_2 + \eta_3 = 1 \tag{6.29}$$

The relationships between coordinates x, y and η_1, η_2, η_3 can be written as follows:

$$\begin{bmatrix} \eta_1 \\ \eta_2 \\ \eta_3 \end{bmatrix} = \begin{bmatrix} a_1 & b_1 & c_1 \\ a_2 & b_2 & c_2 \\ a_3 & b_3 & c_3 \end{bmatrix} \begin{bmatrix} 1 \\ x \\ y \end{bmatrix} \tag{6.30}$$

$$\begin{bmatrix} 1 \\ x \\ y \end{bmatrix} = \begin{bmatrix} 1 & 1 & 1 \\ x_1 & x_2 & x_3 \\ y_1 & y_2 & y_3 \end{bmatrix} \begin{bmatrix} \eta_1 \\ \eta_2 \\ \eta_3 \end{bmatrix} \tag{6.31}$$

Equation (6.30) is the matrix form of three equations in Eq. (6.28), and Eq. (6.31) is the inverse relation of Eq. (6.30). The area coordinate η_1–η_3 is a one-to-one correspondent to the x–y coordinate. Namely, the pair (x, y) can be determined by the combination of (η_1, η_2, η_3).

The differentiation of the area coordinate can be formulated as in the following formula, for the first-order polynomial

$$\frac{\partial}{\partial x} \begin{bmatrix} \eta_1 \\ \eta_2 \\ \eta_3 \end{bmatrix} = \begin{bmatrix} b_1 \\ b_2 \\ b_3 \end{bmatrix} \tag{6.32}$$

$$\frac{\partial}{\partial y} \begin{bmatrix} \eta_1 \\ \eta_2 \\ \eta_3 \end{bmatrix} = \begin{bmatrix} c_1 \\ c_2 \\ c_3 \end{bmatrix} \tag{6.33}$$

and for the second-order polynomial

$$\frac{\partial}{\partial x} \begin{bmatrix} \eta_1{}^2 \\ \eta_2{}^2 \\ \eta_3{}^2 \\ \eta_2\eta_3 \\ \eta_3\eta_1 \\ \eta_1\eta_2 \end{bmatrix} = \begin{bmatrix} 2b_1 & & \\ & 2b_2 & \\ & & 2b_3 \\ & b_3 & b_2 \\ b_3 & & b_1 \\ b_2 & b_1 & \end{bmatrix} \begin{bmatrix} \eta_1 \\ \eta_2 \\ \eta_3 \end{bmatrix} \tag{6.34}$$

$$\frac{\partial}{\partial y}\begin{bmatrix} \eta_1{}^2 \\ \eta_2{}^2 \\ \eta_3{}^2 \\ \eta_2\eta_3 \\ \eta_3\eta_1 \\ \eta_1\eta_2 \end{bmatrix} = \begin{bmatrix} 2c_1 & & \\ & 2c_2 & \\ & & 2c_3 \\ & c_3 & c_2 \\ c_3 & & c_1 \\ c_2 & c_1 & \end{bmatrix}\begin{bmatrix} \eta_1 \\ \eta_2 \\ \eta_3 \end{bmatrix} \tag{6.35}$$

Those formula are helpful for the differentiation of interpolation function. The advantage of the area coordinate is that the integral formula over the triangular area is given by explicit form as follows:

$$\frac{1}{2\boldsymbol{\Delta}}\int(\eta_1{}^l\eta_2{}^m\eta_3{}^n)dxdy = \frac{l!m!n!}{(l+m+n+2)!} \tag{6.36}$$

where l, m, n are arbitrary integer numbers and $\boldsymbol{\Delta}$ is the area of a triangle.

6.6 Interpolation Function

The interpolation function can be formulated using the area coordinate introduced in Sect. 6.5. As was stated in the previous Sect. 6.4, the mixed interpolation should be used; namely, the interpolation of velocity should be one order or more higher than that of pressure. The pressure inside an element is assumed to be given by a linear function as

$$p = \alpha_1\eta_1 + \alpha_2\eta_2 + \alpha_3\eta_3 \tag{6.37}$$

where α_1, α_2, α_3 are unknown constants. The pressure p given by Eq. (6.37) should coincide with the pressure p_1, p_2, p_3 at nodes 1, 2, 3 of the triangular element. The three nodes are expressed as $(1,0,0)$, $(0,1,0)$, $(0,0,1)$ by the area coordinate. Substituting these values into Eq. (6.37)

$$\begin{bmatrix} p_1 \\ p_2 \\ p_3 \end{bmatrix} = \begin{bmatrix} 1 & & \\ & 1 & \\ & & 1 \end{bmatrix}\begin{bmatrix} \alpha_1 \\ \alpha_2 \\ \alpha_3 \end{bmatrix} \tag{6.38}$$

Then, the unknown constants are

$$\begin{bmatrix} \alpha_1 \\ \alpha_2 \\ \alpha_3 \end{bmatrix} = \begin{bmatrix} p_1 \\ p_2 \\ p_3 \end{bmatrix} \tag{6.39}$$

Equation (6.37) can be reformulated as

$$p = [\eta_1\ \eta_2\ \eta_3] \begin{bmatrix} \alpha_1 \\ \alpha_2 \\ \alpha_3 \end{bmatrix} = [\eta_1\ \eta_2\ \eta_3] \begin{bmatrix} p_1 \\ p_2 \\ p_3 \end{bmatrix} \tag{6.40}$$

Comparing Eq. (6.40) with Eq. (6.19), we obtain

$$\begin{bmatrix} \Psi_1 \\ \Psi_2 \\ \Psi_3 \end{bmatrix} = \begin{bmatrix} \eta_1 \\ \eta_2 \\ \eta_3 \end{bmatrix} \tag{6.41}$$

The shape function of the interpolation function of pressure is the area coordinate itself.

For the velocity, there are two components u and v. However, because the interpolation is the same, we only consider the interpolation of velocity u. It is assumed that the velocity u inside an element is given by the second-order polynomial as

$$u = \beta_1 {\eta_1}^2 + \beta_2 {\eta_2}^2 + \beta_3 {\eta_3}^2 + \beta_4 \eta_2 \eta_3 + \beta_5 \eta_3 \eta_1 + \beta_6 \eta_1 \eta_2 \tag{6.42}$$

where $\beta_1, \ldots, \beta_6$ are unknown constants. As shown in Fig. 6.3a, we use three nodes and three mid nodes, we call them six nodes as a whole, of which coordinates are expressed as

$$(1,0,0),\ (0,1,0),\ (0,0,1),\ \left(0, \frac{1}{2}, \frac{1}{2}\right),\ \left(\frac{1}{2}, 0, \frac{1}{2}\right),\ \left(\frac{1}{2}, \frac{1}{2}, 0\right)$$

by the area coordinate. Substituting these values into Eq. (6.42), we obtain the following:

$$\begin{bmatrix} u_1 \\ u_2 \\ u_3 \\ u_4 \\ u_5 \\ u_6 \end{bmatrix} = \begin{bmatrix} 1 & & & & & \\ & 1 & & & & \\ & & 1 & & & \\ & \frac{1}{4} & \frac{1}{4} & \frac{1}{4} & & \\ \frac{1}{4} & & \frac{1}{4} & & \frac{1}{4} & \\ \frac{1}{4} & \frac{1}{4} & & & & \frac{1}{4} \end{bmatrix} \begin{bmatrix} \beta_1 \\ \beta_2 \\ \beta_3 \\ \beta_4 \\ \beta_5 \\ \beta_6 \end{bmatrix} \tag{6.43}$$

Inverting Eq. (6.43), the unknown constants can be obtained.

$$\begin{bmatrix} \beta_1 \\ \beta_2 \\ \beta_3 \\ \beta_4 \\ \beta_5 \\ \beta_6 \end{bmatrix} = \begin{bmatrix} 1 & & & & & \\ & 1 & & & & \\ & & 1 & & & \\ & -1 & -1 & 4 & & \\ -1 & & -1 & & 4 & \\ -1 & -1 & & & & 4 \end{bmatrix} \begin{bmatrix} u_1 \\ u_2 \\ u_3 \\ u_4 \\ u_5 \\ u_6 \end{bmatrix} \tag{6.44}$$

Using Eq. (6.44), Eq. (6.42) is reformulated as

$$\begin{aligned} u &= \left[\eta_1{}^2 \; \eta_2{}^2 \; \eta_3{}^2 \; \eta_2\eta_3 \; \eta_3\eta_1 \; \eta_1\eta_2 \right] \begin{bmatrix} \beta_1 \\ \beta_2 \\ \beta_3 \\ \beta_4 \\ \beta_5 \\ \beta_6 \end{bmatrix} \\ &= \left[\eta_1{}^2 \; \eta_2{}^2 \; \eta_3{}^2 \; \eta_2\eta_3 \; \eta_3\eta_1 \; \eta_1\eta_2 \right] \begin{bmatrix} 1 & & & & & \\ & 1 & & & & \\ & & 1 & & & \\ & -1 & -1 & 4 & & \\ -1 & & -1 & & 4 & \\ -1 & -1 & & & & 4 \end{bmatrix} \begin{bmatrix} u_1 \\ u_2 \\ u_3 \\ u_4 \\ u_5 \\ u_6 \end{bmatrix} \end{aligned} \tag{6.45}$$

Equation (6.45) is the interpolation relation for the velocity u. The velocity v is also assumed to be given by

$$v = \gamma_1\eta_1{}^2 + \gamma_2\eta_2{}^2 + \gamma_3\eta_3{}^2 + \gamma_4\eta_2\eta_3 + \gamma_5\eta_3\eta_1 + \gamma_6\eta_1\eta_2 \tag{6.46}$$

where $\gamma_1, \ldots, \gamma_6$ are unknown constants. The same procedure leads to the same interpolation relation. Therefore, the shape function of the interpolation function of velocity can be represented as follows:

$$\begin{bmatrix} \Phi_1 \\ \Phi_2 \\ \Phi_3 \\ \Phi_4 \\ \Phi_5 \\ \Phi_6 \end{bmatrix} = \begin{bmatrix} 1 & & & & -1 & -1 \\ & 1 & & -1 & & -1 \\ & & 1 & -1 & -1 & \\ & & & 4 & & \\ & & & & 4 & \\ & & & & & 4 \end{bmatrix} \begin{bmatrix} {\eta_1}^2 \\ {\eta_2}^2 \\ {\eta_3}^2 \\ \eta_2\eta_3 \\ \eta_3\eta_1 \\ \eta_1\eta_2 \end{bmatrix} \tag{6.47}$$

Making use of Eq. (6.47), the interpolation relations for velocity u and v are as follows:

$$u = \Phi_\alpha u_\alpha \tag{6.48}$$

$$v = \Phi_\alpha v_\alpha \tag{6.49}$$

Both Eqs. (6.48) and (6.49) can be written as Eq. (6.18). The shape function Eq. (6.41) can be written as

$$\Psi_\lambda = I_{\lambda\mu} f_\mu \tag{6.50}$$

where $I_{\lambda\mu}$ is unit matrix and f_μ consists of first-order area coordinate as follows:

$$I_{\lambda\mu} = \begin{bmatrix} 1 & & \\ & 1 & \\ & & 1 \end{bmatrix} \tag{6.51}$$

$$f_\mu = \begin{bmatrix} \eta_1 \\ \eta_2 \\ \eta_3 \end{bmatrix} \tag{6.52}$$

In the same manner, Eq. (6.47) can be written as

$$\Phi_\alpha = \mathfrak{A}_{\alpha\xi} g_\xi \tag{6.53}$$

where

$$\mathfrak{A}_{\alpha\xi} = \begin{bmatrix} 1 & & & & -1 & -1 \\ & 1 & & -1 & & -1 \\ & & 1 & -1 & -1 & \\ & & & 4 & & \\ & & & & 4 & \\ & & & & & 4 \end{bmatrix} \tag{6.54}$$

$$g_\xi = \begin{bmatrix} {\eta_1}^2 \\ {\eta_2}^2 \\ {\eta_3}^2 \\ \eta_2\eta_3 \\ \eta_3\eta_1 \\ \eta_1\eta_2 \end{bmatrix} \tag{6.55}$$

The first- and second-order interpolation functions are expressed as Eqs. (6.50) and (6.53) by the area coordinate.

6.7 Concrete Forms of Finite Element Equation

We have used the indicial notation x_i $(i = 1, 2)$ for coordinate (x, y) until Sect. 6.4. However, we do not use the indicial notation to compute the concrete forms of Eqs. (6.25) and (6.26) to avoid confusion. The summation convention is used for subscripts ξ, η, λ, μ to express the matrix multiplication. We start by differentiating Eq. (6.47) with respect to x and y. Using Eqs. (6.34) and (6.35), we can write

$$\Phi_{\alpha,x} = \begin{bmatrix} 1 & & & & -1 & -1 \\ & 1 & & -1 & & -1 \\ & & 1 & -1 & -1 & \\ & & & 4 & & \\ & & & & 4 & \\ & & & & & 4 \end{bmatrix} \begin{bmatrix} 2b_1 & & \\ & 2b_2 & \\ & & 2b_3 \\ & b_3 & b_2 \\ b_3 & & b_1 \\ b_2 & b_1 & \end{bmatrix} \begin{bmatrix} \eta_1 \\ \eta_2 \\ \eta_3 \end{bmatrix} \tag{6.56}$$

$$\Phi_{\alpha,y} = \begin{bmatrix} 1 & & & & -1 & -1 \\ & 1 & & -1 & & -1 \\ & & 1 & -1 & -1 & \\ & & & 4 & & \\ & & & & 4 & \\ & & & & & 4 \end{bmatrix} \begin{bmatrix} 2c_1 & & \\ & 2c_2 & \\ & & 2c_3 \\ & c_3 & c_2 \\ c_3 & & c_1 \\ c_2 & c_1 & \end{bmatrix} \begin{bmatrix} \eta_1 \\ \eta_2 \\ \eta_3 \end{bmatrix} \tag{6.57}$$

where subscripted $,x$ and $,y$ mean differentiation with respect to x and y. Equations (6.56) and (6.57) can be rewritten as follows:

$$\Phi_{\alpha,x} = \mathfrak{A}_{\alpha\xi} B_{\xi\eta} f_\eta \tag{6.58}$$

$$\Phi_{\alpha,y} = \mathfrak{A}_{\alpha\xi} C_{\xi\eta} f_\eta \tag{6.59}$$

where

$$B_{\xi\eta} = \begin{bmatrix} 2b_1 & & \\ & 2b_2 & \\ & & 2b_3 \\ & b_3 & b_2 \\ b_3 & & b_1 \\ b_2 & b_1 & \end{bmatrix} \tag{6.60}$$

$$C_{\xi\eta} = \begin{bmatrix} 2c_1 & & \\ & 2c_2 & \\ & & 2c_3 \\ & c_3 & c_2 \\ c_3 & & c_1 \\ c_2 & c_1 & \end{bmatrix} \tag{6.61}$$

Consider the concrete form of Eq. (6.24). Equation (6.41) means that the shape function Ψ_λ is f_λ itself. Making use of Eqs. (6.50), (6.58), and (6.59), the following equations can be obtained, replacing x_i $(i = 1, 2)$ by (x, y):

$$\begin{aligned} H_{\lambda\beta}{}^{x} &= \int_V (\Psi_\lambda \Phi_{\beta,x}) dxdy \\ &= \mathfrak{A}_{\beta\xi} B_{\xi\eta} \int_V (f_\eta f_\lambda) dxdy = \mathfrak{A}_{\beta\xi} B_{\xi\eta} G_{\eta\lambda} \end{aligned} \tag{6.62}$$

$$\begin{aligned} H_{\lambda\beta}{}^{y} &= \int_V (\Psi_\lambda \Phi_{\beta,y}) dxdy \\ &= \mathfrak{A}_{\beta\xi} C_{\xi\eta} \int_V (f_\eta f_\lambda) dxdy = \mathfrak{A}_{\beta\xi} C_{\xi\eta} G_{\eta\lambda} \end{aligned} \tag{6.63}$$

where

$$G_{\eta\lambda} = \int_V (f_\eta f_\lambda) dxdy = \frac{\mathbf{\Delta}}{12} \begin{bmatrix} 2 & 1 & 1 \\ 1 & 2 & 1 \\ 1 & 1 & 2 \end{bmatrix} \tag{6.64}$$

Matrix forms of Eq. (6.26) are expressed as follows:

$$
H_{\beta\lambda}{}^{x} = \begin{bmatrix} 1 & & & & -1 & -1 \\ & 1 & & -1 & & -1 \\ & & 1 & -1 & -1 & \\ & & & 4 & & \\ & & & & 4 & \\ & & & & & 4 \end{bmatrix} \begin{bmatrix} 2b_1 & & \\ & 2b_2 & \\ & & 2b_3 \\ & b_3 & b_2 \\ b_3 & & b_1 \\ b_2 & b_1 & \end{bmatrix} \frac{\boldsymbol{\Delta}}{12} \begin{bmatrix} 2 & 1 & 1 \\ 1 & 2 & 1 \\ 1 & 1 & 2 \end{bmatrix} \tag{6.65}
$$

$$
H_{\beta\lambda}{}^{y} = \begin{bmatrix} 1 & & & & -1 & -1 \\ & 1 & & -1 & & -1 \\ & & 1 & -1 & -1 & \\ & & & 4 & & \\ & & & & 4 & \\ & & & & & 4 \end{bmatrix} \begin{bmatrix} 2c_1 & & \\ & 2c_2 & \\ & & 2c_3 \\ & c_3 & c_1 \\ c_3 & & c_1 \\ c_2 & c_1 & \end{bmatrix} \frac{\boldsymbol{\Delta}}{12} \begin{bmatrix} 2 & 1 & 1 \\ 1 & 2 & 1 \\ 1 & 1 & 2 \end{bmatrix} \tag{6.66}
$$

Consider the concrete form of Eq. (6.25). Replacing x_i $(i = 1, 2)$ by (x, y), the matrices can be written as follows:

$$S_{\alpha\beta}{}^{xx} = 2\mu\mathfrak{M}_{\alpha\beta}{}^{xx} + \mu\mathfrak{M}_{\alpha\beta}{}^{yy} \tag{6.67}$$

$$S_{\alpha\beta}{}^{xy} = \mu\mathfrak{M}_{\alpha\beta}{}^{xy} \tag{6.68}$$

$$S_{\alpha\beta}{}^{yx} = \mu\mathfrak{M}_{\alpha\beta}{}^{yx} \tag{6.69}$$

$$S_{\alpha\beta}{}^{yy} = \mu\mathfrak{M}_{\alpha\beta}{}^{xx} + 2\mu\mathfrak{M}_{\alpha\beta}{}^{yy} \tag{6.70}$$

where

$$\mathfrak{M}_{\alpha\beta}{}^{xx} = \int_V (\Phi_{\alpha,x}\Phi_{\beta,x})dxdy \tag{6.71}$$

$$\mathfrak{M}_{\alpha\beta}{}^{xy} = \int_V (\Phi_{\alpha,y}\Phi_{\beta,x})dxdy \tag{6.72}$$

$$\mathfrak{M}_{\alpha\beta}{}^{yx} = \int_V (\Phi_{\alpha,x}\Phi_{\beta,y})dxdy \tag{6.73}$$

$$\mathfrak{M}_{\alpha\beta}{}^{yy} = \int_V (\Phi_{\alpha,y}\Phi_{\beta,y})dxdy \tag{6.74}$$

Making use of Eqs. (6.56) and (6.57), and carrying out the integration, those are

$$\mathfrak{M}_{\alpha\beta}{}^{xx} = \mathfrak{A}_{\alpha\xi}B_{\xi\eta}\mathfrak{A}_{\beta\lambda}B_{\lambda\mu}G_{\eta\mu} \tag{6.75}$$

$$\mathfrak{M}_{\alpha\beta}{}^{xy} = \mathfrak{A}_{\alpha\xi}C_{\xi\eta}\mathfrak{A}_{\beta\lambda}B_{\lambda\mu}G_{\eta\mu} \tag{6.76}$$

$$\mathfrak{M}_{\alpha\beta}{}^{yx} = \mathfrak{A}_{\alpha\xi} B_{\xi\eta} \mathfrak{A}_{\beta\lambda} C_{\lambda\mu} G_{\eta\mu} \tag{6.77}$$

$$\mathfrak{M}_{\alpha\beta}{}^{yy} = \mathfrak{A}_{\alpha\xi} C_{\xi\eta} \mathfrak{A}_{\beta\lambda} C_{\lambda\mu} G_{\eta\mu} \tag{6.78}$$

Matrix forms of Eq. (6.75)–(6.78) can be determined as in the following equations:

$$\mathfrak{M}_{\alpha\beta}{}^{xx} = \begin{bmatrix} 1 & & & & -1 & -1 \\ & 1 & & -1 & & -1 \\ & & 1 & -1 & -1 & \\ & & & 4 & & \\ & & & & 4 & \\ & & & & & 4 \end{bmatrix} \begin{bmatrix} 2b_1 & & \\ & 2b_2 & \\ & & 2b_3 \\ & b_3 & b_2 \\ b_3 & & b_1 \\ b_2 & b_1 & \end{bmatrix} \frac{\Delta}{12} \begin{bmatrix} 2 & 1 & 1 \\ 1 & 2 & 1 \\ 1 & 1 & 2 \end{bmatrix} \cdot \begin{bmatrix} 2b_1 & & & & b_3 & b_2 \\ & 2b_2 & & b_3 & & b_1 \\ & & 2b_3 & b_2 & b_1 & \end{bmatrix} \begin{bmatrix} 1 & & & & & \\ & 1 & & & & \\ & & 1 & & & \\ & -1 & -1 & 4 & & \\ -1 & & -1 & & 4 & \\ -1 & -1 & & & & 4 \end{bmatrix} \tag{6.79}$$

$$\mathfrak{M}_{\alpha\beta}{}^{xy} = \begin{bmatrix} 1 & & & & -1 & -1 \\ & 1 & & -1 & & -1 \\ & & 1 & -1 & -1 & \\ & & & 4 & & \\ & & & & 4 & \\ & & & & & 4 \end{bmatrix} \begin{bmatrix} 2c_1 & & \\ & 2c_2 & \\ & & 2c_3 \\ & c_3 & c_2 \\ c_3 & & c_1 \\ c_2 & c_1 & \end{bmatrix} \frac{\Delta}{12} \begin{bmatrix} 2 & 1 & 1 \\ 1 & 2 & 1 \\ 1 & 1 & 2 \end{bmatrix} \cdot \begin{bmatrix} 2b_1 & & & & b_3 & b_2 \\ & 2b_2 & & b_3 & & b_1 \\ & & 2b_3 & b_2 & b_1 & \end{bmatrix} \begin{bmatrix} 1 & & & & & \\ & 1 & & & & \\ & & 1 & & & \\ & -1 & -1 & 4 & & \\ -1 & & -1 & & 4 & \\ -1 & -1 & & & & 4 \end{bmatrix} \tag{6.80}$$

$$
\mathfrak{M}_{\alpha\beta}{}^{yx} = \begin{bmatrix} 1 & & & & -1 & -1 \\ & 1 & & -1 & & -1 \\ & & 1 & -1 & -1 & \\ & & & 4 & & \\ & & & & 4 & \\ & & & & & 4 \end{bmatrix} \begin{bmatrix} 2b_1 & & \\ & 2b_2 & \\ & & 2b_3 \\ & b_3 & b_2 \\ b_3 & & b_1 \\ b_2 & b_1 & \end{bmatrix} \frac{\Delta}{12} \begin{bmatrix} 2 & 1 & 1 \\ 1 & 2 & 1 \\ 1 & 1 & 2 \end{bmatrix}
\cdot \begin{bmatrix} 2c_1 & & & & c_3 & c_2 \\ & 2c_2 & & c_3 & & c_1 \\ & & 2c_3 & c_2 & c_1 & \end{bmatrix} \begin{bmatrix} 1 & & & & & \\ & 1 & & & & \\ & & 1 & & & \\ & -1 & -1 & 4 & & \\ -1 & & -1 & & 4 & \\ -1 & -1 & & & & 4 \end{bmatrix} \tag{6.81}
$$

$$
\mathfrak{M}_{\alpha\beta}{}^{yy} = \begin{bmatrix} 1 & & & & -1 & -1 \\ & 1 & & -1 & & -1 \\ & & 1 & -1 & -1 & \\ & & & 4 & & \\ & & & & 4 & \\ & & & & & 4 \end{bmatrix} \begin{bmatrix} 2c_1 & & \\ & 2c_2 & \\ & & 2c_3 \\ & c_3 & c_2 \\ c_3 & & c_1 \\ c_2 & c_1 & \end{bmatrix} \frac{\Delta}{12} \begin{bmatrix} 2 & 1 & 1 \\ 1 & 2 & 1 \\ 1 & 1 & 2 \end{bmatrix}
\cdot \begin{bmatrix} 2c_1 & & & & c_3 & c_2 \\ & 2c_2 & & c_3 & & c_1 \\ & & 2c_3 & c_2 & c_1 & \end{bmatrix} \begin{bmatrix} 1 & & & & & \\ & 1 & & & & \\ & & 1 & & & \\ & -1 & -1 & 4 & & \\ -1 & & -1 & & 4 & \\ -1 & -1 & & & & 4 \end{bmatrix} \tag{6.82}
$$

In Eqs. (6.48)–(6.50), (6.53), (6.58), (6.59), (6.62), (6.63), (6.75)–(6.78), summation convention is used for subscripts ξ, η, λ, μ. Employing Eqs. (6.65), (6.66), and (6.79)–(6.82), coefficients of Eq. (6.24) can be formulated. Based on coefficients of

Eq. (6.24) for all finite elements in the flow field, the local finite element equation is derived. Following the procedures described in Chaps. 2–5, the global finite element equation is obtained.

6.8 Matrix Form of the Finite Element Equation

As we have already described in Sect. 6.4, the local finite element equation is expressed in Eq. (6.24). To clarify the meaning of Eq. (6.24), it is helpful to show the matrix form of Eq. (6.24). First, Eq. (6.24) can be rewritten as follows:

$$S_{\alpha i\beta j}v_{\beta_j} - H_{\alpha i\lambda}p_\lambda = \Omega_{\alpha i} \tag{6.83}$$

$$H_{\mu\beta j}v_{\beta j} = 0 \tag{6.84}$$

Then, the matrix form of the first term of Eq. (6.83) can be expressed as in Eq. (6.85).

$$S_{\alpha i\beta j}v_{\beta j}$$

$$= \left[\begin{array}{cccccc|cccccc}
S_{1111} & S_{1121} & S_{1131} & S_{1141} & S_{1151} & S_{1161} & S_{1112} & S_{1122} & S_{1132} & S_{1142} & S_{1152} & S_{1162} \\
S_{2111} & S_{2121} & S_{2131} & S_{2141} & S_{2151} & S_{2161} & S_{2112} & S_{2122} & S_{2132} & S_{2142} & S_{2152} & S_{2162} \\
S_{3111} & S_{3121} & S_{3131} & S_{3141} & S_{3151} & S_{3161} & S_{3112} & S_{3122} & S_{3132} & S_{3142} & S_{3152} & S_{3162} \\
S_{4111} & S_{4121} & S_{4131} & S_{4141} & S_{4151} & S_{4161} & S_{4112} & S_{4122} & S_{4132} & S_{4142} & S_{4152} & S_{4162} \\
S_{5111} & S_{5121} & S_{5131} & S_{5141} & S_{5151} & S_{5161} & S_{5112} & S_{5122} & S_{5132} & S_{5142} & S_{5152} & S_{5162} \\
S_{6111} & S_{6121} & S_{6131} & S_{6141} & S_{6151} & S_{6161} & S_{6112} & S_{6122} & S_{6132} & S_{6142} & S_{6152} & S_{6162} \\
\hline
S_{1211} & S_{1221} & S_{1231} & S_{1241} & S_{1251} & S_{1261} & S_{1212} & S_{1222} & S_{1232} & S_{1242} & S_{1252} & S_{1262} \\
S_{2211} & S_{2221} & S_{2231} & S_{2241} & S_{2251} & S_{2261} & S_{2212} & S_{2222} & S_{2232} & S_{2242} & S_{2252} & S_{2262} \\
S_{3211} & S_{3221} & S_{3231} & S_{3241} & S_{3251} & S_{3261} & S_{3212} & S_{3222} & S_{3232} & S_{3242} & S_{3252} & S_{3262} \\
S_{4211} & S_{4221} & S_{4231} & S_{4241} & S_{4251} & S_{4261} & S_{4212} & S_{4222} & S_{4232} & S_{4242} & S_{4252} & S_{4262} \\
S_{5211} & S_{5221} & S_{5231} & S_{5241} & S_{5251} & S_{5261} & S_{5212} & S_{5222} & S_{5232} & S_{5242} & S_{5252} & S_{5262} \\
S_{6211} & S_{6221} & S_{6231} & S_{6241} & S_{6251} & S_{6261} & S_{6212} & S_{6222} & S_{6232} & S_{6242} & S_{6252} & S_{6262}
\end{array}\right] \cdot \left[\begin{array}{c}
v_{11} \\ v_{21} \\ v_{31} \\ v_{41} \\ v_{51} \\ v_{61} \\ \hline v_{12} \\ v_{22} \\ v_{32} \\ v_{42} \\ v_{52} \\ v_{62}
\end{array}\right] \tag{6.85}$$

The matrix form of the second term of Eq. (6.83) is in Eq. (6.86).

$$H_{\alpha i\lambda}p_\lambda = \left[\begin{array}{ccc}
H_{111} & H_{211} & H_{311} \\
H_{121} & H_{221} & H_{321} \\
H_{131} & H_{231} & H_{331} \\
H_{141} & H_{241} & H_{341} \\
H_{151} & H_{251} & H_{351} \\
H_{161} & H_{261} & H_{361} \\
\hline
H_{112} & H_{212} & H_{312} \\
H_{122} & H_{222} & H_{322} \\
H_{132} & H_{232} & H_{332} \\
H_{142} & H_{242} & H_{342} \\
H_{152} & H_{252} & H_{352} \\
H_{162} & H_{262} & H_{362}
\end{array}\right] \cdot \left[\begin{array}{c} p_1 \\ p_2 \\ p_3 \end{array}\right] \tag{6.86}$$

The matrix form of left side of Eq. (6.84) is Eq. (6.87).

$$H_{\mu\beta j}v_{\beta j} = \left[\begin{array}{cccccc|cccccc} H_{111} & H_{121} & H_{131} & H_{141} & H_{151} & H_{161} & H_{112} & H_{122} & H_{132} & H_{142} & H_{152} & H_{162} \\ H_{211} & H_{221} & H_{231} & H_{241} & H_{251} & H_{261} & H_{212} & H_{222} & H_{232} & H_{242} & H_{252} & H_{262} \\ H_{311} & H_{321} & H_{331} & H_{341} & H_{351} & H_{361} & H_{312} & H_{322} & H_{332} & H_{342} & H_{352} & H_{362} \end{array}\right] \left[\begin{array}{c} v_{11} \\ v_{21} \\ v_{31} \\ v_{41} \\ v_{51} \\ v_{61} \\ \hline v_{12} \\ v_{22} \\ v_{32} \\ v_{42} \\ v_{52} \\ v_{62} \end{array}\right] \tag{6.87}$$

The global finite element equation can be derived by the superposition procedure precisely described in Chap. 4. As is stated in Chap. 4, the global finite element equation can be described by the same form of Eq. (6.24). The matrix form of Eq. (6.24) can also be expressed as in the following equation:

$$\begin{bmatrix} \boldsymbol{S}_{xx} & \boldsymbol{S}_{xy} & -\boldsymbol{H}_x \\ \boldsymbol{S}_{yx} & \boldsymbol{S}_{yy} & -\boldsymbol{H}_y \\ \boldsymbol{H}_x^T & \boldsymbol{H}_y & \boldsymbol{0} \end{bmatrix} \begin{bmatrix} \boldsymbol{u} \\ \boldsymbol{v} \\ \boldsymbol{p} \end{bmatrix} = \begin{bmatrix} \boldsymbol{\Omega}_x \\ \boldsymbol{\Omega}_y \\ \boldsymbol{0} \end{bmatrix} \tag{6.88}$$

where $\boldsymbol{u}$, $\boldsymbol{v}$, $\boldsymbol{p}$ are two velocity components and pressure on the whole nodes in the flow field. Coefficient matrices of Eq. (6.88) can be obtained by the superposition procedures based on Eqs. (6.85), (6.86), and (6.87). Solving the global form of Eq. (6.24) or Eq. (6.88), we can obtain velocity $v_{\beta j}$ and pressure p_λ, or velocity $\boldsymbol{u}$, $\boldsymbol{v}$ and pressure $\boldsymbol{p}$. The computer program of the creeping flow can be constructed based on Eq. (6.24) or Eq. (6.88), which is implemented in Sect. 6.9.

6.9 Computer Program creeping_flow

In this section, computer program to implement the idea described in this chapter is shown. The name of the program is creeping_flow. The principal theme of the program is the mixed interpolation, which is reflected in Eq. (6.88). Therefore, essentially, the purpose of this program is to solve Eq. (6.88). The characteristic point of this equation is that the degrees of freedom of velocity and pressure are different. Total number of main nodes is nx, and total number of whole nodes is nax. The degrees of freedom of velocity are $2 \times nax$, and degrees of freedom of pressure are nx. Thus, total degrees of freedom of Eq. (6.88) is $ns = 2 \times nax + nx$.

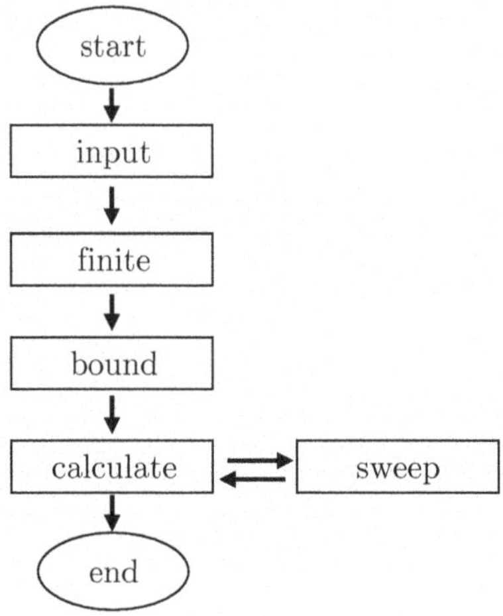

Fig. 6.5 Flow chart of the program creeping flow

The maximum number of array dimension is limited for use of the maximum 50 main nodes. The flow chart of the program is shown in Fig. 6.5. After data is fed into the subroutine input, coefficient matrix of Eq. (6.88) is coded in the subroutine finite. In the subroutine bound, the known terms of the right side in Eq. (6.88) is programed, and the boundary condition is imposed. The subroutine calcul calculates the velocity and pressure by the subroutine sweep. The computed results are output in the subroutine results.

6.9.1 Subroutine Input

To use the mixed interpolation, we will use main nodes and subnodes of triangular element. The main nodes consist of three nodes of triangle and the subnodes consist of mid nodes of triangle. The total number of main nodes is nx, and the sum of total number of main and subnodes is nax. The input data are listed in Table 6.1. The configulation of the channel, which is expressed by the coordinate of nodes and the connectivity relations between nodes and elements, constitute the input. In this computer program, the connectivity for six nodes of a triangle should be given, although three mid nodes can be automatically generated. The boundary conditions of both components of velocity and pressure are specified. Those input data are output for confirmation.

Table 6.1 Input data

Identifier	Maximum	Explanation
nx	1	Total number of main nodes
nax	1	Total number of nodes
mx	1	Total number of elements
iux	1	Total number of nodes on which velocity *u* is given
ivy	1	Total number of nodes on which velocity *v* is given
iet	1	Total number of nodes on which pressure is given
$xx(i)$	*nx*	Coordinate *x* of main node *i*
$yy(i)$	*nx*	Coordinate *y* of main node *i*
$ie(i, j)$	6, *mx*	Connectivity index of element *j*
$jux(i)$	*iux*	Number of nodes on which velocity *u* is given
$ufx(i)$	*iux*	Specified value of velocity *u*
$jvy(i)$	*ivy*	Number of nodes on which velocity *v* is given
$vfy(i)$	*ivy*	Specified value of velocity *v*
$jet(i)$	*iet*	Number of nodes on which pressure *p* is given
$wet(i)$	*iet*	Specified value of pressure *p*

6.9.2 *Subroutine Finite*

The role of the subroutine finite is to code the global coefficient matrix of Eq. (6.88) based on Eq. (6.24), i.e., the global finite element equation. The main identifiers are listed in Table 6.2. The main index is $im = 1$ to mx, which is the number of element. The global coefficient matrix is made at every finite element. The local coefficient matrix is in the array $sa(i, j)$, $i = 1$ to 15 and $j = 1$ to 15. The total degrees of freedom 15 is 2 components of velocity u and v on 6 nodes plus 1 pressure on 3 nodes. The matrix components are $aa(i, j)$, $i = 1$ to 6, and $j = 1$ to 6, which is $\mathfrak{A}_{\alpha\xi}$ in Eq. (6.54), $bb(i, j)$, $i = 1$ to 6, and $j = 1$ to 3, which is $B_{\xi\eta}$ in Eq. (6.60), $cc(i, j)$, $i = 1$ to 6, and $j = 1$ to 3, which is $C_{\xi\eta}$ in Eq. (6.61) and $gg(i, j)$, $i = 1$ to 3, and $j = 1$ to 3, which is $G_{\eta\lambda}$ in Eq. (6.64). Intermediate matrices are $\mathfrak{M}_{\alpha\beta}{}^{xx}$ in Eq. (6.79), $\mathfrak{M}_{\alpha\beta}{}^{xy}$ in Eq. (6.80), $\mathfrak{M}_{\alpha\beta}{}^{yx}$in Eq. (6.81), and $\mathfrak{M}_{\alpha\beta}{}^{yy}$ in Eq. (6.82). Those matrices are not made explicitly in the program. The local coefficient matrices $S_{\alpha\beta}{}^{xx}$ in Eq. (6.67), $S_{\alpha\beta}{}^{xy}$ in Eq. (6.68), $S_{\alpha\beta}{}^{yx}$ in Eq. (6.69), $S_{\alpha\beta}{}^{yy}$ in Eq. (6.70), $H^{x}_{\lambda\beta}$ in Eq. (6.65), and $H_{\lambda\beta}{}^{y}$ in Eq. (6.66) are coded in $sa(i, j)$ at once. The global coefficient matrix of Eq. (6.88) can be superimposed in the array $sk(i, j)$, $i = 1$ to ns, and $j = 1$ to ns, where ns is the total number of degrees of freedom in the whole flow channel. The algorithm of the superposition is the extension of the one written in Sect. 4.1.2 in Chap. 4.

Table 6.2 Main identifiers in subroutines finite and bound

Identifier	Maximum	Explanation
$aa(i,j)$	6, 6	$\mathfrak{A}_{\alpha\xi}$ in Eq. (6.54)
$bb(i,j)$	6, 3	$B_{\xi\eta}$ in Eq. (6.60)
$cc(i,j)$	6, 3	$C_{\xi\eta}$ in Eq. (6.61)
$gg(i,j)$	3, 3	$G_{\eta\lambda}$ in Eq. (6.64)
$sa(i,j)$	15, 15	$2\mu\mathfrak{M}_{\alpha\beta}{}^{xx} + \mu\mathfrak{M}_{\alpha\beta}{}^{yy}$
		$\mu\mathfrak{M}_{\alpha\beta}{}^{xy}$
		$\mu\mathfrak{M}_{\alpha\beta}{}^{yx}$
		$\mu\mathfrak{M}_{\alpha\beta}{}^{xx} + 2\mu\mathfrak{M}_{\alpha\beta}{}^{yy}$
		in Eqs. (6.67)–(6.70)
		$H_{\lambda\beta}{}^{x}$ in Eq. (6.65)
		$H_{\lambda\beta}{}^{y}$ in Eq. (6.66)
ns	1	$2 \times nax + nx$
$sk(i,j)$	ns, ns	Global coefficient matrix of Eq. (6.88)
$dmy(i)$	ns	Known term of Eq. (6.88) and calculated velocity and pressure obtained by the subroutine sweep
sweep		Solution of simultaneous equation

6.9.3 Subroutine Bound

The boundary condition is assigned in the coefficient matrix of Eq. (6.88). The known term $\Omega_{\alpha i}$ is coded in the array $dmy(i)$, $i = 1$ to ns. The algorithm is the natural extension of the one written in Sect. 4.13.5 in Chap. 4. In this computer program, surface force $\hat{t}_i$ in Eq. (6.6) is assumed to be zero on the boundary S_2. In the trial computation shown in Fig. 6.7, CD is the S_2 boundary. The boundary condition on CD corresponds to the one for which the flow is going out to a vacuum across the boundary.

6.9.4 Subroutine Calcul

In the subroutine calcul, the velocity and pressure on whole nodes are calculated. The subroutine sweep is used in the subroutine to solve the simultaneous equation based on the coefficient matrix $sk(i,j)$, $i = 1$ to ns, and $j = 1$ to ns and the known term $dmy(i)$, $i = 1$ to ns. By the computation of the subroutine sweep, the solutions, i.e., velocity and pressure, are obtained in the array $dmy(i)$, $i = 1$ to ns. The method of solution is the standard sweep out method.

6.9.5 *Subroutine Result*

The computed velocity and pressure are output in the subroutine result.

6.9.6 *Program List of creeping_flow*

The computer program list is shown in Fig. 6.6.

6.9.7 *Trial Computation*

The trial computation is carried out using the channel shown in Fig. 6.7. On the boundary AB, the velocity is given as parabolic distribution. On the boundaries AC and BD, both components of velocity are zero. The pressure is not specified at any nodes because the surface force on the outlet CD is assumed to be zero. The input data are shown in Fig. 6.8. The computed results of velocity and pressure are listed and plotted in Fig. 6.9. The numerical results of velocity show the uniform parabolic distribution, and results of pressure show the linear distribution. Those results express a uniform channel flow.

6.10 An Example of Mixed Interpolation

In this section, we will examine the mixed interpolation using a simple one-dimensional equation. By denoting velocity u and pressure p, the one-dimensional basic equation is expressed as

$$-\mu\frac{d^2u}{dx^2}+\frac{dp}{dx}=0 \quad \text{in} \quad [0,L] \tag{6.89}$$

$$\frac{du}{dx}=0 \quad \text{in} \quad [0,L] \tag{6.90}$$

$$(u)_{x=0}=1 \tag{6.91}$$

where μ is viscosity and L is total channel length, as shown in Fig. 6.10. The solution of Eqs. (6.89)—(6.91) is $u=1$ and $p=0$, which has no practical meaning. However, we just use for explanation of the mixed interpolation. Using the weighting functions u^* and p^*, the weighted residual equation can be described as follows:

```
!*************************************************************************
!*************************************************************************
!**********                                                   ***********
!**********     finite element analysis for creeping flow     ***********
!**********     coded by Wakui                                ***********
!**********                                                   ***********
!*************************************************************************
!*************************************************************************
!
!      nx---------------total number of main nodes
!      nax--------------total number of nodes
!      mx---------------total number of elements
!      iux--------------total number of boundary u's
!      ivx--------------total number of boundary v's
!      iet--------------total number of boundary e's
!      xx,yy------------node coordinates
!      ie---------------conection index for elements
!      al---------------viscosity coefficient
!      jux,jvy,jet-----node numbers for u,v,e on the boundery
!                      respectively
!      fux,vfy,wet-----specified values for u,v,e on the boudery
!                      respectively
!
!*************************************************************************
!*************************************************************************


!*************************************************************************
!****              ******************************************************
!****  main program ******************************************************
!****              ******************************************************
!*************************************************************************

      program creeping_flow

      implicit real(8) (a-h,o-z)
      implicit integer(4) (i-n)

      dimension xx(50),yy(50),ie(6,60)
      dimension aa(6,6),bb(6,3),cc(6,3)
      dimension sk(300,300),sa(15,15),gg(3,3)
      dimension uu(150),vv(150),et(50)
      dimension jux(50),jvy(50),jet(50)
      dimension ufx(50),vfy(50),wet(50)
      dimension dmy(300)

      open(10,file='input_creep.txt')
      open(20,file='check_input_creep.txt')
      open(30,file='output_creep.txt')

      call input
     &   (nx,nax,mx,ns,xx,yy,ie,al,iux,ivy,iet,
     &    jux,jvy,jet,ufx,vfy,wet              )
      call finite
     &   (nax,xx,yy,ie,aa,bb,cc,gg,sa,sk,al,mx)
      call bound (ns,nax,iux,ivy,iet,ufx,vfy,
     &    wet,jux,jvy,jet,dmy,sk               )
      call calcul (nx,ns,nax,iux,ivy,iet,jux,
     &    jvy,jet,ufx,vfy,wet,dmy,sk,uu,vv,et )
      call result (nx,nax,uu,vv,et)

      close(10)
      close(20)
      close(30)
```

Fig. 6.6 Program list of creeping flow

```
      stop
      end program creeping_flow

!***********************************************************************


!***********************************************************************
!****             ******************************************************
!****  sub program ******************************************************
!****             ******************************************************
!***********************************************************************

      subroutine input
     &    (nx,nax,mx,ns,xx,yy,ie,al,iux,ivy,iet,
     &     jux,jvy,jet,ufx,vfy,wet              )

      implicit real(8) (a-h,o-z)
      implicit integer(4) (i-n)

      dimension xx(50),yy(50),ie(6,60)
      dimension jux(50),jvy(50),jet(50)
      dimension ufx(50),vfy(50),wet(50)

!***********************************************************************
!****              *****************************************************
!****  file input  *****************************************************
!****              *****************************************************
!***********************************************************************

      read(10,*) nx,nax,mx,iux,ivy,iet,al
      read(10,*) (i,xx(i),yy(i),j=1,nx)
      read(10,*) (i,ie(1,i),ie(2,i),ie(3,i),
     &                ie(4,i),ie(5,i),ie(6,i),j=1,mx)
      if(iux/=0) then
        read(10,*) (jux(i),ufx(i),i=1,iux)
      end if
      if(ivy/=0) then
        read(10,*) (jvy(i),vfy(i),i=1,ivy)
      end if
      if(iet/=0)then
        read(10,*) (jet(i),wet(i),i=1,iet)
      end if

!***********************************************************************

      ns = 2 * nax + nx

!***********************************************************************
!****                        *******************************************
!****  input data confirmation *****************************************
!****                        *******************************************
!***********************************************************************

      write(20,*) '*********  input data check  *********************'
      write(20,*)
      write(20,*) 'total number of main nodes    nx  = ',nx
      write(20,*) 'total number of nodes         nax = ',nax
      write(20,*) 'total number of ele,emts      mx  = ',mx
      write(20,*) 'total number of boundary u's iux = ',iux
      write(20,*) 'total number of boundary v's ivy = ',ivy
      write(20,*) 'total number of boundary e's iet = ',iet
      write(20,*) 'viscosity coefficient         al  = ', al
      write(20,*)
      write(20,*)'*********  X,Y cordinate  *************************'
```

Fig. 6.6 (continued)

```
      write(20,*)
      write(20,*)'No.          X(m/s)          Y(m/s)'
      write(20,'(i5,3x,2f16.10)') (i,xx(i),yy(i),i=1,nx)
      write(20,*)
      write(20,*)'*********  node connectivity  **********************'
      write(20,*)
      write(20,*)'No.'
      write(20,'(7i5)') (i,ie(1,i),ie(2,i),ie(3,i),
     &               ie(4,i),ie(5,i),ie(6,i),i=1,mx)
      write(20,*)
      write(20,*)'*********  boundary condition  *********************'
      if(iux/=0) then
        write(20,*)
        write(20,*)'No.          UU(m/s)'
        write(20,'(2i5,f16.10)') (i,jux(i),ufx(i),i=1,iux)
      end if
      if(ivy/=0) then
        write(20,*)
        write(20,*)'No.          VV(m/s)'
        write(20,'(2i5,f16.10)') (i,jvy(i),ufx(i),i=1,ivy)
      end if
      if(iet/=0) then
        write(20,*)
        write(20,*)'No.          PPM(m/s)'
        write(20,'(2i5,f16.10)') (i,jet(i),wet(i),i=1,iet)
      end if

      return
      end subroutine input

!**********************************************************************


!**********************************************************************
!****                     ********************************************
!****  subroutine finite  ********************************************
!****                     ********************************************
!**********************************************************************

      subroutine finite
     &    (nax,xx,yy,ie,aa,bb,cc,gg,sa,sk,al,mx)

      implicit real(8) (a-h,o-z)
      implicit integer(4) (i-n)

      dimension xx(50),yy(50),ie(6,60)
      dimension aa(6,6),bb(6,3),cc(6,3)
      dimension gg(3,3),sa(15,15),sk(300,300)

      do i = 1, 300
        do j = 1, 300
          sk(i,j) = 0.0d0
        end do
      end do

      do i = 1, 15
        do j = 1, 15
          sa(i,j) = 0.0d0
        end do
      end do

      do i = 1, 6
        do j = 1, 6
          aa(i,j) =0.0d0
        end do
```

Fig. 6.6 (continued)

```
          do k = 1, 3
            bb(i,k) = 0.0d0
            cc(i,k) = 0.0d0
        end do
        end do
        do i = 1, 3
          do j = 1, 3
            gg(i,j) = 0.0d0
          end do
        end do

!**************************************************************************

        aa(1,1) = 1.0d0
        aa(2,2) = 1.0d0
        aa(3,3) = 1.0d0
        aa(4,4) = 4.0d0
        aa(5,5) = 4.0d0
        aa(6,6) = 4.0d0
        aa(1,5) = -1.0d0
        aa(1,6) = -1.0d0
        aa(2,4) = -1.0d0
        aa(2,6) = -1.0d0
        aa(3,4) = -1.0d0
        aa(3,5) = -1.0d0

        do im = 1, mx
          ia = ie(1,im)
          ib = ie(2,im)
          ic = ie(3,im)
          xa = xx(ia)
          xb = xx(ib)
          xc = xx(ic)
          ya = yy(ia)
          yb = yy(ib)
          yc = yy(ic)
          area = ((xb-xa)*(yc-ya)-(xc-xa)*(yb-ya))*0.5d0
          rarea = 0.5d0 / (area)
          area12 = area / 12.0d0
          area06 = area / 6.0d0
          b1 = (yb-yc)*rarea
          b2 = (yc-ya)*rarea
          b3 = (ya-yb)*rarea
          c1 = (xc-xb)*rarea
          c2 = (xa-xc)*rarea
          c3 = (xb-xa)*rarea

!**************************************************************************

          bb(1,1) = 2.0d0*b1
          bb(2,2) = 2.0d0*b2
          bb(3,3) = 2.0d0*b3
          bb(4,2) = b3
          bb(4,3) = b2
          bb(5,1) = b3
          bb(5,3) = b1
          bb(6,1) = b2
          bb(6,2) = b1
          cc(1,1) = 2.0d0*c1
          cc(2,2) = 2.0d0*c2
          cc(3,3) = 2.0d0*c3
          cc(4,2) = c3
          cc(4,3) = c2
          cc(5,1) = c3
```

Fig. 6.6 (continued)

```
        cc(5,3) = c1
        cc(6,1) = c2
        cc(6,2) = c1

!************************************************************************

        gg(1,1) = area06
        gg(2,2) = area06
        gg(3,3) = area06
        gg(1,2) = area12
        gg(1,3) = area12
        gg(2,1) = area12
        gg(2,3) = area12
        gg(3,1) = area12
        gg(3,2) = area12

!************************************************************************
!****  oueal matrix sa(i,j), i = 1 to 6, j = 1 to 6  *******************

        do ia = 1, 6
          do ib = 1, 6
            sii = 0.0d0
            sjj = 0.0d0
            sij = 0.0d0
            sji = 0.0d0
            do i = 1, 6
              do j = 1, 3
                do k = 1, 3
                  do l = 1, 6

!************************************************************************

      sii = sii+aa(ia,i)*bb(i,j)*gg(j,k)*bb(l,k)*aa(ib,l)
      sjj = sjj+aa(ia,i)*cc(i,j)*gg(j,k)*cc(l,k)*aa(ib,l)
      sij = sij+aa(ia,i)*bb(i,j)*gg(j,k)*cc(l,k)*aa(ib,l)
      sji= s ji+aa(ia,i)*cc(i,j)*gg(j,k)*bb(l,k)*aa(ib,l)
                  end do
                end do
              end do
            end do
            ic = ia+6
            id = ib+6

!************************************************************************

            sa(ia,ib)  = 2.0d0*sii*al+sjj*al
            sa(ia,id)  = sji*al
            sa(ic,ib)  = sij*al
            sa(ic,id)  = 2.0d0*sjj*al+sii*al
          end do
        end do

        do ia = 1, 6
          do ib = 1, 3
            hij = 0.0d0
            hji = 0.0d0
            do i = 1 ,6
              do j = 1, 3

!************************************************************************

                hij = hij + aa(ia,i)*cc(i,j)*gg(j,ib)
                hji = hji + aa(ia,i)*bb(i,j)*gg(j,ib)
              end do
            end do
```

Fig. 6.6 (continued)

```
            ic = ia + 6
            id = ib + 12
            sa(ia,id) = -hji
            sa(ic,id) = -hij
            sa(id,ia) = hji
            sa(id,ic) = hij
          end do
        end do

        do i = 1, 6
          do j = 1, 6
            ii = ie(i,im)
            jj = ie(j,im)
            kk = i + 6
            ll = j + 6
            mm = nax+ii
            nn = nax+jj
            sk(ii,jj) = sk(ii,jj) + sa(i,j)
            sk(ii,nn) = sk(ii,nn) + sa(i,ll)
            sk(mm,jj) = sk(mm,jj) + sa(kk,j)
            sk(mm,nn) = sk(mm,nn) + sa(kk,ll)
          end do
        end do

!*************************************************************************
!****  global  matrix sk(i,j) = 1 to ns, j= 1 to ns                 *******

        do i = 1, 6
          do j = 1, 3
            ii = ie(i,im)
            jj = ie(j,im)
            kk = i + 6
            ll = j + 12
            mm = nax+ii
            nn = 2*nax + jj
            sk(ii,nn) = sk(ii,nn)+sa(i,ll)
            sk(mm,nn) = sk(mm,nn)+sa(kk,ll)
            sk(nn,ii) = sk(nn,ii)+sa(ll,i)
            sk(nn,mm) = sk(nn,mm)+sa(ll,kk)
          end do
        end do
      end do

      return
      end subroutine finite

!*************************************************************************


!*************************************************************************
!****                     ***********************************************
!****  subroutine bound   ***********************************************
!****                     ***********************************************
!*************************************************************************

      subroutine bound (ns,nax,iux,ivy,iet,ufx,vfy,
     &          wet,jux,jvy,jet,dmy,sk                    )

      implicit real(8) (a-h,o-z)
      implicit integer(4) (i-n)

      dimension jux(50),jvy(50),jet(50)
      dimension ufx(50),vfy(50),wet(50)
      dimension sk(300,300),dmy(300)
```

Fig. 6.6 (continued)

```
      do i = 1, 300
        dmy(i) = 0.0d0
      end do

!****  known term for velocity u and boundary condition treatment  *****

      if(iux/=0) then
        do i = 1, iux
          ii = jux(i)
          do j = 1, ns
            dmy(j) = dmy(j) - sk(j,ii)*ufx(i)
            sk(ii,j) = 0.0d0
            sk(j,ii) = 0.0d0
          end do
          sk(ii,ii) = 1.0d0
        end do
      end if

!****  known term for velocity v and boundary condition treatment  *****

      if(ivy/=0) then
        do i = 1,ivy
          ii = nax + jvy(i)
          do j = 1, ns
            dmy(j) = dmy(j) - sk(j,ii)*vfy(i)
            sk(ii,j) = 0.0d0
            sk(j,ii) = 0.0d0
          end do
          sk(ii,ii) = 1.0d0
        end do
      end if

!****  known term for pressure and boundary treatment  *****************

      if(iet/=0) then
        do i = 1, iet
          ii = 2*nax + jet(i)
          do j = 1, ns
            dmy(j) = dmy(j) - sk(j,ii)*wet(i)
            sk(ii,j) = 0.0d0
            sk(j,ii) = 0.0d0
          end do
          sk(ii,ii) = 1.0d0
        end do
      end if

      return
      end subroutine bound

!***********************************************************************


!***********************************************************************
!****                      *********************************************
!****  subroutine calcul   *********************************************
!****                      *********************************************
!***********************************************************************

      subroutine calcul (nx,ns,nax,iux,ivy,iet,jux,
     &     jvy,jet,ufx,vfy,wet,dmy,sk,uu,vv,et       )

      implicit real(8) (a-h,o-z)
      implicit integer(4) (i-n)

      dimension sk(300,300),dmy(300)
```

Fig. 6.6 (continued)

```
      dimension jux(50),jvy(50),jet(50)
      dimension ufx(50),vfy(50),wet(50)
      dimension uu(150),vv(150),et(150)

!****  calculation of sweep out  ***************************************

      call sweep(ns,sk,dmy)

!****  treatment of boundary condition  ********************************

      if(iux/=0) then
        do i = 1, iux
          ii = jux(i)
          dmy(ii) = ufx(i)
        end do
      end if
      if(ivy/=0) then
        do i = 1, ivy
          ii = nax + jvy(i)
          dmy(ii) = vfy(i)
        end do
      end if
      if(iet/=0) then
        do i = 1, iet
          ii = 2*nax + jet(i)
          dmy(ii) = wet(i)
        end do
      end if

      do i = 1, nax
        uu(i) = dmy(i)
        ii = nax + i
        vv(i) = dmy(ii)
      end do
      do i = 1, nx
        ii = 2*nax + i
        et(i) = dmy(ii)
      end do

      return
      end subroutine calcul

!***********************************************************************


!***********************************************************************
!****                     **********************************************
!****  subroutine result  **********************************************
!****                     **********************************************
!***********************************************************************

      subroutine result (nx,nax,uu,vv,et)

      implicit real(8) (a-h,o-z)
      implicit integer(4) (i-n)

      dimension uu(150),vv(150),et(150)

      write(30,*)'**  output results  ********************************'
      write(30,*)
      write(30,*)'**  computed velocity and pressure on main nodes  **'
      write(30,*)
      write(30,*)'No.        UU(m/s)          VV(m/s)          PPM(m/s)'
      write(30,'(i5,3x,3f16.10)') (i,uu(i),vv(i),et(i),i=1,nx)
```

Fig. 6.6 (continued)

```
      write(30,*)
      write(30,*)'**  computed velocity and pressure on sub nodes  ***'
      write(30,*)
      write(30,*)'No.        UU(m/s)          VV(m/s)'
      write(30,'(i5,3x,2f16.10)') (i,uu(i),vv(i),i=nx+1,nax)

      return
      end subroutine result

!*************************************************************************


!*************************************************************************
!****                  **********************************************
!****  subroutine sweep  **********************************************
!****                  **********************************************
!*************************************************************************

      subroutine sweep(n,a,u)

      implicit real(8) (a-h,o-z)
      implicit integer(4) (i-n)

      dimension a(300,300),u(300)
      do i = 1, n
        b = abs(a(i,i))
        if(b.gt.0.1d-10) then
          p = 1.d0/a(i,i)
          do j = i, n
            a(i,j) = a(i,j)*p
          end do
          u(i)     = u(i)   *p
          do k = i + 1, n
            q = a(k,i)
            do j = 1, n
              a(k,j) = a(k,j) - q*a(i,j)
            end do
            u(k) = u(k) - u(i)*q
          end do
        else
          write(*,*) i,i,'diagonal element is zero'
        end if
      end do
      do i = n-1, 1, -1
        do j = i+1, n
          u(i) = u(i) - u(j)*a(i,j)
        end do
      end do

      return
      end subroutine sweep

!*************************************************************************
```

Fig. 6.6 (continued)

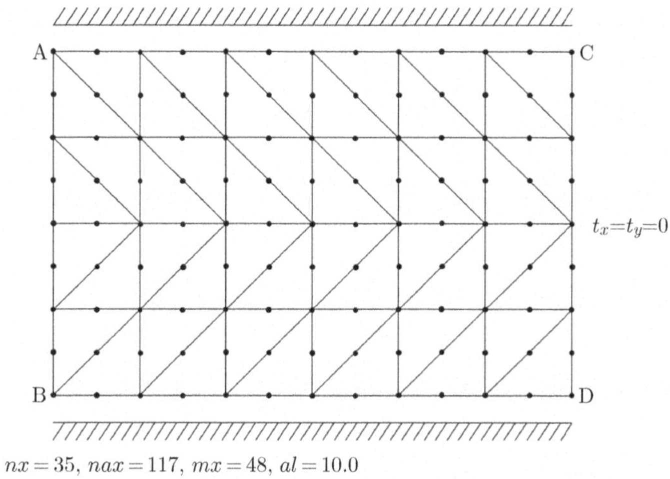

Fig. 6.7 Channel used for trial computation

$$\mu \int_0^l \left(\frac{du^*}{dx} \frac{du}{dx} \right) dx - \int_0^l \left(\frac{du^*}{dx} p \right) dx = u_a^* F_a + u_b^* F_b \tag{6.92}$$

$$\int_0^l \left(p^* \frac{du}{dx} \right) dx = 0 \tag{6.93}$$

where l is the element length as shown in Fig. 6.10, and

$$u_a^* = (u^*)_{x=0} \tag{6.94}$$

$$u_b^* = (u^*)_{x=l} \tag{6.95}$$

$$F_a = - \left[-p + \mu \frac{du}{dx} \right]_{x=0} \tag{6.96}$$

$$F_b = \left[-p + \mu \frac{du}{dx} \right]_{x=l} \tag{6.97}$$

Consider next the normal interpolation, i.e., the same interpolations are used for velocity and pressure

$$u = \left(1 - \frac{x}{l} \right) u_a + \left(\frac{x}{l} \right) u_b \tag{6.98}$$

$$p = \left(1 - \frac{x}{l} \right) p_a + \left(\frac{x}{l} \right) p_b \tag{6.99}$$

where u_a, u_b, p_a, p_b are velocities on nodes a, b and pressures on nodes a, b, respectively. The weighting functions are

```
*********  input data check  *********************

total number of main nodes   nx  = 35
total number of nodes        nax = 117
total number of ele,emts     mx  = 48
total number of boundary u's iux = 33
total number of boundary v's ivy = 40
total number of boundary e's iet = 0
viscosity coefficient        al  = 10.00000000000000

*********  X,Y cordinate  *************************

No.        X(m/s)          Y(m/s)
   1       0.0000000000     0.0000000000
   2       0.0000000000     5.0000000000
   3       0.0000000000    10.0000000000
   4       0.0000000000    15.0000000000
   5       0.0000000000    20.0000000000
   6       5.0000000000     0.0000000000
   7       5.0000000000     5.0000000000
   8       5.0000000000    10.0000000000
   9       5.0000000000    15.0000000000
  10       5.0000000000    20.0000000000
  11      10.0000000000     0.0000000000
  12      10.0000000000     5.0000000000
  13      10.0000000000    10.0000000000
  14      10.0000000000    15.0000000000
  15      10.0000000000    20.0000000000
  16      15.0000000000     0.0000000000
  17      15.0000000000     5.0000000000
  18      15.0000000000    10.0000000000
  19      15.0000000000    15.0000000000
  20      15.0000000000    20.0000000000
  21      20.0000000000     0.0000000000
  22      20.0000000000     5.0000000000
  23      20.0000000000    10.0000000000
  24      20.0000000000    15.0000000000
  25      20.0000000000    20.0000000000
  26      25.0000000000     0.0000000000
  27      25.0000000000     5.0000000000
  28      25.0000000000    10.0000000000
  29      25.0000000000    15.0000000000
  30      25.0000000000    20.0000000000
  31      30.0000000000     0.0000000000
  32      30.0000000000     5.0000000000
  33      30.0000000000    10.0000000000
  34      30.0000000000    15.0000000000
  35      30.0000000000    20.0000000000

*********  node connectivity  *********************

No.
   1    6    7    1   41   40   49
   2    2    1    7   41   42   36
   3    7    8    2   43   42   50
   4    3    2    8   43   44   37
   5    3    8    4   45   38   44
   6    9    4    8   45   51   46
   7    4    9    5   47   39   46
   8   10    5    9   47   52   48
   9   15   10   14   60   65   61
  10    9   14   10   60   52   59
  11   14    9   13   58   64   59
  12    8   13    9   58   51   57
```

Fig. 6.8 Input data

```
    13     8     7    13    56    57    50
    14    12    13     7    56    55    63
    15     7     6    12    54    55    49
    16    11    12     6    54    53    62
    17    16    17    11    67    66    75
    18    12    11    17    67    68    62
    19    17    18    12    69    68    76
    20    13    12    18    69    70    63
    21    13    18    14    71    64    70
    22    19    14    18    71    77    72
    23    14    19    15    73    65    72
    24    20    15    19    73    78    74
    25    25    20    24    86    91    87
    26    19    24    20    86    78    85
    27    24    19    23    84    90    85
    28    18    23    19    84    77    83
    29    18    17    23    82    83    76
    30    22    23    17    82    81    89
    31    17    16    22    80    81    75
    32    21    22    16    80    79    88
    33    26    27    21    93    92   101
    34    22    21    27    93    94    88
    35    27    28    22    95    94   102
    36    23    22    28    95    96    89
    37    23    28    24    97    90    96
    38    29    24    28    97   103    98
    39    24    29    25    99    91    98
    40    30    25    29    99   104   100
    41    35    30    34   112   117   113
    42    29    34    30   112   104   111
    43    34    29    33   110   116   111
    44    28    33    29   110   103   109
    45    28    27    33   108   109   102
    46    32    33    27   108   107   115
    47    27    26    32   106   107   101
    48    31    32    26   106   105   114

*********  boundary condition  *********************

No.          UU(m/s)
     1     1    0.0000000000
     2     2    0.7500000000
     3     3    1.0000000000
     4     4    0.7500000000
     5     5    0.0000000000
     6     6    0.0000000000
     7    10    0.0000000000
     8    11    0.0000000000
     9    15    0.0000000000
    10    16    0.0000000000
    11    20    0.0000000000
    12    21    0.0000000000
    13    25    0.0000000000
    14    26    0.0000000000
    15    30    0.0000000000
    16    31    0.0000000000
    17    35    0.0000000000
    18    36    0.4375000000
    19    37    0.9375000000
    20    38    0.9375000000
    21    39    0.4375000000
    22    40    0.0000000000
    23    48    0.0000000000
    24    53    0.0000000000
```

Fig. 6.8 (continued)

```
  25   61    0.0000000000
  26   66    0.0000000000
  27   74    0.0000000000
  28   79    0.0000000000
  29   87    0.0000000000
  30   92    0.0000000000
  31  100    0.0000000000
  32  105    0.0000000000
  33  113    0.0000000000

No.          VV(m/s)
   1    1    0.0000000000
   2    2    0.7500000000
   3    3    1.0000000000
   4    4    0.7500000000
   5    5    0.0000000000
   6    6    0.0000000000
   7   10    0.0000000000
   8   11    0.0000000000
   9   15    0.0000000000
  10   16    0.0000000000
  11   20    0.0000000000
  12   21    0.0000000000
  13   25    0.0000000000
  14   26    0.0000000000
  15   30    0.0000000000
  16   31    0.0000000000
  17   32    0.0000000000
  18   33    0.4375000000
  19   34    0.9375000000
  20   35    0.9375000000
  21   36    0.4375000000
  22   37    0.0000000000
  23   38    0.0000000000
  24   39    0.0000000000
  25   40    0.0000000000
  26   48    0.0000000000
  27   53    0.0000000000
  28   61    0.0000000000
  29   66    0.0000000000
  30   74    0.0000000000
  31   79    0.0000000000
  32   87    0.0000000000
  33   92    0.0000000000
  34  100    0.0000000000
  35  105    0.0000000000
  36  113    0.0000000000
  37  114    0.0000000000
  38  115    0.0000000000
  39  116    0.0000000000
  40  117    0.0000000000
```

Fig. 6.8 (continued)

$$u^* = \left(1 - \frac{x}{l}\right) u_a^* + \left(\frac{x}{l}\right) u_b^* \tag{6.100}$$

$$p^* = \left(1 - \frac{x}{l}\right) p_a^* + \left(\frac{x}{l}\right) p_b^* \tag{6.101}$$

where u_a^*, u_b^*, p_a^*, p_b^* are nodal values of u^* and v^*, and those are arbitrary constants.

Substituting Eqs. (6.98)–(6.101) into Eqs. (6.92) and (6.93), and rearranging the terms, the local finite element equation for the element ⓘ in Fig. 6.10 can be obtained as follows:

```
**   output results   *******************************

**   computed velocity and pressure on main nodes   **

No.         UU(m/s)          VV(m/s)          PPM(m/s)
   1        0.0000000000     0.0000000000     6.0000000000
   2        0.7500000000     0.0000000000     6.0000000000
   3        1.0000000000     0.0000000000     6.0000000000
   4        0.7500000000     0.0000000000     6.0000000000
   5        0.0000000000     0.0000000000     6.0000000000
   6        0.0000000000     0.0000000000     5.0000000000
   7        0.7500000000     0.0000000000     5.0000000000
   8        1.0000000000     0.0000000000     5.0000000000
   9        0.7500000000     0.0000000000     5.0000000000
  10        0.0000000000     0.0000000000     5.0000000000
  11        0.0000000000     0.0000000000     4.0000000000
  12        0.7500000000     0.0000000000     4.0000000000
  13        1.0000000000     0.0000000000     4.0000000000
  14        0.7500000000     0.0000000000     4.0000000000
  15        0.0000000000     0.0000000000     4.0000000000
  16        0.0000000000     0.0000000000     3.0000000000
  17        0.7500000000     0.0000000000     3.0000000000
  18        1.0000000000     0.0000000000     3.0000000000
  19        0.7500000000     0.0000000000     3.0000000000
  20        0.0000000000     0.0000000000     3.0000000000
  21        0.0000000000     0.0000000000     2.0000000000
  22        0.7500000000     0.0000000000     2.0000000000
  23        1.0000000000     0.0000000000     2.0000000000
  24        0.7500000000     0.0000000000     2.0000000000
  25        0.0000000000     0.0000000000     2.0000000000
  26        0.0000000000     0.0000000000     1.0000000000
  27        0.7500000000     0.0000000000     1.0000000000
  28        1.0000000000     0.0000000000     1.0000000000
  29        0.7500000000     0.0000000000     1.0000000000
  30        0.0000000000     0.0000000000     1.0000000000
  31        0.0000000000     0.0000000000     0.0000000000
  32        0.7500000000     0.0000000000     0.0000000000
  33        1.0000000000     0.0000000000     0.0000000000
  34        0.7500000000     0.0000000000     0.0000000000
  35        0.0000000000     0.0000000000     0.0000000000

**   computed velocity and pressure on sub nodes   ***

No.         UU(m/s)          VV(m/s)
  36        0.4375000000     0.0000000000
  37        0.9375000000     0.0000000000
  38        0.9375000000     0.0000000000
  39        0.4375000000     0.0000000000
  40        0.0000000000     0.0000000000
  41        0.4375000000     0.0000000000
  42        0.7500000000     0.0000000000
  43        0.9375000000     0.0000000000
  44        1.0000000000     0.0000000000
  45        0.9375000000     0.0000000000
  46        0.7500000000     0.0000000000
  47        0.4375000000     0.0000000000
  48        0.0000000000     0.0000000000
  49        0.4375000000     0.0000000000
  50        0.9375000000     0.0000000000
  51        0.9375000000     0.0000000000
  52        0.4375000000     0.0000000000
  53        0.0000000000     0.0000000000
  54        0.4375000000     0.0000000000
  55        0.7500000000     0.0000000000
```

Fig. 6.9 Output results

```
  56     0.9375000000    0.0000000000
  57     1.0000000000    0.0000000000
  58     0.9375000000    0.0000000000
  59     0.7500000000    0.0000000000
  60     0.4375000000    0.0000000000
  61     0.0000000000    0.0000000000
  62     0.4375000000    0.0000000000
  63     0.9375000000    0.0000000000
  64     0.9375000000    0.0000000000
  65     0.4375000000    0.0000000000
  66     0.0000000000    0.0000000000
  67     0.4375000000    0.0000000000
  68     0.7500000000    0.0000000000
  69     0.9375000000    0.0000000000
  70     1.0000000000    0.0000000000
  71     0.9375000000    0.0000000000
  72     0.7500000000    0.0000000000
  73     0.4375000000    0.0000000000
  74     0.0000000000    0.0000000000
  75     0.4375000000    0.0000000000
  76     0.9375000000    0.0000000000
  77     0.9375000000    0.0000000000
  78     0.4375000000    0.0000000000
  79     0.0000000000    0.0000000000
  80     0.4375000000    0.0000000000
  81     0.7500000000    0.0000000000
  82     0.9375000000    0.0000000000
  83     1.0000000000    0.0000000000
  84     0.9375000000    0.0000000000
  85     0.7500000000    0.0000000000
  86     0.4375000000    0.0000000000
  87     0.0000000000    0.0000000000
  88     0.4375000000    0.0000000000
  89     0.9375000000    0.0000000000
  90     0.9375000000    0.0000000000
  91     0.4375000000    0.0000000000
  92     0.0000000000    0.0000000000
  93     0.4375000000    0.0000000000
  94     0.7500000000    0.0000000000
  95     0.9375000000    0.0000000000
  96     1.0000000000    0.0000000000
  97     0.9375000000    0.0000000000
  98     0.7500000000    0.0000000000
  99     0.4375000000    0.0000000000
 100     0.0000000000    0.0000000000
 101     0.4375000000    0.0000000000
 102     0.9375000000    0.0000000000
 103     0.9375000000    0.0000000000
 104     0.4375000000    0.0000000000
 105     0.0000000000    0.0000000000
 106     0.4375000000    0.0000000000
 107     0.7500000000    0.0000000000
 108     0.9375000000    0.0000000000
 109     1.0000000000    0.0000000000
 110     0.9375000000    0.0000000000
 111     0.7500000000    0.0000000000
 112     0.4375000000    0.0000000000
 113     0.0000000000    0.0000000000
 114     0.4375000000    0.0000000000
 115     0.9375000000    0.0000000000
 116     0.9375000000    0.0000000000
 117     0.4375000000    0.0000000000
```

Fig. 6.9 (continued)

Fig. 6.9 (continued)

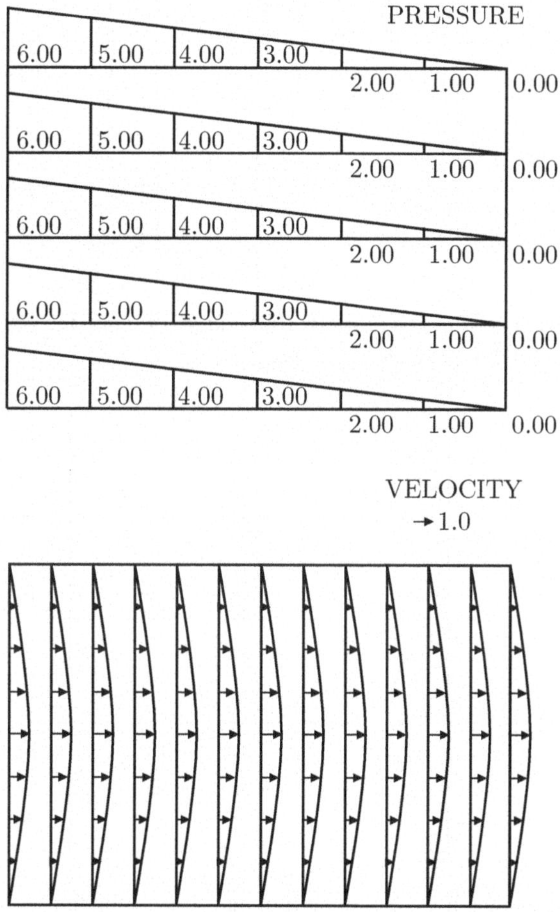

Fig. 6.10 Finite element mesh of the normal interpolation

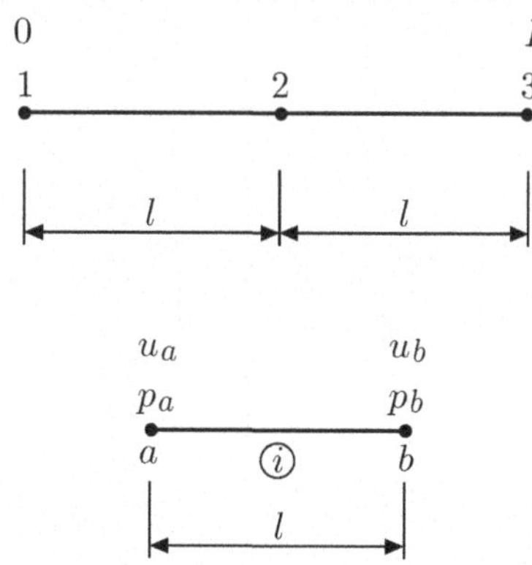

$$\begin{bmatrix} \frac{\mu}{l} & -\frac{\mu}{l} \\ -\frac{\mu}{l} & \frac{\mu}{l} \end{bmatrix} \begin{bmatrix} u_a \\ u_b \end{bmatrix} - \begin{bmatrix} -\frac{1}{2} & -\frac{1}{2} \\ \frac{1}{2} & \frac{1}{2} \end{bmatrix} \begin{bmatrix} p_a \\ p_b \end{bmatrix} = \begin{bmatrix} F_a \\ F_b \end{bmatrix} \tag{6.102}$$

$$\begin{bmatrix} -\frac{1}{2} & \frac{1}{2} \\ -\frac{1}{2} & \frac{1}{2} \end{bmatrix} \begin{bmatrix} u_a \\ u_b \end{bmatrix} = \begin{bmatrix} 0 \\ 0 \end{bmatrix} \tag{6.103}$$

By superimposing Eqs. (6.102) and (6.103) into the whole channel (Fig. 6.10), which consists of two elements, we get

$$\left[\begin{array}{ccc|ccc} \frac{\mu}{l} & -\frac{\mu}{l} & & \frac{1}{2} & \frac{1}{2} & \\ -\frac{\mu}{l} & \left(\frac{\mu}{l}+\frac{\mu}{l}\right) & -\frac{\mu}{l} & -\frac{1}{2} & & \frac{1}{2} \\ & -\frac{\mu}{l} & \frac{\mu}{l} & & -\frac{1}{2} & -\frac{1}{2} \\ \hline -\frac{1}{2} & \frac{1}{2} & & & & \\ -\frac{1}{2} & & \frac{1}{2} & & & \\ & -\frac{1}{2} & \frac{1}{2} & & & \end{array}\right] \left[\begin{array}{c} u_1 \\ u_2 \\ u_3 \\ \hline p_1 \\ p_2 \\ p_3 \end{array}\right] = \left[\begin{array}{c} 0 \\ 0 \\ 0 \\ \hline 0 \\ 0 \\ 0 \end{array}\right] \tag{6.104}$$

Using the boundary condition that $\hat{u}_1 = 1$, the final form of the global finite element equation based on the normal interpolation is as follows:

$$\left[\begin{array}{cc|ccc} 2 & -1 & -\frac{1}{2} & & \frac{1}{2} \\ -1 & 1 & & -\frac{1}{2} & -\frac{1}{2} \\ \hline \frac{1}{2} & & & & \\ & \frac{1}{2} & & & \\ -\frac{1}{2} & \frac{1}{2} & & & \end{array}\right] \left[\begin{array}{c} u_2 \\ u_3 \\ \hline p_1 \\ p_2 \\ p_3 \end{array}\right] = \left[\begin{array}{c} 1 \\ \\ \hline \frac{1}{2} \\ \frac{1}{2} \\ \\ \end{array}\right] \tag{6.105}$$

where we put $l = 1$ and $\mu = 1$ for simplicity.

The solution of Eq. (6.105) may lead to velocities u_2, u_3 and pressures p_1, p_2, p_3. However, it is impossible to solve Eq. (6.105) by the normal manner because the determinant of Eq. (6.105) is zero. From the third equation, we get $u_2 = 1$, and from the fourth equation, we also get $u_3 = 1$. Thus, the last equation is superfluous. One of the last three equations should be subtracted from Eq. (6.105); otherwise, we cannot solve Eq. (6.105).

Now, let us consider the mixed interpolation, i.e., the quadratic interpolation for velocity and the linear interpolation for pressure are used as

$$u = \left(1 - \frac{3x}{l} + \frac{2x^2}{l^2}\right) u_a + \left(-\frac{x}{l} + \frac{2x^2}{l^2}\right) u_b + \left(\frac{4x}{l} - \frac{4x^2}{l^2}\right) u_c \tag{6.106}$$

$$p = \left(1 - \frac{x}{l}\right) p_a + \left(\frac{x}{l}\right) p_b \tag{6.107}$$

where u_a, u_b, u_c are velocities on nodes a, b, c and p_a, p_b are pressures on nodes a, b. Accordingly, the corresponding weighting functions are

$$u^* = \left(1 - \frac{3x}{l} + \frac{2x^2}{l^2}\right) u_a^* + \left(-\frac{x}{l} + \frac{2x^2}{l^2}\right) u_b^* + \left(\frac{4x}{l} - \frac{4x^2}{l^2}\right) u_c^* \tag{6.108}$$

$$p^* = \left(1 - \frac{x}{l}\right) p_a^* + \left(\frac{x}{l}\right) p_b^* \tag{6.109}$$

where u_a^*, u_b^*, u_c^* are arbitrary constants corresponding to nodes a, b, c and p_a^*, p_b^* are also constants corresponding to nodes a, b. Substituting Eqs. (6.106)–(6.109) into Eqs. (6.92) and (6.93), and rearranging the terms, the following local finite element equation for element ⓘ in Fig. 6.11 can be obtained:

$$\begin{bmatrix} \frac{7\mu}{3l} & \frac{\mu}{3l} & -\frac{8\mu}{3l} & \frac{5}{6} & \frac{1}{6} \\ \frac{\mu}{3l} & \frac{7\mu}{3l} & -\frac{8\mu}{3l} & -\frac{1}{6} & -\frac{5}{6} \\ -\frac{8\mu}{3l} & -\frac{8\mu}{3l} & \frac{16\mu}{3l} & -\frac{2}{3} & \frac{2}{3} \end{bmatrix} \begin{bmatrix} u_a \\ u_b \\ u_c \\ p_a \\ p_b \end{bmatrix} = \begin{bmatrix} F_a \\ F_b \\ F_c \end{bmatrix} \tag{6.110}$$

$$\begin{bmatrix} -\frac{5}{6} & \frac{1}{6} & \frac{2}{3} \\ -\frac{1}{6} & \frac{5}{6} & -\frac{2}{3} \end{bmatrix} \begin{bmatrix} u_a \\ u_b \\ u_c \end{bmatrix} = \begin{bmatrix} 0 \\ 0 \end{bmatrix} \tag{6.111}$$

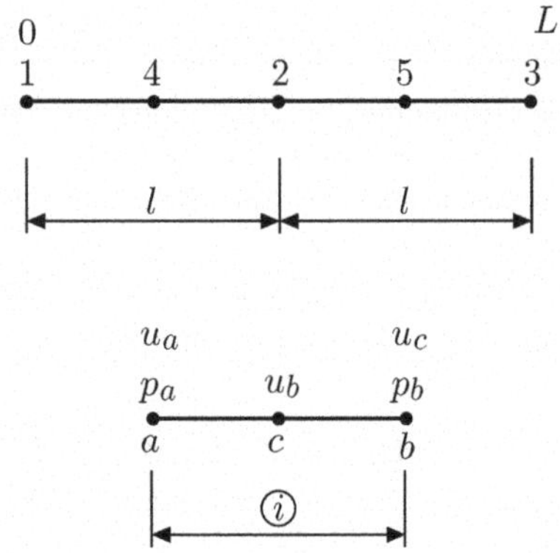

Fig. 6.11 Finite element mesh of the mixed interpolation

Superposing Eq. (6.110) and (6.111) into the whole channel shown in Fig. 6.11, which consists of two elements, we get

$$
\left[\begin{array}{ccccc:ccc}
\left(\frac{7\mu}{3l}\right) & \left(\frac{\mu}{3l}\right) & & \left(-\frac{8\mu}{3l}\right) & & \left(\frac{5}{6}\right) & \left(\frac{1}{6}\right) & \\
\left(\frac{\mu}{3l}\right) & \left(\frac{7\mu}{3l}+\frac{7\mu}{3l}\right) & \left(\frac{\mu}{3l}\right) & \left(-\frac{8\mu}{3l}\right) & \left(-\frac{8\mu}{3l}\right) & \left(-\frac{1}{6}\right) & \left(-\frac{5}{6}+\frac{5}{6}\right) & \left(\frac{1}{6}\right) \\
 & \left(\frac{\mu}{3l}\right) & \left(\frac{7\mu}{3l}\right) & & \left(-\frac{8\mu}{3l}\right) & & \left(-\frac{1}{6}\right) & \left(-\frac{5}{6}\right) \\
\left(-\frac{8\mu}{3l}\right) & \left(-\frac{8\mu}{3l}\right) & & \left(\frac{16\mu}{3l}\right) & & \left(-\frac{2}{3}\right) & \left(\frac{2}{3}\right) & \\
 & \left(-\frac{8\mu}{3l}\right) & \left(-\frac{8\mu}{3l}\right) & & \left(\frac{16\mu}{3l}\right) & & \left(-\frac{2}{3}\right) & \left(\frac{2}{3}\right) \\
\hdashline
\left(-\frac{5}{6}\right) & \left(\frac{1}{6}\right) & & \left(\frac{2}{3}\right) & & & & \\
\left(-\frac{1}{6}\right) & \left(\frac{5}{6}-\frac{5}{6}\right) & \left(\frac{1}{6}\right) & \left(-\frac{2}{3}\right) & \left(\frac{2}{3}\right) & & & \\
 & \left(-\frac{1}{6}\right) & \left(\frac{5}{6}\right) & & \left(-\frac{2}{3}\right) & & &
\end{array}\right]
\begin{bmatrix} u_1 \\ u_2 \\ u_3 \\ u_4 \\ u_5 \\ \hdashline p_1 \\ p_2 \\ p_3 \end{bmatrix}
= \begin{bmatrix} 0 \\ 0 \\ 0 \\ 0 \\ 0 \\ \hdashline 0 \\ 0 \\ 0 \end{bmatrix} \tag{6.112}
$$

Using the boundary condition that $\hat{u}_1 = 1$, the final form of the global finite element equation based on the mixed interpolation is described as follows:

$$
\left[\begin{array}{cccc|ccc}
\frac{14}{3} & \frac{1}{3} & -\frac{8}{3} & -\frac{8}{3} & -\frac{1}{6} & & -\frac{1}{6} \\
\frac{1}{3} & \frac{7}{3} & & -\frac{8}{3} & & -\frac{1}{6} & -\frac{5}{6} \\
-\frac{8}{3} & & \frac{16}{3} & & -\frac{2}{3} & \frac{2}{3} & \\
-\frac{8}{3} & -\frac{8}{3} & & \frac{16}{3} & & -\frac{2}{3} & \frac{2}{3} \\
\hline
\frac{1}{6} & & \frac{2}{3} & & & & \\
& \frac{1}{6} & -\frac{2}{3} & \frac{2}{3} & & & \\
-\frac{1}{6} & \frac{5}{6} & & -\frac{2}{3} & & &
\end{array}\right]
\begin{bmatrix} u_2 \\ u_3 \\ u_4 \\ u_5 \\ p_1 \\ p_2 \\ p_3 \end{bmatrix}
=
\left[\begin{array}{c}
-\frac{1}{3} \\ \\ \frac{8}{3} \\ \\ \hline \frac{5}{6} \\ \frac{1}{6} \\ \\
\end{array}\right]
\tag{6.113}
$$

where we put $l = 1$ and $\mu = 1$. The coefficient matrix of Eq. (6.108) is

$$
\begin{bmatrix}
4.6667 & 0.3333 & -2.6667 & -2.6667 & 0.1667 & 0.0 & -0.1667 \\
0.3333 & 2.3333 & 0.0 & -2.6667 & 0.0 & 0.1667 & 0.8333 \\
-2.6667 & 0.0 & 5.3333 & 0.0 & 0.6667 & -0.6667 & 0.0 \\
-2.6667 & -2.6667 & 0.0 & 5.3333 & 0.0 & 0.6667 & -0.6667 \\
0.1667 & 0.0 & 0.6667 & 0.0 & 0.0 & 0.0 & 0.0 \\
0.0 & 0.1667 & -0.6667 & 0.6667 & 0.0 & 0.0 & 0.0 \\
-0.1667 & 0.8333 & 0.0 & -0.6667 & 0.0 & 0.0 & 0.0
\end{bmatrix}
\tag{6.114}
$$

The inverse matrix of the coefficient matrix of Eq. (6.114) is computed as

$$
\begin{bmatrix}
0.1250 & 0.0 & -0.0312 & -0.0312 & 1.2500 & 0.5000 & -0.2500 \\
-0.0000 & 0.0000 & 0.0000 & 0.0000 & 1.0000 & 1.0000 & 1.0000 \\
-0.0312 & 0.0000 & 0.0078 & 0.0078 & 1.1875 & -0.1250 & 0.0625 \\
-0.0313 & 0.0000 & 0.0078 & 0.0078 & 0.9375 & 1.1250 & -0.1875 \\
1.2500 & 1.0000 & 1.1875 & 0.9375 & -3.5000 & 1.0000 & -0.5000 \\
0.5000 & 1.0000 & -0.1250 & 1.1250 & 1.0000 & -2.0000 & 1.0000 \\
-0.2500 & 1.0000 & 0.0625 & -0.1875 & -0.5000 & 1.0000 & -3.5000
\end{bmatrix}
\tag{6.115}
$$

The right side of Eq. (6.113) is

$$\begin{bmatrix} -0.333 \\ 0.0 \\ 2.667 \\ 0.0 \\ 0.833 \\ 0.167 \\ 0.0 \end{bmatrix}. \tag{6.116}$$

Multiplying the matrix in Eqs. (6.115) and (6.116), we obtain the results as follows:

$$\begin{bmatrix} 1.000 \\ 1.000 \\ 1.000 \\ 1.000 \\ 0.000 \\ 0.000 \\ 0.000 \end{bmatrix} \tag{6.117}$$

The first four are velocities u_2, u_3, u_4, u_5 and the last three are the pressures p_1, p_2, p_3. We can see that by the finite element method based on the mixed interpolation of quadratic polynomial for velocity and linear polynomial for pressure, the exact solution can be obtained.

6.11 Summary and Conclusion

In this chapter, the finite element analysis of creeping flow, which has high viscosity and slow velocity, is presented. Although the problem is linear and time independent, a plenty of standard techniques for the finite element method of the viscous flows are included. The weighted residual equation is formulated from the basic equation, which includes the natural boundary condition. Consequently, the pressure on the whole nodes can be computed without any treatment on the boundary when the surface force on the boundary is zero. For the interpolation of velocity and pressure, the mixed interpolation is used. The interpolation function based on the six nodes of the triangular element is employed for velocity, and the function based on the three nodes of the element is for pressure. The formulations are relatively complex; however, all coefficient matrices can be obtained by matrix multiplications. Many combinations of the interpolation for velocity and pressure have been presented in the literature. An example is the combination of the bubble function for velocity and linear function for pressure. Matrix form is described in addition to indicial notation form to clarify the meaning of the finite element method. We avoid to reiterate the explanation of description by the indicial notation in the later chapters. The introductory computer program is shown to clarify the algorithm of the mixed interpolation. The computer program is more complex than that based on the normal interpolation.

In this chapter, we treated the steady linear flow. However, actual flows are transient and nonlinear phenomena. In the following chapters, we will discuss the nonlinear time-dependent fluid flows. The fundamental techniques of the analyses of those flows are the extended techniques presented in Chaps. 4–6.

Part II
Computational Methods and Applications of Finite Element Method in Fluid Flows

Chapter 7
Continuum Mechanics of Fluid Flows

7.1 Introduction

Although numerous theories in continuum mechanics in fluid flows have been proposed in the past decades, only theories related to the later chapters will be presented in this chapter. First, definition of description and concepts of deformation, displacement, velocity, and acceleration will be discussed. Then, the main fundamental concepts of the continuum mechanics, i.e., the conservation laws of mass, momentum, and energy, are defined. We will provide the global and local forms of the conservation laws. The global forms are given by the integral equations, whereas the local forms are expressed in terms of the differential equations. Mutual relations among conservation laws are discussed. We then introduce stress as flux of momentum, and followed by the conservation low of momentum leads to the equilibrium equation, i.e., mass times acceleration is equal to force. As the constitutive equation, linear relation between stress and deformation rate is employed. We will witness that the conservation of energy is equivalent to the thermal transport equation. As consequences of the Clausius–Duhem inequality, restrictions on the thermal conduction and viscosity coefficient will be imposed. Introducing Helmholtz free energy, the conservation of energy can be transformed into the simplified form whose physical meaning is more clarified. The concept of ideal gas in the adiabatic state is described. There are many good books in the field of continuum mechanics. Among them, Eringen (1967), Leigh (1968), Sedov (1971), Mase and Mase (1999), Bonet and Wood (2008), etc. are helpful for the readers for further reading.

Except for the explanation of material description, the equations in this chapter are expressed using the spatial description. In this chapter, we denote $\dfrac{D}{Dt}$ as superscript dot $\dot{}$.

M. Kawahara, *Finite Element Methods in Incompressible, Adiabatic, and Compressible Flows*, Mathematics for Industry 19,
DOI 10.1007/978-4-431-55450-9_7

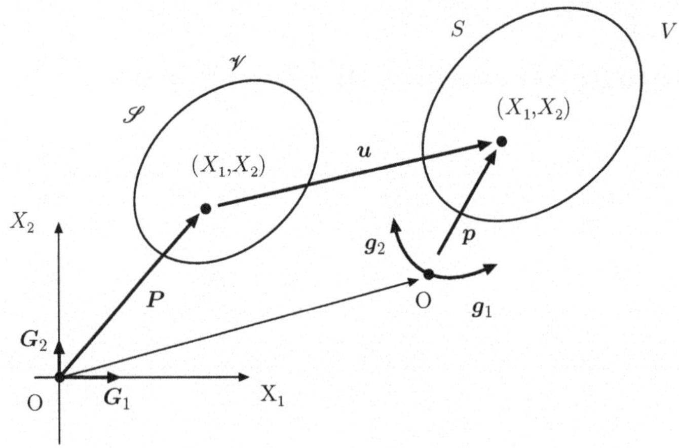

Fig. 7.1 Material description

7.2 Description of Deformation

Let a body B consist of material points and have the material volume $\mathscr{V}$ surrounded by its surface $\mathscr{S}$. The position of a material point is defined by position vector $\boldsymbol{P}$ as shown in Fig. 7.1.

$$\boldsymbol{P} = X_i \boldsymbol{G}_i \tag{7.1}$$

where X_i is coordinate and $\boldsymbol{G}_i$ is a base vector. A coordinate system consists of two elements, one is coordinate index, henceforth simply called coordinate, and the other is coordinate axis, called axis. Because the base vector expresses the unit scale of the axis,

$$\boldsymbol{G}_i = \boldsymbol{P}_{,i} \tag{7.2}$$

We use the Cartesian coordinate axis, whose base vector is $\boldsymbol{G}_i$.

After deformation takes place, the volume of the body is V and its surface is S. Let X_i be the coordinate before deformation, the position of the material point is expressed by

$$\boldsymbol{p} = X_i \boldsymbol{g}_i \tag{7.3}$$

where $\boldsymbol{g}_i$ is the base vector of the axis of coordinate in the deformed state, which can be derived as

$$\boldsymbol{g}_i = \boldsymbol{p}_{,i} \tag{7.4}$$

The deformation can be expressed by the displacement $\boldsymbol{u}$:

$$\boldsymbol{u} = u_i \boldsymbol{G}_i \tag{7.5}$$

where u_i is the component of displacement. In Fig. 7.1, the following equation holds:

$$\boldsymbol{p} + \boldsymbol{C} = \boldsymbol{P} + \boldsymbol{u} \tag{7.6}$$

where $\boldsymbol{C}$ is the movement of the origin of the coordinate system. Differentiating both sides of Eq. (7.6), we get

$$\boldsymbol{p}_{,i} = \boldsymbol{P}_{,i} + \boldsymbol{u}_{,i} \tag{7.7}$$

Thus, the relation

$$\boldsymbol{g}_i = (\delta_{ji} + u_{j,i})\boldsymbol{G}_j \tag{7.8}$$

can be derived. This expression is referred to as the "material description." Using the material description, the deformation can be described in terms of the relation between $\boldsymbol{g}_i$ and $\boldsymbol{G}_i$. The material coordinate X_i is not changed after the deformation.

When the coordinate system is fixed on the space, the position vector after deformation is expressed as follows:

$$\boldsymbol{p} = x_i \boldsymbol{G}_i \tag{7.9}$$

where x_i represents components of position vector $\boldsymbol{p}$ after deformation. If we take coordinate system after deformation $\boldsymbol{g}_i$, $\boldsymbol{g}_i$ and $\boldsymbol{G}_i$ can be the same. Then

$$\boldsymbol{p} = x_i \boldsymbol{g}_i \tag{7.10}$$

Introducing the displacement vector $\boldsymbol{u}$,

$$\boldsymbol{u} = u_i \boldsymbol{G}_i \tag{7.11}$$

where u_i is the components of the displacement in the axis $\boldsymbol{G}_i$, and u_i is also the components of the displacement in $\boldsymbol{g}_i$. In Fig. 7.2, the displacement can be expressed by the position vectors:

$$\boldsymbol{P} + \boldsymbol{u} = \boldsymbol{p} \tag{7.12}$$

From Eqs. (7.1), (7.9), and (7.11), we obtain

$$x_i = X_i + u_i \tag{7.13}$$

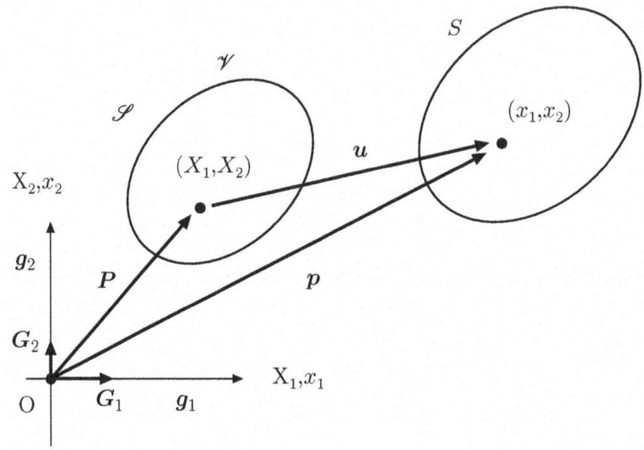

Fig. 7.2 Spatial description

This expression is called as the "spatial description." In this case, the deformation can be understood as the difference in the positions of the material points between before and after the deformation takes place. In the fluid mechanics, the spatial description is usually used.

7.3 Motion, Velocity, and Acceleration

A motion of a body B is a continuous sequence of configuration in time, i.e.,

$$\boldsymbol{p} = \boldsymbol{p}(\boldsymbol{P}(X_i, t)) \tag{7.14}$$

or

$$x_i = x_i(X_i, t) \tag{7.15}$$

where $\boldsymbol{p}$ is a position at a material point at time t, and x_i is the components of $\boldsymbol{p}$. The position vector before motion is $\boldsymbol{P}$, which is expressed in Eq. (7.1). Equation (7.14) is the spatial description. As stated in Chap. 1, there are two notation methods: vectorial and componential notations. Equation (7.14) uses the vectorial notation and Eq. (7.15) uses the componential notation. Recently, the vectorial notation is more commonly used. However, for the sake of exposition, we use the componential notation. Multiplying base vector to both sides of Eq. (7.15), we get Eq. (7.14).

Velocity $\boldsymbol{v}$ is defined as derivative of displacement with respect to time fixing material point $\boldsymbol{P}$. In the material description, displacement is

$$\boldsymbol{u} = \boldsymbol{u}(X_i, t) = u_i(X_i, t)\boldsymbol{G}_i \tag{7.16}$$

From Eq. (7.16), we define

$$\boldsymbol{v} = \left.\frac{d\boldsymbol{u}}{dt}\right|_{\boldsymbol{P}} \tag{7.17}$$

where vertical bar with subscripted $\boldsymbol{P}$ means partial derivative fixing material point $\boldsymbol{P}$. The component of velocity v_k is

$$\boldsymbol{v} = v_k \boldsymbol{G}_k \tag{7.18}$$

and

$$v_k = \left.\frac{du_k}{dt}\right|_{X_i} = \frac{\partial u_k}{\partial t}(X_i, t) \tag{7.19}$$

where vertical bar subscripted by X_i represents partial derivative fixing material point X_i.

In the spatial description, the velocity is defined by Eq. (7.17) and displacement is defined by Eq. (7.12) or (7.13):

$$\boldsymbol{u} = \boldsymbol{p} - \boldsymbol{P} \tag{7.20}$$

or

$$u_i = x_i - X_i \tag{7.21}$$

thus

$$\boldsymbol{v} = \left.\frac{d\boldsymbol{u}}{dt}\right|_{\boldsymbol{P}} = \left.\frac{d\boldsymbol{p}}{dt}\right|_{\boldsymbol{P}} = \left.\frac{du_k}{dt}\right|_{X_i} \boldsymbol{G}_k \tag{7.22}$$

which leads to

$$v_k = \left.\frac{du_k}{dt}\right|_{X_i} = \frac{\partial u_k}{\partial t}(x_i, t) \tag{7.23}$$

Acceleration $\boldsymbol{a}$ is derivative of velocity with respect to time fixing material point $\boldsymbol{P}$:

$$\boldsymbol{a} = \left.\frac{d\boldsymbol{v}}{dt}\right|_{\boldsymbol{P}} \tag{7.24}$$

In the material description, the velocity is

$$\boldsymbol{v} = v_k(X_i, t)\boldsymbol{G}_k \tag{7.25}$$

thus,

$$\boldsymbol{a} = a_k\boldsymbol{G}_k = \left.\frac{dv_k}{dt}\right|_{X_k}\boldsymbol{G}_k = \frac{\partial v_k}{\partial t}\boldsymbol{G}_k = \frac{\partial^2 u_k}{\partial t^2}\boldsymbol{G}_k \tag{7.26}$$

In the spatial description, the velocity is written as

$$\boldsymbol{v} = v_k(x_i, t)\boldsymbol{G}_k \tag{7.27}$$

thus,

$$\begin{aligned}\boldsymbol{a} = a_k\boldsymbol{G}_k &= \left.\frac{dv_k}{dt}(x_i, t)\right|_{X_i}\boldsymbol{G}_k \\ &= \left(\frac{\partial v_k}{\partial t} + \frac{\partial x_i}{\partial t}\frac{\partial v_k}{\partial x_i}\right)\boldsymbol{G}_k \\ &= \frac{Dv_k}{Dt}\boldsymbol{G}_k\end{aligned} \tag{7.28}$$

namely,

$$a_k = \frac{Dv_k}{Dt} = \frac{\partial v_k}{\partial t} + v_i\frac{\partial v_k}{\partial x_i} \tag{7.29}$$

where

$$\frac{D}{Dt} = \frac{\partial}{\partial t} + \frac{\partial x_i}{\partial t}\frac{\partial}{\partial x_i} = \frac{\partial}{\partial t} + v_i\frac{\partial}{\partial x_i} \tag{7.30}$$

The material time rate of change of a function $\boldsymbol{f}$ is defined as

$$\frac{d\boldsymbol{f}}{dt} = \left.\frac{d\boldsymbol{f}}{dt}\right|_{\boldsymbol{P}} \tag{7.31}$$

If $\boldsymbol{f}$ is a material function,

$$\boldsymbol{f} = \boldsymbol{f}(\boldsymbol{P}, t) \equiv f_k(X_i, t)\boldsymbol{G}_k \tag{7.32}$$

then

$$\frac{d\boldsymbol{f}}{dt} = \frac{\partial f_k}{\partial t}(X_i, t)\boldsymbol{G}_k \tag{7.33}$$

If $\boldsymbol{f}$ is a spatial function,

$$\boldsymbol{f} = \boldsymbol{f}(\boldsymbol{p}, t) \equiv f_k(x_i, t)\boldsymbol{G}_k \tag{7.34}$$

then

$$\begin{aligned}\frac{d\boldsymbol{f}}{dt} &= \left(\left.\frac{\partial f_k}{\partial t}\right|_{x_i} + \frac{\partial f_k}{\partial x_i}\frac{\partial x_i}{\partial t} \right)\boldsymbol{G}_k \\ &= \frac{Df_k}{Dt}\boldsymbol{G}_k \end{aligned} \tag{7.35}$$

Note that in Eq. (7.35), the differentiation has been applied while fixing $\boldsymbol{P}$, i.e., X_i, thus x_i is not fixed.

7.4 Conservation Law

Here and henceforth until declared otherwise, we will use the spatial description. Consider physical quantity ϕ flows with velocity v_i, as exemplified by Fig. 7.3.

Let ϕ be a quantity per unit mass, which may be a scalar, vector, or tensor. The conservation law for a volume V surrounded by a surface S is stated as follows: the time rate of total amount of ϕ must be equal to the sum of a quantity flown in and out across the surface and a quantity produced inside the volume. Therefore, in Fig. 7.3, the conservation law can be formulated in the global form. Because ϕ is a quantity per unit mass and $\rho\phi$ is a quantity per unit volume, where ρ is density, the time rate of total amount of ϕ over the volume V is as follows:

$$\frac{D}{Dt}\int_V (\rho\phi)dV$$

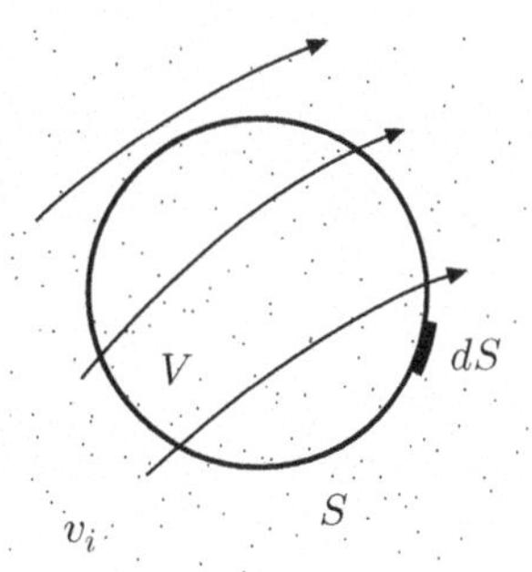

Fig. 7.3 Conservation law

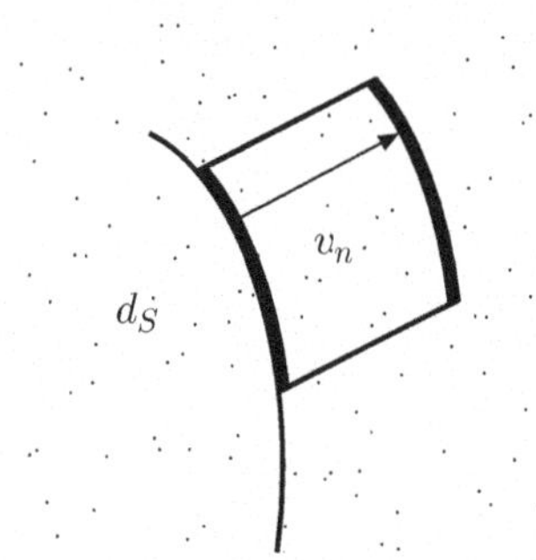

Fig. 7.4 Infinitesimal area per unit time

An infinitesimal area per unit time on the surface is considered as rectangle or cuboid, whose volume is $v_n dS$ as shown in Fig. 7.4. Here v_n means $v_i n_i$, which is normal velocity, denoting n_i component of unit outward normal. Thus, the time rate of total amount of ϕ flown in and out across the surface S is expressed as

$$\int_S (\rho\phi v_i n_i) dS$$

Let production per unit mass per unit time be Q, total amount of production is

$$\int_V (\rho Q) dV$$

Thus, the conservation law can be derived as follows:

$$\frac{D}{Dt}\int_V (\rho\phi) dV = \int_S (\rho\phi v_i n_i) dS + \int_V (\rho Q) dV \tag{7.36}$$

Equation (7.36) is called a global form of conservation equation, which can be reformulated into the local form. The first term is

$$\frac{D}{Dt}\int_V (\rho\phi) dV = \int_V \frac{D}{Dt}(\rho\phi) dV + \int_V \rho\phi \frac{D}{Dt}(dV) \tag{7.37}$$

Using the well-known relation

$$\frac{D}{Dt}(dV) = v_{i,i} dV \tag{7.38}$$

we obtain

$$\frac{D}{Dt}\int_V (\rho\phi) dV = \int_V \left\{ \frac{D}{Dt}(\rho\phi) + v_{i,i} \right\} dV \tag{7.39}$$

The second term is

$$\int_S (\rho\phi v_i n_i) dS = \int_S (q_i n_i) dS = \int_V (q_{i,i}) dV \tag{7.40}$$

where q_i is called flux, which is

$$q_i = \rho\phi v_i \tag{7.41}$$

Based on Eqs. (7.39) and (7.40), Eq. (7.36) can be written as

$$\int_V \left\{ \frac{D}{Dt}(\rho\phi) + \rho\phi v_{i,i} - q_{i,i} - \rho Q \right\} dV = 0 \tag{7.42}$$

Equation (7.42) is valid over any arbitrary volume. Thus

$$\frac{D}{Dt}(\rho\phi) + \rho\phi v_{i,i} - q_{i,i} - \rho Q = 0 \tag{7.43}$$

Equation (7.43) is referred to as a local form of the conservation equation. In Eq. (7.40), the surface integral is converted into the volume integral using the Green–Gauss theorem.

7.5 Conservation of Mass

The conservation equation of mass can be derived by substituting $\phi = 1$ in Eq. (7.36). It is assumed here that the mass does not change across the surface nor is produced inside the volume V. Equation (7.36) then becomes

$$\frac{D}{Dt} \int_V \rho dV = 0 \tag{7.44}$$

This equation is the global form of conservation of mass. The local form of conservation of mass can be obtained from Eq. (7.44) as

$$\frac{D\rho}{Dt} + \rho v_{i,i} = 0 \tag{7.45}$$

Equation (7.45) can also be derived by putting

$$q_i = 0 \quad \text{and} \quad Q = 0 \tag{7.46}$$

in Eq. (7.43). Equation (7.45) is transformed into the following form:

$$\frac{\partial \rho}{\partial t} + (\rho v_i)_{,i} = 0 \tag{7.47}$$

If we assume incompressibility, which is defined as

$$\frac{D\rho}{Dt} = 0 \tag{7.48}$$

then Eq. (7.45) leads to the following equation.

$$v_{i,i} = 0 \tag{7.49}$$

Equation (7.49) is usually used as equation of continuity of the incompressible fluid flows.

7.6 Conservation of Substance Transport

We can obtain the global form of conservation equation of substance transport replacing ϕ in Eq. (7.36) with $\mathfrak{c}$:

$$\frac{D}{Dt}\int_V (\rho\mathfrak{c})dV = \int_S (\rho\mathfrak{c}v_i n_i)dS + \int_V (\rho Q)dV \tag{7.50}$$

where $\mathfrak{c}$ is concentration of substance per unit mass and Q is the production rate of substance per unit mass per unit time. The substance flux is defined by

$$N_i = \rho\mathfrak{c}v_i \tag{7.51}$$

Substituting Eq. (7.51) into Eq. (7.50), we get

$$\frac{D}{Dt}\int_V (\rho\mathfrak{c})dV = \int_S (N_i n_i)dS + \int_V (\rho Q)dV \tag{7.52}$$

Equation (7.52) is referred to as the global form of conservation equation of substance.

Replacing ϕ and q_i in Eq. (7.43) with $\mathfrak{c}$ and N_i, respectively, we obtain

$$\frac{D}{Dt}(\rho\mathfrak{c}) + \rho\mathfrak{c}v_{i,i} - N_{i,i} - \rho Q = 0 \tag{7.53}$$

Transforming Eq. (7.53) and rearranging the terms, the following equation can be derived:

$$\left(\frac{D\rho}{Dt} + \rho v_{i,i}\right) c + \rho \frac{Dc}{Dt} - N_{i,i} - \rho Q = 0 \tag{7.54}$$

Applying the conservation equation of mass, Eq. (7.45), the local form of conservation of substance can be described as follows:

$$\rho \frac{Dc}{Dt} = N_{i,i} + \rho Q \tag{7.55}$$

Assume that the flux N_i is proportional to the gradient of substance:

$$N_i = kc_{,i} \tag{7.56}$$

where k is referred to as diffusion coefficient. Substituting Eq. (7.56) into Eq. (7.55), the governing equation of substance transport can be obtained:

$$\frac{\partial c}{\partial t} + v_i c_{,i} - kc_{,ii} + Q = 0 \tag{7.57}$$

Equation (7.57) corresponds to Eq. (5.1) in Chap. 5. With boundary and initial conditions, the substance transport can be solved.

7.7 Conservation of Momentum

The global form of conservation equation of momentum can be derived by replacing ϕ and Q in Eq. (7.36) with velocity v_i and body force f_i, respectively, as

$$\frac{D}{Dt} \int_V (\rho v_i) dV = \int_S (\rho v_j v_i n_j) dS + \int_V (\rho f_i) dV \tag{7.58}$$

In Eq. (7.58), the left side term represents the time rate of momentum, and the last term on the right side represents the momentum rate created by the body force. For the flux of momentum, we posit

$$\sigma_{ji} = \rho v_j v_i \tag{7.59}$$

Introducing Eq. (7.59) into Eq. (7.58), the following equation can be obtained:

$$\frac{D}{Dt} \int_V (\rho v_i) dV = \int_S (\sigma_{ji} n_j) dS + \int_V (\rho f_i) dV \tag{7.60}$$

Replacing the first term on the right side of Eq. (7.60) with

$$t_i = \sigma_{ij} n_j \tag{7.61}$$

the global form of the conservation equation of momentum can be derived as follows:

$$\frac{D}{Dt}\int_V (\rho v_i) dV = \int_S (t_i) dS + \int_V (\rho f_i) dV \tag{7.62}$$

If we consider t_i as traction force, the right side of Eq. (7.62) expresses the sum of forces applied on the surface S and produced inside the volume V. Therefore, Eq. (7.62) states that the time rate of the momentum of the body is equal to the time rate of total force applied to the body. This is the conservation law of momentum. Considering Eq. (7.61), the traction force t_i is characterized by its components σ_{ij}. Namely, σ_{ij} defined by Eq. (7.59) is stress. Thus, it can be stated that the stress is the flux of momentum.

Replacing ϕ, q_j, and Q in Eq. (7.43) with v_i, σ_{ji}, and f_i, respectively, we obtain

$$\frac{D}{Dt}(\rho v_i) + \rho v_i v_{j,j} - \sigma_{ji,j} - \rho f_i = 0 \tag{7.63}$$

Equation (7.63) can be transformed as

$$\left(\frac{D\rho}{Dt} + \rho v_{j,j}\right) v_i + \rho \frac{Dv_i}{Dt} - \sigma_{ji,j} - \rho f_i = 0 \tag{7.64}$$

Using the conservation of mass, Eq. (7.45), the first term of Eq. (7.64) is zero. The local form of the conservation equation of momentum can be obtained as follows:

$$\rho \frac{Dv_i}{Dt} = \sigma_{ji,j} + \rho f_i \tag{7.65}$$

Equation (7.65) is also referred to as the equation of motion. Namely, mass times acceleration is equal to the force per unit mass per unit time.

We now take a closer look at Eq. (7.62) from a different perspective. The traction vector is defined as

$$\boldsymbol{t} = t_i \boldsymbol{G}_i \tag{7.66}$$

where $\boldsymbol{G}_i$ is the base vector. The traction is force. For simplicity, we set aside acceleration and body force for now, and then Eq. (7.62) is reduced to the following equation:

$$\int_S \boldsymbol{t} dS = 0 \tag{7.67}$$

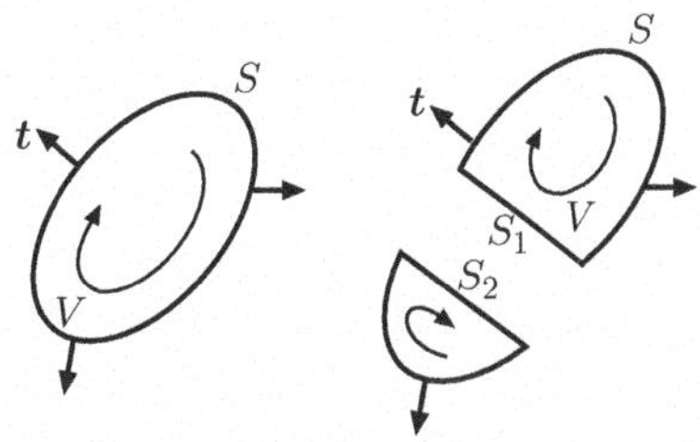

Fig. 7.5 Equilibrium of force

Equation (7.67) represents static equilibrium. Namely, if we introduce stress as the flux of momentum, as in Eq. (7.59), then static equilibrium is deduced as in Eq. (7.67). In the literature, Eq. (7.67) is often assumed first and then the stress is deduced as in Eq. (7.61). In this book, we emphasize the equation of motion can be derived from the conservation law and the stress corresponds to the flux of momentum, which is in Eq. (7.59).

It is understood that Eq. (7.67) is important because the stress equilibrium can be explained by this equation. If a body V with its surface S divided into two parts, for instance, as shown in Fig. 7.5, Eq. (7.67) is valid over two parts with cutting planes S_1 and S_2. The integration directions along two sides at the cutting plane are definitely inverse on S_1 and S_2. Thus,

$$\int_{S_1} \boldsymbol{t} dS + \int_{S_2} \boldsymbol{t} dS = 0 \tag{7.68}$$

This relation is normally represented as

$$\boldsymbol{t}_1 + \boldsymbol{t}_2 = 0 \tag{7.69}$$

where

$$\boldsymbol{t}_1 = \int_{S_1} \boldsymbol{t} dS$$
$$\boldsymbol{t}_2 = \int_{S_2} \boldsymbol{t} dS$$

Equation (7.69) is referred to as the equilibrium of internal force.

7.8 Constitutive Equation

The constitutive equations are used to define the continua, i.e., solids, fluids, mixtures, etc. Although a number of constitutive equations have been presented, we restrict here only necessary equations in fluids to be used in the following chapters. The components of stress are defined in Eq. (7.59). From the equilibrium of moment of momentum, the stress must be symmetric, i.e.,

$$\sigma_{ij} = \sigma_{ji} \tag{7.70}$$

In general, solids are defined as continua whose stress is expressed by a function of deformation gradient of displacement, whereas fluids are defined as continua whose stress is expressed by a function of deformation rate d_{ij}, which is represented as

$$d_{ij} = \frac{1}{2}(v_{i,j} + v_{j,i}) \tag{7.71}$$

where v_i is velocity.

Reiner and Rivlin introduced the following equation as the constitutive equation of fluid:

$$\sigma_{ij} = -p\delta_{ij} + \alpha_0\delta_{ij} + \alpha_1 d_{ij} + \alpha_2 d_{ik}d_{kj} \tag{7.72}$$

where p is pressure and

$$\alpha_0,\ \alpha_1,\ \alpha_2 = \alpha(\mathrm{I}_d,\ \mathrm{II}_d,\ \mathrm{III}_d) \tag{7.73}$$

in which α is some function and $\mathrm{I}_d, \mathrm{II}_d, \mathrm{III}_d$ are referred to as invariants of deformation rate, i.e.,

$$\begin{aligned}
\mathrm{I}_d &= d_{ii} \\
\mathrm{II}_d &= \frac{1}{2}(d_{ii}d_{jj} - d_{ij}d_{ij}) \\
\mathrm{III}_d &= \det|d_{ij}|
\end{aligned}$$

We sometimes assume the linear relation, i.e.,

$$\sigma_{ij} = -p\delta_{ij} + \lambda d_{kk}\delta_{ij} + 2\mu d_{ij} \tag{7.74}$$

where λ and μ are constants.

Equation (7.74) can be divided into two parts: conservative stress and dissipative stress τ_{ij}, i.e.,

$$\sigma_{ij} = -p\delta_{ij} + \tau_{ij} \tag{7.75}$$

where

$$\tau_{ij} = \lambda d_{kk}\delta_{ij} + 2\mu d_{ij} \tag{7.76}$$

From Eq. (7.76), we have

$$d_{kk} = \frac{\tau_{kk}}{3\lambda + 2\mu} \tag{7.77}$$

then the inverse relation can be written as follows:

$$d_{ij} = \frac{\tau_{ij}}{2\mu} + \frac{\lambda}{2\mu(3\lambda + 2\mu)}\tau_{kk}\delta_{ij} \tag{7.78}$$

Therefore, the coefficients must satisfy the following relations:

$$\mu \neq 0, \quad \text{and} \quad (3\lambda + 2\mu) \neq 0 \tag{7.79}$$

From Eq. (7.74), we can write

$$\sigma_{kk} = -3p + (3\lambda + 2\mu)d_{kk} \tag{7.80}$$

Therefore, σ_{kk} is not equal to hydrostatic pressure. Practically, we sometimes use the relation

$$3\lambda + 2\mu = 0 \tag{7.81}$$

and σ_{kk} is considered to be hydrostatic pressure. However, these relations are valid only in case of the incompressible flow, i.e., $d_{kk} = 0$.

7.9 Conservation of Energy

The equation for the conservation of energy can be derived by introducing the following quantity into the conservation equation Eq. (7.36):

$$\rho\phi = \rho e = \frac{\rho}{2}v_i v_i + \rho\epsilon \tag{7.82}$$

where e is total energy and ϵ is internal energy. The first term on the right side of Eq. (7.82) is kinetic energy. Here and throughout, we use the term "energy" to mean "energy density." Equation (7.82) means that total energy consists of kinetic and internal energies. Flux of energy and internal production of energy are expressed by the following equations:

$$\rho\phi v_j = v_i\sigma_{ij} + q_j \tag{7.83}$$

$$\rho Q = \rho f_i v_i + \rho h \tag{7.84}$$

where q_j and h are heat flux and heat source, respectively, and ρf_i is body force. Substituting Eqs. (7.82), (7.83), and (7.84) into Eq. (7.36), we get

$$\frac{D}{Dt}\int_V \rho\left(\frac{1}{2}v_i v_i\right)dV + \frac{D}{Dt}\int_V (\rho\epsilon)dV = \int_S (v_i\sigma_{ij}n_j)dS + \int_S (q_j n_j)dS + \int_V (\rho f_i v_i)dV + \int_V (\rho h)dV \quad (7.85)$$

Equation (7.85) can be written as

$$\dot{\mathscr{K}} + \dot{\mathscr{E}} = W + \Sigma \quad (7.86)$$

where

$$\mathscr{K} = \int_V \frac{\rho}{2}(v_i v_i)dV$$
$$\mathscr{E} = \int_V (\rho\epsilon)dV$$
$$W = \int_V (\rho f_i v_i)dV + \int_S (v_i\sigma_{ij}n_j)dS$$
$$\Sigma = \int_V (\rho h)dV + \int_S (q_i n_i)dS$$

Equation (7.86) is the global form of conservation equation of energy. The sum of time rate of kinetic energy $\mathscr{K}$ and internal energy $\mathscr{E}$ is equal to rate of mechanical power W and heat input Σ. In this chapter, overdot $\dot{}$ stands for material time derivative, i.e., $\frac{D}{Dt}$. In general, Eq. (7.86) is known as the first law of thermodynamics.

Here, let us reconsider the flux of energy equation Eq. (7.83). Using Eq. (7.82), the flux of energy can be written as

$$\rho\phi v_j = \rho e v_j = \frac{\rho}{2}v_i v_i v_j + \rho\epsilon v_j \quad (7.87)$$

The internal energy is assumed to be

$$\rho\epsilon = \frac{\rho}{2}v_i v_i + \rho\xi \quad (7.88)$$

where ξ is the internal energy caused by heat. Namely, the internal energy is assumed to be the sum of the kinetic energy and the internal energy by heat. introducing Eq. (7.88) into Eq. (7.87) and rearranging the terms, we obtain

$$\rho\phi v_j = \frac{\rho}{2}v_i v_i v_j + \frac{1}{2}\rho v_i v_i v_j + \rho\xi v_j \quad (7.89)$$

using Eq. (7.59) and putting

$$q_j = \rho \xi v_j \tag{7.90}$$

the flux of energy can be expressed as in Eq. (7.83).

Let us derive the local form of the conservation equation of energy. Left side of Eq. (7.85) is

$$\begin{aligned}\frac{D}{Dt}\int_V (\rho e)dV &= \int_V \frac{D}{Dt}(\rho e)dV + \int_V (\rho e)\frac{D}{Dt}(dV)\\ &= \int_V \left\{\frac{D}{Dt}(\rho e) + \rho e v_{k,k}\right\}dV\\ &= \int_V \left\{\frac{\partial}{\partial t}(\rho e) + v_j(\rho e)_{,j} + \rho e v_{j,j}\right\}dV\\ &= \int_V \left\{\frac{\partial}{\partial t}(\rho e) + (\rho e v_i)_{,i}\right\}dV\end{aligned} \tag{7.91}$$

using Eq. (7.38). The right side terms of Eq. (7.85) can be transformed into the following forms converting the surface integral into the volume integral using the Green–Gauss theorem:

$$W = \int_V \{(\rho f_i v_i) + (\sigma_{ij} v_i)_{,j}\}\, dV \tag{7.92}$$

$$\Sigma = \int_V (\rho h + q_{i,i})dV \tag{7.93}$$

Using Eqs. (7.47), (7.91), (7.92), and (7.93), we obtain the following equation from Eq. (7.85):

$$\rho\frac{De}{Dt} - (\sigma_{ij} v_i)_{,j} - q_{i,i} - \rho f_i v_i - \rho h = 0 \tag{7.94}$$

Introducing Eq. (7.82) into Eq. (7.94) and rearranging the terms, we get

$$v_i\left(\rho\frac{Dv_i}{Dt} - \sigma_{ij,j} - \rho f_i\right) + (\rho\dot{\epsilon} - \sigma_{ij} d_{ij} - q_{i,i} - \rho h) = 0 \tag{7.95}$$

in which

$$\sigma_{ij} d_{ij} = \sigma_{ij}\frac{1}{2}(v_{i,j} + v_{j,i}) = \sigma_{ij} v_{i,j} \tag{7.96}$$

because of Eq. (7.70). Introducing conservation equation of momentum, Eq. (7.65), the local form of conservation equation of energy can be expressed by the following equation:

$$\rho \frac{D\epsilon}{Dt} = \sigma_{ij} d_{ij} + q_{i,i} + \rho h \tag{7.97}$$

The first term on the right side of Eq. (7.97) is referred to as mechanical power, where d_{ij} is deformation rate defined in Eq. (7.71). Equation (7.97) expresses the conservation law in which the internal energy rate per unit time is due to the mechanical power and heat energy.

For continua without heat transfer, the conservation equation of energy can be simplified. We consider the case in which

$$q_i = 0 \quad \text{and} \quad h = 0 \tag{7.98}$$

then Eq. (7.97) is reduced to

$$\rho \frac{D\epsilon}{Dt} = \sigma_{ij} d_{ij} \tag{7.99}$$

Adding conservation equation of mass Eq. (7.45), we can posit

$$\epsilon \left(\frac{D\rho}{Dt} + \rho v_{k,k} \right) + \rho \frac{D\epsilon}{Dt} = \sigma_{ij} d_{ij} \tag{7.100}$$

Integrating Eq. (7.100) over the volume V and using Eq. (7.38), the following equation can be derived:

$$\frac{D}{Dt} \int_V (\rho\epsilon) dV = \int_V (\sigma_{ij} d_{ij}) dV \tag{7.101}$$

Equation (7.101) means that time rate of internal energy occurred only by the mechanical power. In this case, conservation equation of energy becomes

$$\frac{D\mathscr{K}}{Dt} + \frac{D\mathscr{E}}{Dt} = W \tag{7.102}$$

which is

$$\begin{aligned} \frac{D}{Dt} \int_V \frac{1}{2} \rho (v_i v_i) dV &+ \int_V (\sigma_{ij} v_{i,j}) dV \\ &= \int_S (\sigma_{ij} v_i n_j) dS + \int_V (\rho v_i f_i) dV \end{aligned} \tag{7.103}$$

Transforming Eq. (7.103) leads to the following equation:

$$\rho\frac{Dv_i}{Dt} = \sigma_{ij,j} + \rho f_i \tag{7.104}$$

Equation (7.104) is equivalent to Eq. (7.65). Namely, the conservation equation of energy is the conservation equation of momentum in case of continua without heat transfer.

7.10 Clausius–Duhem Inequality

To express the irreversible process of deformation of continua, it is well known that the second law of thermodynamics, also known as Clausius–Duhem inequality, can be used. The axiom states that the time rate of change of the total entropy H is never smaller than the sum of the flux of entropy $\mathfrak{S}$ and the entropy supplied by the internal source B, which can be described as follows:

$$\frac{DH}{Dt} - B - \mathfrak{S} \geqq 0 \tag{7.105}$$

where

$$\begin{aligned} H &= \int_V (\rho\eta)dV \\ B &= \int_V \rho\left(\frac{h}{T}\right)dV \\ \mathfrak{S} &= \int_S \left(\frac{q_i}{T}\right)n_i dS \end{aligned}$$

in which η is entropy, q_i is flux of entropy across the boundary S of the volume V, and T is temperature. Using the Green–Gauss theorem, Eq. (7.105) is transformed into the local form:

$$\rho T\dot{\eta} - \rho h - q_{i,i} + \frac{q_i T_{,i}}{T} \geqq 0 \tag{7.106}$$

Let the internal production γ be

$$\rho T\gamma = \rho T\dot{\eta} - \rho h - q_{i,i} + \frac{q_i T_{,i}}{T} \tag{7.107}$$

Equation (7.106) can be written as

$$\rho T\gamma \geqq 0 \tag{7.108}$$

In order to deduce the final consequence of the Clausius–Duhem inequality, equation of the internal energy ϵ, Eq. (7.97), is transformed into simplified form using Eq. (7.75) as follows:

$$\begin{aligned}\rho\dot{\epsilon} &= \sigma_{ij}d_{ij} + q_{i,i} + \rho h \\ &= -pd_{ii} + \tau_{ij}d_{ij} + q_{i,i} + \rho h\end{aligned} \tag{7.109}$$

Internal dissipation σ is now defined as

$$\sigma = \rho T\dot{\eta} - q_{i,i} - \rho h \tag{7.110}$$

If $\sigma = 0$ then the process is called as reversible process. In this case, the second term of the right side of Eq. (7.109) is zero. Thus, the internal energy in the reversible process is derived as

$$\rho\dot{\epsilon} = -pd_{ii} + \rho T\dot{\eta} \tag{7.111}$$

Based on Eqs. (7.109) and (7.111), the entropy can be deduced in the irreversible process:

$$\rho T\dot{\eta} = \tau_{ij}d_{ij} + q_{i,i} + \rho h \tag{7.112}$$

Substituting Eq. (7.112) into Eq. (7.108):

$$\rho T\gamma = \tau_{ij}d_{ij} + \frac{q_i T_{,i}}{T} \tag{7.113}$$

We can see in Eq. (7.113) that the entropy production consists of two parts, one is due to the dissipative stress and the other is due to the thermal effects.

Using Eq. (7.110), Eq. (7.107) is transformed into the following form:

$$\rho T\gamma = \sigma + \frac{q_i T_{,i}}{T} \geqq 0 \tag{7.114}$$

Assuming the Fourier's law for the thermal flux of the form,

$$q_i = \kappa T_{,i} \tag{7.115}$$

where κ is the thermal conduction coefficient, Eq. (7.114) with Eq. (7.115) leads to the following inequality:

$$\sigma + \frac{\kappa T_{,i} T_{,i}}{T} \geqq 0 \tag{7.116}$$

Equation (7.116) should hold even if $\sigma = 0$, i.e., in the reversible process. Therefore,

$$\kappa \geqq 0 \tag{7.117}$$

On the other hand, even if thermal effect is zero, i.e., $q_i = 0$, Eq. (7.113) should also be positive:

$$(\lambda \delta_{ij} + \mu d_{ij}) d_{ij} \geqq 0 \tag{7.118}$$

Equation (7.118) can be transformed as

$$(3\lambda + 2\mu) d_{ii} d_{kk} + 2\mu d'_{ij} d'_{ij} \geqq 0 \tag{7.119}$$

where

$$d'_{ij} = d_{ij} - \frac{1}{3} d_{kk} \delta_{ij} \tag{7.120}$$

which is referred to as deviatoric deformation rate. From Eq. (7.119), we obtain

$$(3\lambda + 2\mu) \geqq 0 \quad \text{and} \quad 2\mu \geqq 0 \tag{7.121}$$

Equations (7.117) and (7.121) are final consequences derived from the Clausius–Duhem inequality.

7.11 Helmholtz Free Energy

Introducing the Helmholtz free energy, the conservation equation of energy can be transformed into the thermal conduction equation. Let the Helmholtz free energy be ψ, entropy η, and temperature T, then the internal energy ε can be expressed as

$$\epsilon = \psi + \eta T \tag{7.122}$$

Substituting Eq. (7.122) into Eq. (7.97), we get

$$\rho(\dot{\psi} + \dot{\eta} T + \eta \dot{T}) = q_{i,i} + \rho h + \sigma_{ij} d_{ij} \tag{7.123}$$

Assume

$$\psi = \psi(T, \nu) \tag{7.124}$$

where ν is the specific volume,

$$\nu = \frac{1}{\rho} \tag{7.125}$$

and also assume

$$\eta = -\frac{\partial \psi}{\partial T} \tag{7.126}$$

then the following equations are derived:

$$\dot{\psi} = \frac{\partial \psi}{\partial T}\dot{T} + \frac{\partial \psi}{\partial \nu}\dot{\nu}, \tag{7.127}$$

$$\eta \dot{T} = -\frac{\partial \psi}{\partial T}\dot{T} = -\dot{\psi} + \frac{\partial \psi}{\partial \nu}\dot{\nu} \tag{7.128}$$

Substituting Eq. (7.128) into Eq. (7.123) and rearranging the terms, we get

$$\rho T \dot{\eta} + \rho \frac{\partial \psi}{\partial \nu}\dot{\nu} = -p d_{ii} + \tau_{ij} d_{ij} + q_{i,i} + \rho h \tag{7.129}$$

Assuming that

$$p = -\frac{\partial \psi}{\partial \nu} \tag{7.130}$$

and using the conservation equation of mass, i.e., Eq. (7.45):

$$\rho \dot{\nu} = d_{ii} \tag{7.131}$$

Equation (7.129) is transformed as follows:

$$\rho T \dot{\eta} = \tau_{ij} d_{ij} + q_{i,i} + \rho h \tag{7.132}$$

Differentiating both sides of Eq. (7.126) with respect to time t:

$$\dot{\eta} = -\frac{\partial^2 \psi}{\partial T^2}\dot{T} - \frac{\partial^2 \psi}{\partial T \partial \nu}\dot{\nu} \tag{7.133}$$

Substituting Eq. (7.133) into Eq. (7.132), we can obtain the following equation:

$$\rho T \frac{\partial^2 \psi}{\partial T^2}\dot{T} + T \frac{\partial^2 \psi}{\partial T \partial \nu} d_{ii} + q_{i,i} + \rho h + \tau_{ij} d_{ij} = 0 \tag{7.134}$$

Equation (7.134) is sometimes referred to as the "energy transport equation."

From Eqs. (7.122) and (7.126), we get

$$\frac{\partial \epsilon}{\partial T} = \frac{\partial \psi}{\partial T} + \eta + \frac{\partial \eta}{\partial T} T = \frac{\partial \eta}{\partial T} T \tag{7.135}$$

Specific heat at constant volume c_v can be expressed as follows:

$$c_v \equiv \left(\frac{\partial \epsilon}{\partial T} \right)_v = T \left(\frac{\partial \eta}{\partial T} \right)_v = -T \frac{\partial^2 \psi}{\partial T^2} \tag{7.136}$$

where subscript v means that differentiation is carried out by keeping volume constant. Using Eq. (7.136) and the Fourier's law Eq. (7.115), Eq. (7.134) can be transformed as follows:

$$-\rho c_v \dot{T} + T \frac{\partial^2 \psi}{\partial T \partial \nu} d_{ii} + (\kappa T_{,i})_{,i} + \rho h + \lambda d_{ii} d_{kk} + 2\mu d_{ij} d_{ij} = 0 \tag{7.137}$$

The thermal conduction equation is normally Eq. (7.137) without second, fifth, and sixth terms. Namely, the temperature and deformation are uncoupled. In practice the effect of d_{ii} is not often significant. In the incompressible flow, $d_{ii} = 0$, then, Eq. (7.137) can be transformed as follows:

$$-\rho c_v \dot{T} + (\kappa T_{,i})_{,i} + \rho h + 2\mu d'_{ij} d'_{ij} = 0 \tag{7.138}$$

where d'_{ij} is the deviatoric deformation rate, which is defined in Eq. (7.120). The thermal conduction equation is normally Eq. (7.138) without the last term. We should emphasize that the last term is significant in the flow of relatively large shearing deformation rate.

7.12 Ideal Gas

The conservation equations described in the previous sections are conservation of mass discussed in Sect. 7.5, conservation of momentum discussed in Sect. 7.7, and conservation of energy discussed in Sect. 7.9. We also described the constitutive equation in Sect. 7.8. There are five kinds of field variables, i.e., density ρ, velocity v_i, pressure p, total energy e, and heat flux q_i, whereas we have three sets of conservation equations. In those equations, there are some variables that are implicitly included, such as temperature T, internal energy ϵ, etc. To flesh out those implicit variables, we introduce two more equations, i.e., equation of state and equation of thermal flux. For this purpose, we usually introduce the ideal gas assumption.

Although there are several manners in which to define the specific heat c, we use the following definitions here, assuming inviscid flow:

$$c \equiv \frac{\partial}{\partial \dot{T}}(\dot{\epsilon} + p\dot{\nu}) \tag{7.139}$$

where ν is the specific volume defined in Eq. (7.125). We can define the specific heat at constant volume c_v as

$$c_v = \left(\frac{\partial \epsilon}{\partial T}\right)_\nu \tag{7.140}$$

and the specific heat at constant pressure as

$$c_p = \left[\left(\frac{\partial \epsilon}{\partial \nu}\right)_{T,\nu} + p\right]\left(\frac{\partial \nu}{\partial T}\right)_p \tag{7.141}$$

The ratio of the specific heat is defined as

$$\gamma = \frac{c_p}{c_v} \tag{7.142}$$

We have shown in Eq. (7.88) that the internal energy ϵ consists of the deformation part and the thermal part. In case of the specific heat at constant volume defined in Eq. (7.140), we can assume the internal energy as

$$\epsilon = c_v T \tag{7.143}$$

The ideal gas assumption is

$$p = \rho R T \tag{7.144}$$

where R is referred to as the gas constant. Based on Eq. (7.144), equation of state in terms of internal energy ϵ can be derived using the Mayer relation, which is expressed as follows. From Eq. (7.141),

$$\begin{aligned} c_p &= \left(\frac{\partial \epsilon}{\partial \nu}\right)_{T,\nu}\left(\frac{\partial \nu}{\partial T}\right)_p + p\left(\frac{\partial \nu}{\partial T}\right)_p \\ &= \left(\frac{\partial \epsilon}{\partial T}\right)_\nu + p\left(\frac{\partial \nu}{\partial T}\right)_p \end{aligned} \tag{7.145}$$

From Eqs. (7.125) and (7.144), we get

$$\nu p = R T \tag{7.146}$$

Introducing Eqs. (7.140) and (7.146) into Eq. (7.145), the Mayer relation can be formulated as

$$c_p - c_v = R \tag{7.147}$$

To eliminate T from Eq. (7.143), we again use Eq. (7.144) as

$$T = \frac{p\nu}{R} = \frac{p\nu}{c_p - c_v} \tag{7.148}$$

Then, we get

$$\epsilon = c_v T = \frac{c_v}{c_p - c_v} p\nu = \frac{p\nu}{\gamma - 1} \tag{7.149}$$

From Eq. (7.149), we also get

$$p = \rho(\gamma - 1)\epsilon \tag{7.150}$$

Equation (7.150) is the equation of state in terms of internal energy and pressure. The thermal flux equation Eq. (7.115) can be transformed into

$$q_j = \frac{k}{c_v}\epsilon_{,j} \tag{7.151}$$

using Eq. (7.143).

7.13 Adiabatic State

In this section, we introduce the concept of the adiabatic state. Using Eq. (7.111), we can describe the entropy as

$$T\dot{\eta} = \dot{\epsilon} + p\dot{\nu} \tag{7.152}$$

Using Eqs. (7.125), (7.143), and (7.144), Eq. (7.152) is reformulated as

$$\dot{\eta} = c_v \frac{\dot{T}}{T} - R\left(\frac{\dot{\rho}}{\rho}\right) \tag{7.153}$$

From Eq. (7.144), the rate of pressure can be written as

$$\dot{p} = \dot{\rho}RT + \rho R\dot{T} \tag{7.154}$$

Then, we obtain

$$\frac{\dot{p}}{p} = \frac{\dot{\rho}RT}{\rho RT} + \frac{\rho R\dot{T}}{\rho RT} = \frac{\dot{\rho}}{\rho} + \frac{\dot{T}}{T} \tag{7.155}$$

Substituting Eq. (7.155) into Eq. (7.153), the entropy can be reformulated as follows using Eq. (7.147):

$$\dot{\eta} = c_v \left(\frac{\dot{p}}{p}\right) - c_p \left(\frac{\dot{\rho}}{\rho}\right) \tag{7.156}$$

Thus, we can integrate Eq. (7.156) as

$$\frac{\eta - \eta_0}{c_v} = \log\left(\frac{p}{p_0}\right) - \gamma \log\left(\frac{\rho}{\rho_0}\right) \tag{7.157}$$

Equation (7.157) can be transformed into

$$\left(\frac{p}{p_0}\right) = \left(\frac{\rho}{\rho_0}\right)^{\gamma} \cdot \exp\left(\frac{\eta - \eta_0}{c_v}\right) \tag{7.158}$$

If we introduce the assumption that

$$\eta = \eta_0 \tag{7.159}$$

then we obtain

$$\frac{p}{p_0} = \left(\frac{\rho}{\rho_0}\right)^{\gamma} \tag{7.160}$$

Equation (7.160) is referred to as the "pressure density equation," or the "Poisson law."

Acoustic velocity $\mathfrak{a}$ is defined as

$$\mathfrak{a}^2 = \frac{\partial p}{\partial \rho} \tag{7.161}$$

Using Eq. (7.160), we can transform this into the following form:

$$\begin{aligned} \mathfrak{a}^2 &= \frac{\partial p}{\partial \rho} = \frac{p_0}{\rho_0^{\gamma}} \gamma \rho^{\gamma - 1} = \gamma \left(\frac{p_0}{\rho_0}\right)\left(\frac{\rho}{\rho_0}\right)^{\gamma - 1} \\ &= \gamma \frac{p}{\rho} = \gamma RT \end{aligned} \tag{7.162}$$

Therefore, the acoustic velocity is time dependent. Using Eqs. (7.144) and (7.160), we get

$$\left(\frac{T}{T_0}\right) = \left(\frac{p}{p_0}\right)^{\frac{\gamma-1}{\gamma}} \tag{7.163}$$

Equation (7.163) is referred to as the "temperature pressure equation." To express the speed of fluid, the Mach number M_c is defined as

$$M_c = \frac{\mathsf{v}}{\mathfrak{a}} \tag{7.164}$$

where v is the velocity of flow and $\mathfrak{a}$ is its acoustic velocity.

7.14 Interface Condition

Consider the case in which there is discontinuity in density, velocity, or stress in the continuum. In Fig. 7.6 the material volume, surface, and interface are denoted by V, S, and σ, respectively. The interface σ divides V and S into two parts, V^+ and V^-, and S^+ and S^-. The velocity is denoted by v_i. The material derivative of the volume integral of a function φ can be expressed as follows (e.g., Eringen 1967):

$$\frac{D}{Dt}\int_{V-\sigma}\varphi dV = \int_{V-\sigma}\left\{\frac{\partial\varphi}{\partial t} + (\varphi v_i)_{,i}\right\}dV + \int_\sigma [\![\varphi(v_j - \nu_j)n_j]\!]\,dS \tag{7.165}$$

where the symbol $[\![\;]\!]$ represents discontinuity such that

$$[\![A]\!] = A^+ - A^-. \tag{7.166}$$

The velocity of the interface is denoted by ν_j. The conservation law of φ can be derived as follows:

$$\frac{D}{Dt}\int_V \varphi dV = \int_V q_{j,j}dV + \int_\sigma [\![q_j n_j]\!]\,dS \tag{7.167}$$

where q_j denotes the flux through the surface. Based on Eqs. (7.165) and (7.167), the following expression is obtained:

$$\int_{V-\sigma}\left\{\frac{\partial\varphi}{\partial t} + (\varphi v_j)_{,j} - q_{j,j}\right\}dV + \int_\sigma [\![\{\varphi(v_j - \nu_j) - q_j\}n_j]\!]\,dS = 0 \tag{7.168}$$

Because the function φ satisfies the local form of the conservation law, the first term of Eq. (7.168) equals to zero. Thus, the following equation can be obtained:

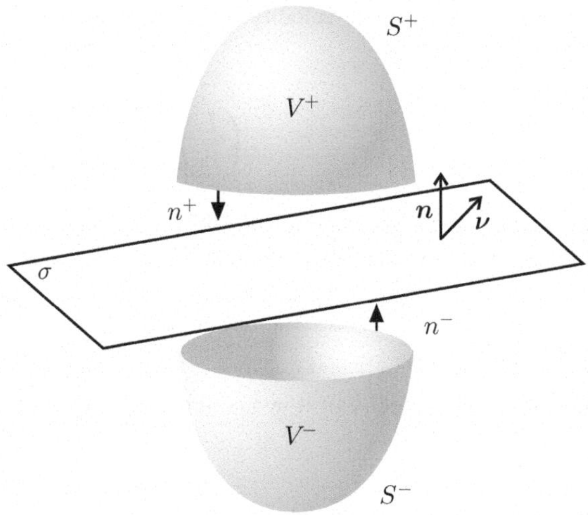

Fig. 7.6 Interface and domains

$$\int_\sigma [\![\{\varphi(v_j - \nu_j) - q_j\} n_j]\!] dS = 0 \tag{7.169}$$

Substituting

$$\varphi = \rho \quad \text{and} \quad q_j = 0$$

in Eq. (7.169), the discontinuity condition of mass is obtained as follows:

$$[\![\rho(v_j - \nu_j) n_j]\!] = 0 \tag{7.170}$$

Substituting

$$\varphi = \rho v_i \boldsymbol{g}_i \quad \text{and} \quad q_j = \sigma_{ji} \boldsymbol{g}_i$$

where $\boldsymbol{g}_i$ is the base vector, Eq. (7.169) leads to the discontinuity condition of momentum:

$$[\![\{\rho v_i (v_j - \nu_j) - \sigma_{ij}\} n_j]\!] = 0 \tag{7.171}$$

Equation (7.171) can be rewritten as

$$[\![\sigma_{ij} n_j]\!] = [\![\rho v_i (v_n - \nu_n)]\!] \tag{7.172}$$

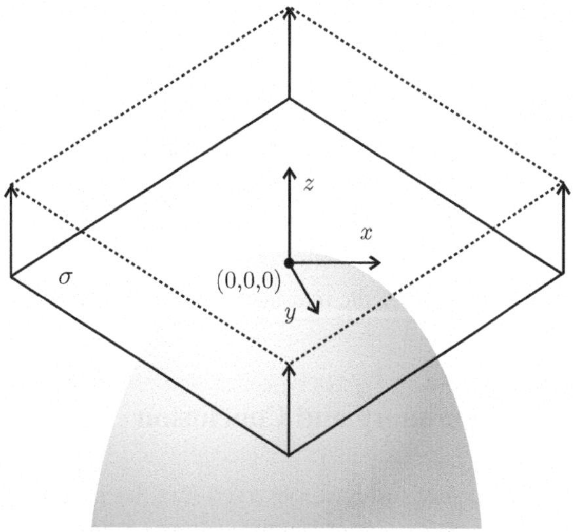

Fig. 7.7 Interface condition

where:

$$v_n = v_j n_j, \tag{7.173}$$

$$\nu_n = \nu_j n_j \tag{7.174}$$

Both v_n and ν_n denote the normal components of velocities of the material points and the interface, respectively. The tangential stress boundary conditions are appropriate for a fluid–fluid interface characterized by an interfacial tension. It is normally assumed that

$$\left[\!\left[\sigma_{ij} n_j\right]\!\right] = \xi \kappa n_i \tag{7.175}$$

where κ denotes the curvature of the interface and ξ denotes the surface tension coefficient.

Based on the local coordinate system shown in Fig. 7.7, the following expression holds:

$$Z = \Phi(x, y, t)$$

where $\Phi(x, y, t)$ is a function which expresses the interface at time t. A point on the interface is still on the interface after deformation takes place. Thus,

$$\frac{D}{Dt}\{Z - \Phi(x, y, t)\} = 0 \tag{7.176}$$

If we put

$$\mathfrak{f} = Z - \Phi(x, y, t) \tag{7.177}$$

Equation (7.176) can be transformed as follows:

$$\frac{D\mathfrak{f}}{Dt} = 0 \tag{7.178}$$

Using Eq. (7.178), the interface position can be determined.

7.15 Summary and Conclusion

In this chapter, we have observed that the governing equations of fluid flows are derived from the conservation equations of mass, momentum, and energy in addition to the constitutive equations. We have clarified the relations between conservation equations.

It is essential that the global forms of conservation equations of mass, momentum, and energy are represented by the integral equation forms. As we have repeatedly emphasized in the past chapters, the finite element method is based on the integral forms of the governing equations. Those are extensions of the global forms of the conservation equation using the weighting functions. Therefore, the finite element method is a numerical method based on a natural extension of conservation laws. Concrete discretization formulations will be presented in the following chapters.

Chapter 8
Analysis of Incompressible Flows

8.1 Introduction

In the past three decades, a number of solution methods have been developed based on the finite element method of incompressible fluid flows. The first paper was presented by Oden (1970). Actually, the majority of the histories of the finite element methods in fluid flows are devoted to the analysis of the incompressible flows. In this chapter, we will discuss two types of the stabilization methods, i.e., the bubble function method and the improved balancing tensor diffusivity method. As shown in Chap. 6, we must use the mixed interpolation to analyze the incompressible flows. The mixed interpolation is summarized in Taylor and Hood (1973), Kawahara et al. (1976), Jackson and Cliffe (1981), Carey and Oden (1986), Zienkiewicz et al. (2014), etc. The simplest mixed interpolation is the combination of the bubble function for velocity and linear function for pressure. Original ideas were discussed by Arnold et al. (1984), and Fortin and Fortin (1985). Many types of bubble functions are presented by, e.g., Franca and Farhat (1995), Pierre (1988, 1995), Baiocchi et al. (1993), Brezzi et al. (1992), Simo et al. (1995), Okumura and Kawahara (2000, 2003), Matsumoto and Kawahara (2001, 2006), Matsumoto et al. (2002), Matsumoto (2005), Yamada (1995, 1998), Nakajima and Kawahara (2000), Kurahashi et al. (2012), etc. We employed so-called "standard bubble function," which is the most popular function. With several assumptions, we can show that the stability of the bubble function method is equivalent to that of the SUPG method, which will be discussed in the next chapter in further depth.

Another stabilization method discussed in this chapter is the improved balancing tensor diffusivity (IBTD) method. If we take into account the second-order terms in the time discretization, which is converted into the terms in the space function, we can derive IBTD method after several transformations of equations. Original ideas of BTD, i.e., balancing tensor diffusivity, method were presented by Donea (1984), Donea et al. (1984), Gresho et al. (1984), Gresho and Sani (2000), etc. This method is

M. Kawahara, *Finite Element Methods in Incompressible, Adiabatic, and Compressible Flows*, Mathematics for Industry 19,
DOI 10.1007/978-4-431-55450-9_8

related to the three-step explicit method by Jiang and Kawahara (1993). We discuss the equivalency of the stability of the IBTD method and the SUPG method.

For the discretization in time, we employ the fractional step method, which is also called the "pressure Poisson equation method." Original ideas are found in Chorin (1968). The first finite element method based on this type of concept is presented by Temam (1968). Practical computations were executed by Ramaswamy and Kawahara (1987a, b), Kawahara and Anju (1988), Hayashi et al. (1991), Donea and Huerta (2003), etc. The stability and convergence are discussed by Guermond and Quartapelle (1998), Bochev and Gunzburger (2004), Corthwaite (2013), etc.

The stabilization methods described in this chapter, i.e., the bubble function method and IBTD method, have some characteristics in common. Namely, both methods employ the linear finite element equation, to which the stabilization terms are added. How to obtain the stabilization terms differ between the two methods. Nevertheless, we can show that the stabilization effects of both methods are equivalent to those of the SUPG method. The stabilization effects of the three methods, i.e., the bubble function method, the IBTD method, and the SUPG method, are almost identical, and stable computations can be carried out by those methods. Several case studies are described to show the adaptability for practical industrial applications. Superposted dot $\dot{}$ means partial differentiation with respect to time in this chapter.

8.2 Governing Equations of Incompressible Flows

The field variables for the computation of incompressible flows are velocity v_i and pressure p. The governing equations can be summarized as follows.

(1) Conservation of mass is

$$v_{i,i} = 0 \tag{8.1}$$

which is repeated from Eq. (7.49). This equation is sometimes referred to as the "equation of continuity."

(2) Conservation of momentum is

$$\rho(\dot{v}_i + v_j v_{i,j}) - \sigma_{ij,j} = \rho f_i \tag{8.2}$$

where ρ, σ_{ij}, and f_i mean density, total stress, and body force, respectively. This equation is referred to as the equation of motion, which is Eq. (7.65).

(3) Constitutive equation is

$$\sigma_{ij} = -p\delta_{ij} + \mu d_{ij} \tag{8.3}$$

which is Eq. (7.75). In Eq. (8.3), μ is called viscosity coefficient. Due to the Clausius–Duhem inequality, Eq. (7.119), the following inequality should hold:

$$\mu \geq 0 \tag{8.4}$$

(4) The deformation rate d_{ij} is given in Eq. (7.71), which is the symmetric part of velocity gradient:

$$d_{ij} = \frac{1}{2}(v_{i,j} + v_{j,i}) \tag{8.5}$$

(5) The boundary conditions are

$$v_i = \hat{v}_i \quad \text{on} \quad S_1 \tag{8.6}$$

$$t_i = \sigma_{ij} n_j = \hat{t}_i \qquad \text{on} \quad S_2 \tag{8.7}$$

where carret ˆ over a symbol means the given quantities on the boundary. In Eq. (8.7), t_i is the surface force and n_j is unit normal to the boundary. We assume that

$$\begin{aligned} S_1 \cup S_2 &= S \\ S_1 \cap S_2 &= \mathbf{0} \end{aligned} \tag{8.8}$$

where S denotes the total boundary and $\mathbf{0}$ is the null set.

(6) The initial conditions are

$$v_i = \hat{v}_i^0 \tag{8.9}$$

at time t_0

$$p = \hat{p}^0 \tag{8.10}$$

where superscript ^{0} means the values at initial time t_0. As the initial conditions, velocity and pressure are given in all flow fields. Substituting Eq. (8.5) into Eq. (8.3), and the derived equation again into Eq. (8.2), and rearranging the terms, the equation of motion can be reformulated as in the following form:

$$\rho(\dot{v}_i + v_j v_{i,j}) + p_{,i} - \mu(v_{i,j} + v_{j,i})_{,j} = \rho f_i \tag{8.11}$$

Equations (8.1) and (8.11) with Eqs. (8.6)–(8.10) make closed equation systems, i.e., we obtain the solution of the unsteady incompressible fluid flows to solve those equations. Substituting Eq. (8.1) into Eq. (8.11), we get

$$\rho(\dot{v}_i + v_j v_{i,j}) + p_{,i} - \mu v_{i,jj} = \rho f_i \tag{8.12}$$

The pairs of Eqs. (8.1) and (8.11) or of Eqs. (8.1) and (8.12) are called the Navier–Stokes equations.

We usually use the nondimensional form. Equation (8.11) can be rewritten as

$$\frac{\partial v_i}{\partial t} + v_j v_{i,j} + \frac{1}{\rho}\frac{\partial p}{\partial x_i} - \nu \frac{\partial}{\partial x_j}\left(\frac{\partial v_i}{\partial x_j} + \frac{\partial v_j}{\partial x_i}\right) = f_i \tag{8.13}$$

The dimension of the terms in Eq. (8.13) has a dimension [L][S^{-2}], which is acceleration. Introducing nondimensional quantities such as

$$\begin{aligned}
&\tilde{x}_i = \tfrac{x_i}{\mathfrak{L}}, \quad \tilde{v}_i = \tfrac{v_i}{\mathcal{V}}, \quad \tilde{t} = \tfrac{\mathcal{V}}{\mathfrak{L}}t \\
&\tilde{p} = \tfrac{p}{\rho \mathcal{V}^2} \\
&\tilde{\mu} = \tfrac{\mu}{\rho \mathcal{V} \mathfrak{L}} \\
&\tilde{f}_i = \tfrac{\mathfrak{L}}{\mathcal{V}^2} f_i
\end{aligned} \tag{8.14}$$

nondimensional equation of Eq. (8.14) can be expressed as

$$\frac{\partial \tilde{v}_i}{\partial t} + \tilde{v}_j \tilde{v}_{i,j} + \frac{\partial \tilde{p}}{\partial \tilde{x}_i} - \frac{\nu}{\mathcal{V}\mathfrak{L}}\frac{\partial}{\partial \tilde{x}_j}\left(\frac{\partial \tilde{v}_i}{\partial \tilde{x}_j} + \frac{\partial \tilde{v}_j}{\partial \tilde{x}_i}\right) = \tilde{f}_i \tag{8.15}$$

We define Reynolds number:

$$Re = \frac{\mathcal{V}\mathfrak{L}}{\nu}, \quad \nu = \frac{\mu}{\rho} \tag{8.16}$$

where ν is referred to as dynamic viscosity. Equation (8.15) is the nondimensional form and we usually write Eq. (8.15) as follows:

$$\dot{v}_i + v_j v_{i,j} + p_{,i} - \mu(v_{i,j} + v_{j,i})_{,j} = f_i \tag{8.17}$$

where

$$\mu = \frac{1}{Re} \tag{8.18}$$

Using Eq. (8.14), the nondimensional form of equation of continuity can also be written as Eq. (8.1). The boundary and initial conditions are also Eqs. (8.6) and (8.7), and Eqs. (8.9) and (8.10), respectively. For the sake of exposition, we will set aside the body force term in Eq. (8.17) in the following sections.

8.3 The Weighted Residual Equation

Multiplying both sides of Eq. (8.1) by the weighting function p^* and integrating over an arbitrary volume V, we get

$$\int_V (p^* v_{i,i}) dV = 0 \tag{8.19}$$

This is the weighted residual equation of continuity. Multiplying both sides of Eq. (8.17) by the weighting function v_i^* and integrating over the volume V and applying the integration by parts, the following equation can be obtained:

$$\int_V (v_i^* \dot{v}_i) dV + \int_V (v_i^* v_j v_{i,j}) dV + \int_V (v_{i,j}^* \sigma_{i,j}) dV = \int_S (v_i^* t_i) dS \tag{8.20}$$

where S is the boundary of the volume V abstracting away from f_i for the sake of simplicity, and

$$t_i = \sigma_{ij} n_j = \{-p\delta_{ij} + \mu(v_{i,j} + v_{j,i})\} n_j \tag{8.21}$$

is the surface force on the boundary S and n_j is the unit normal to the boundary S. The nondimensional stress is denoted by σ_{ij}.

Introducing the constitutive equation Eqs. (8.3) and (8.5) into Eq. (8.20) and arranging the terms, we get

$$\begin{aligned} &\int_V (v_i^* \dot{v}_i) dV + \int_V (v_i^* v_j v_{i,j}) dV - \int_V (v_{i,i}^* p) dV \\ &+ \mu \int_V (v_{i,j}^* v_{i,j}) dV + \mu \int_V (v_{i,j}^* v_{j,i}) dV = \int_S (v_i^* t_i) dS \end{aligned} \tag{8.22}$$

Equations (8.19) and (8.22) with the boundary condition Eq. (8.6) and the initial condition Eqs. (8.9) and (8.10) are the basic equations of the finite element method of the incompressible fluid flows. The boundary condition Eq. (8.7) can be included in Eq. (8.22). If we take V as the volume of the flow field and S as its boundary, then

$$\begin{aligned} &\int_V (v_i^* \dot{v}_i) dV + \int_V (v_i^* v_j v_{i,j}) dV - \int_V (v_{i,i}^* p) dV \\ &+ \mu \int_V (v_{i,j}^* v_{i,j}) dV + \mu \int_V (v_{i,j}^* v_{j,i}) dV = \int_{S_2} (v_i^* t_i) dS \end{aligned} \tag{8.23}$$

To derive Eq. (8.23), we assume

$$v_i^* = 0 \quad \text{on} \quad S_1 \tag{8.24}$$

This assumption is valid for any case, because v_i^* is an arbitrary function. To derive Eq. (8.23), we also use Eq. (8.8). As stated before, Eq. (8.22) is valid for an arbitrary volume V with surface S. Although we limit the volume V with the surface as a finite element and its surface, Eq. (8.22) is also valid. In this case, we use V_e and S_e to make explicit that the integrations of the equations are over the volume of the element V_e and its surface S_e.

8.4 The Finite Element Equation

For the interpolations of velocity v_i and pressure p, we will use the mixed interpolation, i.e., the interpolation for velocity is one or higher order than that for pressure. By denoting the interpolation function for velocity as Φ_α and that for pressure as Ψ_λ, we can write the interpolation relations for an element as follows:

$$v_i = \Phi_\alpha v_{\alpha i} \tag{8.25}$$

$$p = \Psi_\lambda p_\lambda \tag{8.26}$$

where $v_{\alpha i}$ is velocity at node α in the i-direction, and p_λ is pressure at node λ. For the weighting function, the following equations are employed:

$$v_i^* = \Phi_\alpha v_{\alpha i}^* \tag{8.27}$$

$$p^* = \Psi_\lambda p_\lambda^* \tag{8.28}$$

where $v_{\alpha i}^*$ is a constant at node α in the i-direction corresponding to velocity, and p_λ^* is a constant at node λ corresponding to pressure. Introducing Eqs. (8.25)–(8.28) into Eqs. (8.19) and (8.22), and arranging the terms, the local forms of finite element equation can be derived considering $v_{\alpha i}^*$ and p_λ^* are arbitrary constants as follows:

$$M_{\alpha i\beta j}\dot{v}_{\beta j} + K_{\alpha\beta\gamma j}v_{\beta j}v_{\gamma i} - H_{\alpha i\lambda}p_\lambda + S_{\alpha i\beta j}v_{\beta j} = \Omega_{\alpha i} \tag{8.29}$$

$$H_{\alpha i\mu}v_{\alpha i} = 0 \tag{8.30}$$

where

$$M_{\alpha i\beta j} = \int_V (\Phi_\alpha\Phi_\beta)\delta_{ij}dV \tag{8.31}$$

$$K_{\alpha\beta\gamma j} = \int_V (\Phi_\alpha\Phi_\beta\Phi_{\gamma,j})dV \tag{8.32}$$

$$H_{\alpha i\lambda} = \int_V (\Phi_{\alpha,i}\Psi_\lambda)dV \tag{8.33}$$

$$S_{\alpha i\beta j} = \int_V \mu(\Phi_{\alpha,k}\Phi_{\beta,k})\delta_{ij}dV + \int_V \mu(\Phi_{\alpha,i}\Phi_{\beta,j})dV \tag{8.34}$$

$$\Omega_{\alpha i} = \int_S (\Phi_\alpha t_i)dS \tag{8.35}$$

8.5 Bubble Function Finite Element Method

A substantial numbers of the mixed interpolation have been presented in recent decades. The simplest scheme is the mixed interpolation of bubble function for velocity and linear function for pressure. The advantage of the bubble function method is that the stable computation can be carried out using the artificial viscosity added only on the barycentric node. We will discuss the two-dimensional case, although the extension to the three-dimensional case is also straightforward.

8.5.1 Bubble Function Interpolation

Referring to Fig. 8.1, we number three vertices of a triangular finite element 1, 2, 3 and barycenter 4. Because Eqs. (8.29) and (8.30) are the local forms, we must use V_e and S_e instead of V and S in Eqs. (8.31)–(8.35). However, if we consider S as S_2 in Eq. (8.35), and V and S as volume and its surface of whole flow field, then Eqs. (8.29) and (8.30) can also be considered as the global forms based on Eq. (8.23). Therefore, we can take Eqs. (8.29)–(8.35) to be either local forms or global forms. From now on, we will use these expressions unless otherwise specified to avoid duplicate description. Note that in Eqs. (8.30), the time derivative term is not included. This fact is one of the fundamental causes of numerical instability to solve the time-dependent incompressible fluid flows.

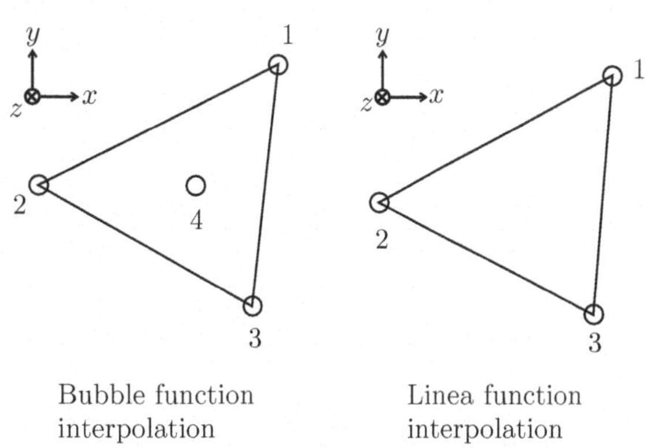

Fig. 8.1 Bubble function and linear function interpolations

The linear interpolation function is the standard interpolation function as defined in Eq. (6.41) in Sect. 6.6. The bubble function interpolation is expressed by the area coordinate η_i ($i = 1$ to 3) presented in Eq. (6.28) in Sect. 6.5. Denoting velocity on vertices of the triangular element as $v_{1i}, v_{2i}, v_{3i}, v_{4i}$ in the i-direction, the interpolation equation of the bubble function for velocity is expressed as follows:

$$v_i = \tilde{\Phi}_1 v_{1i} + \tilde{\Phi}_2 v_{2i} + \tilde{\Phi}_3 v_{3i} + \tilde{\Phi}_4 v_{4i} \tag{8.36}$$

where $\tilde{\Phi}_1$, $\tilde{\Phi}_2$, $\tilde{\Phi}_3$, and $\tilde{\Phi}_4$ are shape functions, which are

$$\begin{aligned} \tilde{\Phi}_1 &= \eta_1 - \frac{27}{3}\eta_1\eta_2\eta_3 \\ \tilde{\Phi}_2 &= \eta_2 - \frac{27}{3}\eta_1\eta_2\eta_3 \\ \tilde{\Phi}_3 &= \eta_3 - \frac{27}{3}\eta_1\eta_2\eta_3 \\ \tilde{\Phi}_4 &= 27\eta_1\eta_2\eta_3 \end{aligned} \tag{8.37}$$

Those functions satisfy the following relations:

$$\sum_{\alpha=1}^{4} \tilde{\Phi}_\alpha = 1, \tag{8.38}$$

and

$$\sum_{\alpha=1}^{4} \tilde{\Phi}_{\alpha,i} = 0 \tag{8.39}$$

Equation (8.36) is simplified as follows:

$$v_i = \eta_1 v_{1i} + \eta_2 v_{2i} + \eta_3 v_{3i} + 27\eta_1\eta_2\eta_3 \bar{v}_{4i} \tag{8.40}$$

where

$$\bar{v}_{4i} = v_{4i} - \frac{1}{3}(v_{1i} + v_{2i} + v_{3i}) \tag{8.41}$$

we can write Eq. (8.40) in the matrix form as

$$v_i = \begin{bmatrix} \eta_1 & \eta_2 & \eta_3 & 27\eta_1\eta_2\eta_3 \end{bmatrix} \begin{bmatrix} v_{1i} \\ v_{2i} \\ v_{3i} \\ \bar{v}_{4i} \end{bmatrix} \tag{8.42}$$

If we posit

$$\Phi_\alpha = \left[\eta_1\ \eta_2\ \eta_3\ 27\eta_1\eta_2\eta_3\right] \tag{8.43}$$

and

$$v_{\alpha i} = \left[v_{1i}\ v_{2i}\ v_{3i}\ \bar{v}_{4i}\right] \tag{8.44}$$

then, Eq. (8.40) can be expressed as follows:

$$v_i = \Phi_\alpha v_{\alpha i} \qquad (\alpha = 1 \text{ to } 4) \tag{8.45}$$

which is the same as Eq. (8.25). Equation (8.36) can also be considered as Eq. (8.45). However, we will use Eq. (8.40) as the bubble function interpolation function, because the first three terms are the same as those of the linear function and those are convenient for later use. Note that the nodal velocity on the barycentric node v_{4i} is computed by Eq. (8.41).

Interpolation function for pressure is the linear interpolation function and is expressed in Eq. (8.26), where

$$\Psi_\lambda = \left[\eta_1\ \eta_2\ \eta_3\right] \tag{8.46}$$

$$p_\lambda = \left[p_1\ p_2\ p_3\right] \tag{8.47}$$

In Eq. (8.47), p_1, p_2, and p_3 denote pressure at three vertices of an element.

8.5.2 Concrete Forms of Coefficients

In this section, we use the expression of coordinates as x and y, and we do not use the summation convention for i and j. We will describe concrete forms of coefficients in Eqs. (8.31)–(8.35) based on Eqs. (8.43) and (8.46). Differentiating both sides of Eqs. (8.43) with respect to coordinates x and y, we get

$$\Phi_{\alpha,x} = \left[b_1\ b_2\ b_3\ d\right] \tag{8.48}$$

$$\Phi_{\alpha,y} = \left[c_1\ c_2\ c_3\ e\right] \tag{8.49}$$

where

$$d = 27(b_1\eta_2\eta_3 + \eta_1 b_2\eta_3 + \eta_1\eta_2 b_3)$$

$$e = 27(c_1\eta_2\eta_3 + \eta_1 c_2\eta_3 + \eta_1\eta_2 c_3)$$

In Eqs. (8.48) and (8.49), b_1, b_2, b_3 and c_1, c_2, c_3 are coefficients expressed in Eq. (4.22) in Sect. 4.4.

Introducing Eq. (8.43) into Eq. (8.31) and performing the integrations by the formula in Eq. (6.36) in Sect. 6.5, Eq. (8.31) can be transformed as follows:

$$M^{xx}_{\alpha\beta} = M^{yy}_{\alpha\beta} = \int_{V_e} (\Phi_\alpha \Phi_\beta) dV = \boldsymbol{\Delta} \begin{bmatrix} \frac{1}{6} & \frac{1}{12} & \frac{1}{12} & \frac{3}{20} \\ \frac{1}{12} & \frac{1}{6} & \frac{1}{12} & \frac{3}{20} \\ \frac{1}{12} & \frac{1}{12} & \frac{1}{6} & \frac{3}{20} \\ \frac{3}{20} & \frac{3}{20} & \frac{3}{20} & \frac{81}{280} \end{bmatrix} \tag{8.50}$$

Introducing Eq. (8.43) into Eq. (8.32), using Eq. (8.48) or (8.49), and performing the integrations by Eq. (6.36), Eq. (8.32) can be transformed as follows:

$$K^x_{\alpha\beta 1} = \int_{V_e} (\Phi_\alpha \Phi_\beta \Phi_{1,x}) dV = b_1 M^{xx}_{\alpha\beta} \tag{8.51}$$

$$K^x_{\alpha\beta 2} = \int_{V_e} (\Phi_\alpha \Phi_\beta \Phi_{2,x}) dV = b_2 M^{xx}_{\alpha\beta} \tag{8.52}$$

$$K^x_{\alpha\beta 3} = \int_{V_e} (\Phi_\alpha \Phi_\beta \Phi_{3,x}) dV = b_3 M^{xx}_{\alpha\beta} \tag{8.53}$$

$$K^x_{\alpha\beta 4} = \int_{V_e} (\Phi_\alpha \Phi_\beta \Phi_{4,x}) dV = \boldsymbol{\Delta} \begin{bmatrix} -\frac{3}{20}(b_1+b_1) & -\frac{3}{20}(b_1+b_2) & -\frac{3}{20}(b_1+b_3) & -\frac{81}{280}b_1 \\ -\frac{3}{20}(b_2+b_1) & -\frac{3}{20}(b_2+b_2) & -\frac{3}{20}(b_2+b_3) & -\frac{81}{280}b_2 \\ -\frac{3}{20}(b_3+b_1) & -\frac{3}{20}(b_3+b_2) & -\frac{3}{20}(b_3+b_3) & -\frac{81}{280}b_3 \\ -\frac{81}{280}b_1 & -\frac{81}{280}b_2 & -\frac{81}{280}b_3 & 0 \end{bmatrix} \tag{8.54}$$

$$K^y_{\alpha\beta 1} = \int_{V_e} (\Phi_\alpha \Phi_\beta \Phi_{1,y}) dV = c_1 M^{yy}_{\alpha\beta} \tag{8.55}$$

$$K^y_{\alpha\beta 2} = \int_{V_e} (\Phi_\alpha \Phi_\beta \Phi_{2,y}) dV = c_2 M^{yy}_{\alpha\beta} \tag{8.56}$$

$$K^y_{\alpha\beta 3} = \int_{V_e} (\Phi_\alpha \Phi_\beta \Phi_{3,y}) dV = c_3 M^{yy}_{\alpha\beta} \tag{8.57}$$

$$K^y_{\alpha\beta 4} = \int_{V_e} (\Phi_\alpha \Phi_\beta \Phi_{4,y}) dV$$
$$= \mathbf{\Delta} \begin{bmatrix} -\frac{3}{20}(c_1+c_1) & -\frac{3}{20}(c_1+c_2) & -\frac{3}{20}(c_1+c_3) & -\frac{81}{280}c_1 \\ -\frac{3}{20}(c_2+c_1) & -\frac{3}{20}(c_2+c_2) & -\frac{3}{20}(c_2+c_3) & -\frac{81}{280}c_2 \\ -\frac{3}{20}(c_3+c_1) & -\frac{3}{20}(c_3+c_2) & -\frac{3}{20}(c_3+c_3) & -\frac{81}{280}c_3 \\ -\frac{81}{280}c_1 & -\frac{81}{280}c_2 & -\frac{81}{280}c_3 & 0 \end{bmatrix} \tag{8.58}$$

Introducing Eqs. (8.46) and (8.48) or Eq. (8.49) into Eq. (8.33), and performing the integrations by Eq. (6.36), Eq. (8.33) can be transformed as follows:

$$H^x_{\alpha\lambda} = \int_{V_e} (\Phi_{\alpha,x} \Psi_\lambda) dV$$
$$= \mathbf{\Delta} \begin{bmatrix} \frac{b_1}{3} & \frac{b_1}{3} & \frac{b_1}{3} \\ \frac{b_2}{3} & \frac{b_2}{3} & \frac{b_2}{3} \\ \frac{b_3}{3} & \frac{b_3}{3} & \frac{b_3}{3} \\ -\frac{9}{20}b_1 & -\frac{9}{20}b_2 & -\frac{9}{20}b_3 \end{bmatrix} \tag{8.59}$$

$$H^y_{\alpha\lambda} = \int_{V_e} (\Phi_{\alpha,y} \Psi_\lambda) dV$$
$$= \mathbf{\Delta} \begin{bmatrix} \frac{c_1}{3} & \frac{c_1}{3} & \frac{c_1}{3} \\ \frac{c_2}{3} & \frac{c_2}{3} & \frac{c_2}{3} \\ \frac{c_3}{3} & \frac{c_3}{3} & \frac{c_3}{3} \\ -\frac{9}{20}c_1 & -\frac{9}{20}c_2 & -\frac{9}{20}c_3 \end{bmatrix} \tag{8.60}$$

To obtain the concrete forms of Eq. (8.34), we formulate the componential coefficients as

$$\mathfrak{M}^{xx}_{\alpha\beta} = \int_{V_e} (\Phi_{\alpha,x} \Phi_{\beta,x}) dV$$
$$= \mathbf{\Delta} \begin{bmatrix} b_1^2 & b_1 b_2 & b_1 b_3 & 0 \\ b_2 b_1 & b_2^2 & b_2 b_3 & 0 \\ b_3 b_1 & b_3 b_2 & b_3^2 & 0 \\ 0 & 0 & 0 & \frac{81}{20}(b_1^2+b_2^2+b_3^2) \end{bmatrix} \tag{8.61}$$

$$\mathfrak{M}^{xy}_{\alpha\beta} = \int_{V_e} (\Phi_{\alpha,x}\Phi_{\beta,y}) dV$$
$$= \mathbf{\Delta} \begin{bmatrix} b_1c_1 & b_1c_2 & b_1c_3 & 0 \\ b_2c_1 & b_2c_2 & b_2c_3 & 0 \\ b_3c_1 & b_3c_2 & b_3c_3 & 0 \\ 0 & 0 & 0 & \frac{81}{20}(b_1c_1 + b_2c_2 + b_3c_3) \end{bmatrix} \tag{8.62}$$

$$\mathfrak{M}^{yx}_{\alpha\beta} = \int_{V} (\Phi_{\alpha,y}\Phi_{\beta,x}) dV$$
$$= \mathbf{\Delta} \begin{bmatrix} c_1b_1 & c_1b_2 & c_1b_3 & 0 \\ c_2b_1 & c_2b_2 & c_2b_3 & 0 \\ c_3b_1 & c_3b_2 & c_3b_3 & 0 \\ 0 & 0 & 0 & \frac{81}{20}(c_1b_1 + c_2b_2 + c_3b_3) \end{bmatrix} \tag{8.63}$$

$$\mathfrak{M}^{yy}_{\alpha\beta} = \int_{V_e} (\Phi_{\alpha,y}\Phi_{\beta,y}) dV$$
$$= \mathbf{\Delta} \begin{bmatrix} c_1^2 & c_1c_2 & c_1c_3 & 0 \\ c_2c_1 & c_2^2 & c_2c_3 & 0 \\ c_3c_1 & c_3c_2 & c_3^2 & 0 \\ 0 & 0 & 0 & \frac{81}{20}(c_1^2 + c_2^2 + c_3^2) \end{bmatrix} \tag{8.64}$$

and we obtain the following results:

$$S^{xx}_{\alpha\beta} = 2\mu\mathfrak{M}^{xx}_{\alpha\beta} + \mu\mathfrak{M}^{yy}_{\alpha\beta} \tag{8.65}$$

$$S^{xy}_{\alpha\beta} = \mu\mathfrak{M}^{xy}_{\alpha\beta} \tag{8.66}$$

$$S^{yx}_{\alpha\beta} = \mu\mathfrak{M}^{yx}_{\alpha\beta} \tag{8.67}$$

$$S^{yy}_{\alpha\beta} = \mu\mathfrak{M}^{xx}_{\alpha\beta} + 2\mu\mathfrak{M}^{yy}_{\alpha\beta} \tag{8.68}$$

To derive Eqs. (8.50)–(8.64), we use the relations

$$b_1 + b_2 + b_3 = 0 \tag{8.69}$$

$$c_1 + c_2 + c_3 = 0 \tag{8.70}$$

and

$$\Phi_4 = 27\eta_1\eta_2\eta_3 = 0 \quad \text{on} \quad S_e \tag{8.71}$$

We can also successfully utilize the integration by parts.

8.5.3 The Global Form of the Finite Element Method

The global form of the finite element method is the superposed forms of Eqs. (8.29) and (8.30). In each element, Eq. (8.29) consists of eight equations with 11 unknown variables, which are two components of velocity on four nodes and pressure on three nodes. Equation (8.30) means three equations with eight unknown variables, which are two components of velocity on four nodes. Total numbers of equations are 11, and total unknown variables are 11. The coefficients of 11 equations consist of Eqs. (8.50)–(8.68). We can write Eqs. (8.29) and (8.30) in the matrix forms as follows:

$$\begin{bmatrix} \mathbf{M_{vv}} & \mathbf{M_{vB}} & \mathbf{0} \\ \mathbf{M_{Bv}} & \mathbf{M_{BB}} & \mathbf{0} \\ \mathbf{0} & \mathbf{0} & \mathbf{0} \end{bmatrix} \begin{bmatrix} \dot{\mathbf{v}} \\ \dot{\mathbf{v}}_\mathbf{B} \\ \dot{\mathbf{p}} \end{bmatrix} + \begin{bmatrix} \mathbf{K_{vv}(v)} & \mathbf{K_{vB}(v)} & \mathbf{0} \\ \mathbf{K_{Bv}(v)} & \mathbf{0} & \mathbf{0} \\ \mathbf{0} & \mathbf{0} & \mathbf{0} \end{bmatrix} \begin{bmatrix} \mathbf{v} \\ \mathbf{v_B} \\ \mathbf{p} \end{bmatrix} + \begin{bmatrix} \mathbf{S} & \mathbf{0} & -\mathbf{H} \\ \mathbf{0} & \mathbf{a} & -\mathbf{h} \\ \mathbf{H}^T & \mathbf{h}^T & \mathbf{0} \end{bmatrix} \begin{bmatrix} \mathbf{v} \\ \mathbf{v_B} \\ \mathbf{p} \end{bmatrix} = \begin{bmatrix} \boldsymbol{\Omega} \\ \mathbf{0} \\ \mathbf{0} \end{bmatrix} \tag{8.72}$$

where $\mathbf{v}$, $\mathbf{v_B}$, and $\mathbf{p}$ denote two components of velocity on three main nodes of the triangular element, two components of velocity on barycentric nodes and pressure on three main nodes, respectively. Altogether, there are 11 degrees of freedom, and we express the velocity on barycentric nodes separately. We do not use summation convention in Eq. (8.72). We show $\mathbf{0}$ elements explicitly. Coefficient $\mathbf{M_{vv}}$ consists of the first 3×3 elements, $\mathbf{M_{vB}}$ of the last 1 column and the first three rows, $\mathbf{M_{Bv}}$ of the last 1 row and the first three columns, and $\mathbf{M_{BB}}$ of the fourth row and fourth column element of $M^{xx}_{\alpha\beta}$ and $M^{yy}_{\alpha\beta}$, respectively. Nonlinear coefficients $\mathbf{K_{vv}(v)}$, $\mathbf{K_{vB}(v)}$, and $\mathbf{K_{Bv}(v)}$ are functions of $\mathbf{v}$, and made by almost the same way. The coefficient $\mathbf{S}$ consists of the first 3×3 elements and $\mathbf{0}$ of the fourth row and fourth column element of $S^{xx}_{\alpha\beta}$, $S^{xy}_{\alpha\beta}$, $S^{yx}_{\alpha\beta}$, $S^{yy}_{\alpha\beta}$, respectively. Coefficients $\mathbf{H}$ and $\mathbf{h}$ consist of the first three rows and the last 1 row of $H^{x}_{\alpha\lambda}$ and $H^{y}_{\alpha\lambda}$, respectively. We can see many zero elements in coefficients in Eq. (8.72), which means there are some non-coupled relations between $\mathbf{v}$, $\mathbf{v_B}$, and $\mathbf{p}$. Especially, values on the barycentric node can be eliminated from Eq. (8.72) if we disregard the term $\mathbf{M_{Bv}}\dot{\mathbf{v}}$. This operation is sometimes referred to as the "static condensation."

The global form of the finite element method is obtained by the superposition procedures applied to Eqs. (8.29) and (8.30), or their matrix form Eq. (8.72). The superposed equation can also be written as Eqs. (8.29) and (8.30), or Eq. (8.72). Thus, we can consider Eqs. (8.29) and (8.30), or Eq. (8.72) as both local and global forms of the finite element method. If we solve Eqs. (8.29) and (8.30), or Eq. (8.72), we obtain the velocity and pressure of the incompressible fluid flows. At this time, it is important to take the stability of computation into consideration. In Sect. 5.8, we presented the stabilized bubble function method. In the following sections, we will discuss the extended stabilized bubble function method to deal with the incompressible fluid flows.

8.5.4 Stabilized Bubble Function Method

In this section, we will describe one of the stabilization methods of the incompressible fluid flows based on the mixed interpolation of the bubble function and linear function. We call this stabilization method the "stabilized bubble function method," because this is the extension of the scheme discussed in Sect. 5.8. We will show that the mixed interpolations based on the bubble function and linear function interpolation are useful, and even more useful when we add simple artificial viscosity for practical computation.

Consider Eqs. (8.19) and (8.22) for the weighted residual equations, and Eqs. (8.25)–(8.28) for the interpolation and weighting functions. Equations (8.25)–(8.28) can be rewritten separating linear interpolation part and bubble function part as follows:

$$v_i = \bar{v}_i + \phi v_{4i} \tag{8.73}$$

$$p = \bar{p} \tag{8.74}$$

$$v_i^* = \bar{v}_i^* + \phi v_{4i}^* \tag{8.75}$$

$$p^* = \bar{p}^* \tag{8.76}$$

where overbar $^{-}$ means the linear interpolation function in this section, i.e., Eq. (8.73) is the separated form of Eq. (8.25), in which the linear part is shown as $\bar{v}_i$ and bubble function part as ϕv_{4i}, and

$$\phi = 27\eta_1\eta_2\eta_3 \tag{8.77}$$

The area coordinate is shown as η_1, η_2, η_3 and v_{4i} is the nodal equivalent velocity on barycentric node 4. Equation (8.75) is the weighting function separating the linear part $\bar{v}_i^*$ and the bubble function part ϕv_{4i}^*, where v_{4i}^* is the weighting constant on the barycentric node. Equations (8.74) and (8.76) mean the interpolation and weighting functions of pressure, which are both linear functions.

Substituting Eqs. (8.73) and (8.76) into Eq. (8.19), we get

$$\int_{V_e} (\bar{p}^* v_{i,i}) dV = 0 \tag{8.78}$$

where V_e means area of a triangular element. Equation (8.78) is transformed into

$$\int_{V_e} (\bar{p}^* \bar{v}_{i,i}) dV + \int_{V_e} (\bar{p}^* \phi_{,i}) dV\, v_{4i} = 0 \tag{8.79}$$

Substituting Eqs. (8.73)–(8.76) into Eq. (8.22), and separating linear and bubble function weighting function parts, we obtain the following two equations:

$$\begin{aligned} &\int_{V_e} (\bar{v}_i^* \dot{v}_i) dV + \int_{V_e} (\bar{v}_i^* v_j v_{i,j}) dV - \int_{V_e} (\bar{v}_{i,i}^* \bar{p}) dV \\ &\quad + \mu \int_{V_e} (\bar{v}_{i,j}^* v_{i,j}) dV + \mu \int_{V_e} (\bar{v}_{i,j}^* v_{j,i}) dV = \int_{S_e} (\bar{v}_i^* t_i) dS \end{aligned} \tag{8.80}$$

$$\begin{aligned} &\int_{V_e} (\phi \dot{v}_i) dV\, v_{4i}^* + \int_{V_e} (\phi v_j v_{i,j}) dV\, v_{4i}^* - \int_{V_e} (\phi_{,i} \bar{p}) dV\, v_{4i}^* \\ &\quad + \mu \int_{V_e} (\phi_{,j} v_{i,j}) dV\, v_{4i}^* + \mu \int_{V_e} (\phi_{,j} v_{j,i}) dV\, v_{4i}^* = \int_{S_e} (\phi t_i) dS\, v_{4i}^* \end{aligned} \tag{8.81}$$

Using the fact that v_{4i}^* are arbitrary constants, and introducing Eq. (8.73) into Eq. (8.81), Eq. (8.81) can be expressed as follows:

$$\begin{aligned} &\int_{V_e} (\phi \dot{v}_i) dV + \int_{V_e} (\phi v_j v_{i,j}) dV + \int_{V_e} (\phi \bar{p}_{,i}) dV + \mu \int_{V_e} (\phi_{,j} \bar{v}_{i,j}) dV \\ &\quad + \mu \int_{V_e} (\phi_{,j} \phi_{,j}) dV\, v_{4i} + \mu \int_{V_e} (\phi_{,j} \bar{v}_{j,i}) dV + \mu \int_{V_e} (\phi_{,j} \phi_{,i}) dV\, v_{4j} = 0 \end{aligned} \tag{8.82}$$

To obtain the stabilization parameter, we assume that the effects of the following terms are zero:

$$\int_{V_e} (\phi \dot{v}_i) dV = 0 \tag{8.83}$$

$$\mu \int_{V_e} (\phi_{,j} \phi_{,i}) dV = 0 \tag{8.84}$$

$$\int_{V_e} (\phi \bar{p}_{,i}) dV = 0 \tag{8.85}$$

Using the equalities,

$$\int_{V_e} (\phi_{,j}\bar{v}_{i,j})dV = 0 \tag{8.86}$$

$$\int_{V_e} (\phi_{,j}\bar{v}_{j,i})dV = 0 \tag{8.87}$$

$$\int_{V_e} (\phi\phi_{,i})dV = 0 \tag{8.88}$$

linearizing the second term of Eq. (8.82), and following the same procedure developed in Sect. 5.8, v_{4i} in Eq. (8.82) can be expressed as follows:

$$v_{4i} = -\frac{\displaystyle\int_{V_e} (U_j\bar{v}_{i,j}\phi)dV}{\mu \displaystyle\int_{V_e} (\phi_{,k}\phi_{,k})dV} \tag{8.89}$$

where U_j is the linearized velocity, which is assumed to be constant in a finite element.

Let us derive the stabilization parameter τ_{eB}, which was discussed in Sect. 5.8. Consider Eq. (8.80). Let the second term of Eq. (8.80) be linearized, setting aside the third term and introducing Eq. (8.73) into the resulted equations, we get

$$\begin{aligned}&\int_{V_e} (\bar{v}_i^*\dot{v}_i)dV + \int_{V_e} (\bar{v}_i^*U_j\bar{v}_{i,j})dV + \int_{V_e} (\bar{v}_i^*U_j\phi_{,j})dV\, v_{4i}\\&+\mu\int_{V_e} (\bar{v}_{i,j}^*v_{i,j})dV + \mu\int_{V_e} (\bar{v}_{i,j}^*v_{j,i})dV = \int_{S_e} (\bar{v}_i^*t_i)dS\end{aligned} \tag{8.90}$$

We reformulate the third term of Eq. (8.90) following the procedures described in Sect. 5.8, and apply Eqs. (8.86) and (8.87) to the fourth and fifth terms. Then, Eq. (8.90) can be described as follows:

$$\begin{aligned}&\int_{V_e} (\bar{v}_i^*\dot{v}_i)dV + \int_{V_e} (\bar{v}_i^*U_j\bar{v}_{i,j})dV + \tau_{eB}\int_{V_e} (\bar{v}_{i,k}^*U_kU_l\bar{v}_{i,l})dV\\&+\mu\int_{V_e} (\bar{v}_{i,j}^*\bar{v}_{i,j})dV + \mu\int_{V_e} (\bar{v}_{i,j}^*\bar{v}_{j,i})dV = \int_{S_e} (\bar{v}_i^*t_i)dS\end{aligned} \tag{8.91}$$

where

$$\tau_{eB} = \frac{\left(\displaystyle\int_{V_e} \phi dV\right)^2}{\mu \displaystyle\int_{V_e} (\phi_{,k}\phi_{,k})dV \cdot \boldsymbol{\Delta}} \tag{8.92}$$

in which $\boldsymbol{\Delta}$ is the area of the triangular finite element. We can see Eq. (8.91) is the similar form of the SUPG scheme based on the linear interpolation function. Because of the third term, the stability of computation is secured. Thus, τ_{eB} is referred to as the "stabilization parameter."

To obtain the final form of the weighted residual equation, let Eqs. (8.73)–(8.76) be introduced into Eq. (8.22):

$$\begin{aligned}
&\int_{V_e} (v_i^* \dot{v}_i) dV + \int_{V_e} v_i^* (\bar{v}_j + \phi v_{4j})(\bar{v}_{i,j} + \phi_{,j} v_{4i}) dV - \int_{V_e} (v_{i,i}^* \bar{p}) dV \\
&\quad + \mu \int_{V_e} (\bar{v}_{i,j}^* + \phi_{,j} v_{4i}^*)(\bar{v}_{i,j} + \phi_{,j} v_{4i}) dV + \mu \int_{V_e} (\bar{v}_{i,j}^* + \phi_{,j} v_{4i}^*)(\bar{v}_{j,i} + \phi_{,i} v_{4j}) dV \\
&\quad = \int_{S_e} (v_i^* t_i) dS
\end{aligned} \tag{8.93}$$

Substituting Eqs. (8.86)–(8.88) and the following equation

$$\int_{V_e} (\phi^2 \phi_{,j}) dV = 0 \tag{8.94}$$

into Eq. (8.93), and rearranging the terms, we obtain

$$\begin{aligned}
&\int_{V_e} (v_i^* \dot{v}_i) dV + \int_{V_e} (v_i^* \bar{v}_j \bar{v}_{i,j}) dV + \int_{V_e} (v_i^* \bar{v}_j \phi_{,j}) dV\, v_{4i} + \int_{V_e} (v_i^* \phi \bar{v}_{i,j}) dV\, v_{4j} \\
&\quad + \int_{V_e} (\bar{v}_i^* \phi \phi_{,j}) dV\, v_{4i} v_{4j} - \int_{V_e} (v_{i,i}^* \bar{p}) dV + \mu \int_{V_e} (\bar{v}_{i,j}^* \bar{v}_{i,j}) dV + \mu \int_{V_e} (\bar{v}_{i,j}^* \bar{v}_{j,i}) dV \\
&\quad + \mu \int_{V_e} (\phi_{,j} \phi_{,j}) dV\, v_{4i}^* v_{4i} + \mu \int_{V_e} (\phi_{,j} \phi_{,i}) dV\, v_{4i}^* v_{4j} \\
&\quad = \int_{S_e} (v_i^* t_i) dS
\end{aligned} \tag{8.95}$$

Substituting the interpolation and weighting functions into Eqs. (8.78) and (8.95), final form of the finite element equation Eq. (8.72) can be derived. Discretizing the time derivative term, i.e., the first term in Eq. (8.95), or Eq. (8.72), and solving Eq. (8.72) we obtain velocity and pressure of the unsteady incompressible viscous fluid flows.

In practical applications, we sometimes find that the computations are not stable using τ_{eB} in Eq. (8.92), i.e., the artificial viscosity in Eq. (8.92) may not be satisfactory for practical computations. To compensate for the artificial viscosity, we can add an additional artificial viscosity, extending the stabilization parameter as follows:

$$\tau_{eB} = \frac{\left(\int_{V_e} \phi dV\right)^2}{(\mu + \mu') \int_{V_e} (\phi_{,k}\phi_{,k}) dV \cdot \boldsymbol{\Delta}} \tag{8.96}$$

where μ' is the additional artificial viscosity. It is known that we can obtain the stable computation by the SUPG scheme, whose stabilization parameter is given as

$$\tau_{eS} = \left\{ \left(\frac{2|U_j|}{h_e}\right)^2 + \left(\frac{4\nu}{h_e^2}\right)^2 + \left(\frac{2}{\Delta t}\right)^2 \right\}^{-1/2} \tag{8.97}$$

where $|U_j|$ is the magnitude of velocity and h_e is the characteristic length:

$$|U_j| = \sqrt{v_i v_i} \tag{8.98}$$

$$h_e = \sqrt{2\boldsymbol{\Delta}}$$

in which v_i is the mean velocity in an element and $\boldsymbol{\Delta}$ is the area of element. Equating Eqs. (8.96) and (8.97), the artificial viscosity can be written as follows:

$$\mu + \mu' = \frac{\left(\int_{V_e} \phi dV\right)^2}{\tau_{eS} \cdot \int_{V_e} (\phi_{,k}\phi_{,k}) dV \cdot \boldsymbol{\Delta}} \tag{8.99}$$

Replacing μ with $\mu + \mu'$ in ninth and tenth terms of Eq. (8.95), we can perform the stable computation. Namely, the final form of the global finite element equation can be expressed based on Eq. (8.95) and separating the terms on the barycentric node:

$$\mathbf{M}_{\alpha\beta}\dot{\mathbf{v}}_\beta + \mathbf{K}_{\alpha\beta}(\mathbf{v}_\gamma)\mathbf{v}_\beta + \mathbf{S}_{\alpha\beta}\mathbf{v}_\beta + \sum_{m=1}^{N_e} (\mu + \mu')\mathbf{a}\mathbf{v}_4 = \boldsymbol{\Omega}_\alpha \tag{8.100}$$

$$\mathbf{a} = [a_{ij}], \qquad a_{ij} = \int_{V_e} (\phi_{,k}\phi_{,k})\delta_{ij} dV + \int_{V_e} (\phi_{,j}\phi_{,i}) dV$$

where $\mathbf{v}_\beta$ denotes both components of velocity on main nodes and on barycentric nodes, and pressure on main nodes, and $\mathbf{v}_4$ means the values on the barycentric nodes. The last term represents that a_{ij} is integrated on each finite element and summed up for $m = 1$ to N_e, N_e is the total number of elements. In Eq. (8.100), the viscosity is on the coefficients only on the barycentric nodes. Advantage of the "bubble function finite element method" is that the stable computation can be performed only by adding an artificial viscosity on the coefficients on barycentric nodes of the flow field. If we eliminate the velocity on the barycentric node from Eq. (8.100), we get the similar equation to the SUPG finite element method. In that sense, the bubble function finite element method is equivalent to the SUPG finite element method. Although Eq. (8.100) is approximately valid in the case of the standard bubble function, we will show in Sect. 8.5.5 that the global finite element equation Eq. (8.100) is satisfied only if we use linear and virtual bubble functions. It should be noted that the last part of Eq. (8.100) is only the function of velocity and the function of pressure is not included. Therefore, only the implicit-type integration method in time can be adaptable to solve Eq. (8.100).

8.5.5 Linear and Virtual Bubble Functions Scheme

We have described the stabilized bubble function method based on the standard bubble function in Sect. 8.5.4. We can carry out stable practical computation using the stabilization parameter, although the parameter is derived on the assumption of the linearized advection term in the case of the unsteady incompressible viscous fluid flows. Matsumoto and Kawahara (2006) derived a stable bubble function scheme for the incompressible flows based on the two different types of bubble functions without assuming the linearized advection.

We can define a number of bubble functions, with nodal values of 1 on the barycentric node, and zero on the side of a finite element. For instance, one sort of bubble functions can be defined as follows:

$$\phi = \begin{cases} 3^{\xi}(1-r-s)^{\xi} & \text{in } \mathbf{\Delta}_1 \\ 3^{\xi}r^{\xi} & \text{in } \mathbf{\Delta}_2 \\ 3^{\xi}s^{\xi} & \text{in } \mathbf{\Delta}_3 \end{cases} \tag{8.101}$$

where ξ is an integer constant and $\mathbf{\Delta}_1$, $\mathbf{\Delta}_2$, and $\mathbf{\Delta}_3$ are areas of triangles as shown in Fig. 8.2. In Eq. (8.101), coordinates r and s are given by the Cartesian coordinates x and y by the following relations:

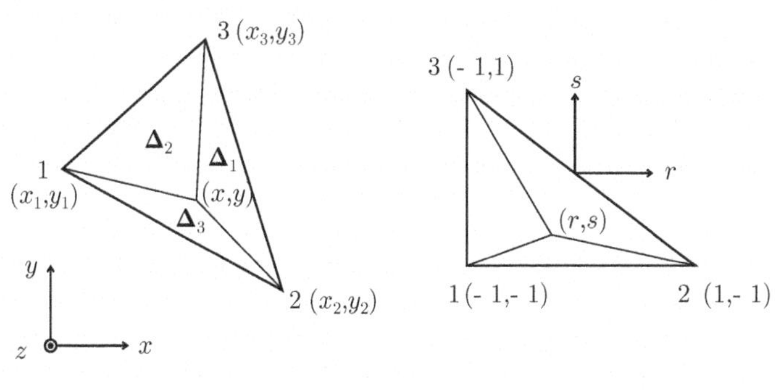

Fig. 8.2 Coordinate system

$$x(r,s) = \frac{1}{2}(1-r-s)x_1 + \frac{1}{2}(1+r)x_2 + \frac{1}{2}(1+s)x_3 \tag{8.102}$$

$$y(r,s) = \frac{1}{2}(1-r-s)y_1 + \frac{1}{2}(1+r)y_2 + \frac{1}{2}(1+s)y_3 \tag{8.103}$$

where (x_1, y_1), (x_2, y_2), and (x_3, y_3) are coordinate values of three vertices of a triangular element as shown in Fig. 8.2.

We will describe in this section the stabilized method using two types of bubble functions, i.e., a bubble function ϕ and a virtual bubble function φ. For the interpolation function v_i and the weighting function $\hat{v}_i^*$, we use

$$v_i = \bar{v}_i + v_i' = \bar{v}_i + \phi v_{4i} \tag{8.104}$$

$$\hat{v}_i^* = \bar{v}_i^* + v_i^{*\prime} + \hat{v}_i^{*\prime} = \bar{v}_i^* + \phi v_{4i}^{*\prime} + \varphi \hat{v}_{4i}^{*\prime} = v_i^* + \hat{v}_i^* \tag{8.105}$$

where v_i and v_i^* are bubble function interpolation and weighting functions, $\bar{v}_i$ and $\bar{v}_i^*$ are the linear interpolation and weighting functions presented in Eqs. (4.24) and (4.27), respectively, $v_i' = \phi v_{4i}'$ and $v_i^{*\prime} = \phi v_{4i}^{*\prime}$ are bubble function interpolation and weighting functions, and $\hat{v}_i^{*\prime} = \varphi \hat{v}_{4i}^{*\prime}$ is the virtual bubble weighting function. In Eqs. (8.104) and (8.105), v_{4i}' represents nodal value of the interpolation function on the barycentric node, and $v_{4i}^{*\prime}$ and $\hat{v}_{4i}^{*\prime}$ are nodal values of the weighting functions on the barycentric node. For the interpolation of pressure, linear interpolation and weighting functions, i.e., $\bar{p}$ and $\bar{p}^*$, are used.

Substituting Eqs. (8.104) and (8.105) into Eqs. (8.19) and (8.22) in Sect. 8.3, we get

$$\int_{V_e} (\bar{p}^* v_{i,i}) dV = 0 \tag{8.106}$$

$$\begin{aligned} &\int_{V_e} (\hat{v}_i^* \dot{v}_i) dV + \int_{V_e} (\hat{v}_i^* v_j v_{i,j}) dV + \int_{V_e} (\hat{v}_i^* p_{,i}) dV \\ &- \mu \int_{V_e} \hat{v}_i^* (v_{i,j} + v_{j,i})_{,j} dV = 0 \end{aligned} \tag{8.107}$$

where V_e represents a finite element. From Eq. (8.107), we can show that

$$\begin{aligned} &\int_{V_e} (v_i^* \dot{v}_i) dV + \int_{V_e} (v_i^* v_i v_{i,j}) dV \\ &+ \int_{V_e} (v_i^* p_{,i}) dV - \mu \int_{V_e} v_i^* (v_{i,j} + v_{j,i})_{,j} dV \\ &+ \mu_a \int_{V_e} (v_{i,j}^{*\prime} v_{i,j}') dV + \mu_b \int_{V_e} (v_{i,j}^{*\prime} v_{j,i}') dV = 0 \end{aligned} \tag{8.108}$$

where μ_a and μ_b are constants, and the last two terms on the left side terms are the stability control term.

Superposing Eqs. (8.106) and (8.108), the global forms of the weighted residual equations can be obtained as

$$\int_V (\bar{p}^* \bar{v}_{i,i}) dV = 0 \tag{8.109}$$

$$\begin{aligned}
&\int_V (v_i^* \dot{v}_i) dV + \int_V (v_i^* v_j v_{i,j}) dV \\
&- \int_V (\bar{v}_{i,i}^* \bar{p}) dV + \mu \int_V \bar{v}_{i,j}^* (\bar{v}_{i,j} + \bar{v}_{j,i}) dV \\
&+ \sum_{e=1}^{N_e} \left\{ (\mu + \mu_a) \int_{V_e} (v_{i,j}^{*\prime} v'_{\hat{i},\hat{j}}) dV + (\mu + \mu_b) \int_{V_e} (v_{i,j}^{*\prime} v'_{\hat{j},\hat{i}}) dV \right\} \\
&= \int_{S_2} (v_i^* t_i) dS
\end{aligned} \tag{8.110}$$

where V represents global flow field, and parameters μ_a and μ_b are additional artificial viscosities, which correspond to μ' in Eq. (8.99) in Sect. 8.5.4. The total number of elements is N_e. Equation (8.110) means that the artificial viscosity is added to the terms only on the barycentric node, and if so the stable computation can be carried out. For the stability control terms, we can use the following relation:

$$(\mu + \mu_a) \int_{V_e} (v_{i,j}^{*\prime} v'_{i,j}) dV + (\mu + \mu_b) \int_{V_e} (v_{i,j}^{*\prime} v'_{j,i}) dV = \frac{\left(\int_{V_e} \phi dV \right)^2}{\int_{V_e} (\phi_{,k} \phi_{,k}) dV} \cdot \boldsymbol{\tau}_{es}^{-1} v'_{4i} v_{4j}^{*\prime} \tag{8.111}$$

where

$$\boldsymbol{\tau}_{eS} = \tau_{eS} \delta_{ij} \tag{8.112}$$

$$\tau_{eS} = \left[\frac{2|v_i|}{h_e} + \left(\frac{4\nu}{h_e^2} \right)^2 + \left(\frac{2}{\Delta t} \right)^2 \right]^{-\frac{1}{2}} \tag{8.113}$$

In Eq. (8.113), h_e is the size of the finite element, $|v_i|$ is the magnitude of velocity, and Δt is the time increment. Using Eqs. (8.111)–(8.113), the effects of the stability control terms become almost identical to those of the SUPG scheme.

Matsumoto and Kawahara (2006) showed that the following equation holds, i.e.,

$$\begin{aligned}&\int_{V_e} \varphi\{\dot{v}_i + v_j v_{i,j} + p_{,i} - \mu(v_{i,j} + v_{j,i})_{,j}\} dV \\ &\quad = \mu \int_{V_e} (\phi_{,i}\phi_{,j}) dV v'_{4j} + \mu \int_{V_e} (\phi_{,i}\phi_{,j}) dV v'_{4j}\end{aligned} \tag{8.114}$$

on the conditions that

$$\int_{V_e} (\phi\varphi) dV = 0 \tag{8.115}$$

$$\int_{V_e} (\Phi_\alpha \varphi) dV = 0 \quad \text{for} \quad \alpha = 1 \text{ to } 3 \tag{8.116}$$

$$\begin{aligned}\int_{V_e} \varphi dV &= \int_{V_e} \left(\sum_{\alpha=1}^{3} \Phi_\alpha \varphi \right) dV \\ &= \sum_{\alpha=1}^{3} \int_{V_e} (\Phi_\alpha \varphi) dV = 0\end{aligned} \tag{8.117}$$

where Φ_α is the ordinary linear interpolation function expressed in Eq. (4.24). We can find the concrete forms of ϕ and φ. We use the linear bubble function for ϕ, which is obtained putting $\xi = 1$ in Eq. (8.101). The linear bubble function is expressed as follows:

$$\phi = \begin{cases} 3(1 - r - s) & \text{in } \mathbf{\Delta}_1 \\ 3r & \text{in } \mathbf{\Delta}_2 \\ 3s & \text{in } \mathbf{\Delta}_3 \end{cases} \tag{8.118}$$

For the precise form of φ, see Matsumoto and Kawahara (2006). The precise form of φ is not necessary for practical computations. Thus, it is referred to as the "virtual bubble function." Equation (8.107) is the extension of Eq. (8.117). Actual practical numerical computations are found in Matsumoto and Kawahara (2006).

8.5.6 Bubble Function Fractional Step Finite Element Method

A number of discretization schemes in time can be adapted to solve the finite element equation Eq. (8.100). For instance, almost all of the schemes described in Chap. 5 are useful. The "fractional step method," which is also referred to as the "pressure Poisson equation method," is one of the standard methods to solve the incompressible viscous flows. In this section, we will describe a time-marching method. For example, the equation of motion discretized in time is solved with the pressure Poisson equation, which is derived by differentiating once the equation of motion.

Denoting velocity and pressure at time point n as v_i^n and p^n, the discretized form of Eq. (8.17) can be expressed as follows by setting aside f_i:

$$\frac{v_i^{n+1} - v_i^n}{\Delta t} + v_j^n v_{i,j}^n + p_{,i}^{n+1} - \mu(v_{i,j}^n + v_{j,i}^n)_{,j} = 0 \tag{8.119}$$

Differentiating both sides of Eq. (8.119) with respect to coordinate x_i, we get

$$\frac{v_{i,i}^{n+1} - v_{i,i}^n}{\Delta t} + (v_j^n v_{i,j}^n)_{,i} + p_{,ii}^{n+1} - V_{ij,ji}^n = 0 \tag{8.120}$$

where

$$V_{ij}^n = (v_{i,j}^n + v_{j,i}^n) \tag{8.121}$$

This term is introduced for the simple exposition. The equation of continuity Eq. (8.1) is discretized implicitly as

$$v_{i,i}^{n+1} = 0 \tag{8.122}$$

Thus, the first term of Eq. (8.120) is zero. From Eq. (8.120), we get

$$p_{,ii}^{n+1} = \frac{v_{i,i}^n}{\Delta t} - (v_j^n v_{i,j}^n)_{,i} \tag{8.123}$$

Equation (8.123) is referred to as the "pressure Poisson equation." In Eq. (8.123), $v_{i,i}^n = 0$ because of Eq. (8.1). However, this term may not be zero because of computational errors. If we use linear interpolation function, the last term in Eq. (8.120) is zero.

The weighted residual equation of Eq. (8.119) can be expressed as

$$\begin{aligned}
&\int_{V_e} v_i^* \left(\frac{v_i^{n+1} - v_i^n}{\Delta t} \right) dV + \int_{V_e} (v_i^* v_j^n v_{i,j}^n) dV \\
&\quad - \int_{V_e} (v_{i,i}^* p^{n+1}) dV + \mu \int_{V_e} (v_{i,j}^* v_{i,j}^n) dV + \mu \int_{V_e} (v_{i,j}^* v_{j,i}^n) dV \\
&\quad = \int_{S_e} (v_i^* t_i^n) dS
\end{aligned} \tag{8.124}$$

where

$$t_i^n = \{-p^{n+1}\delta_{ij} + \mu(v_{i,j}^n + v_{j,i}^n)\} n_j \tag{8.125}$$

and v_i^* is the bubble function weighting function. The weighted residual equation of Eq. (8.123) is as follows:

$$\int_{V_e} (p^*_{,i} p^{n+1}_{,i}) dV = -\frac{1}{\Delta t} \int_{V_e} (p^* v^n_{i,i}) dV - U^n_j \int_{V_e} (p^*_{,i} v^n_{i,j}) dV + U^n_j \int_{S_e} (p^* v^n_{i,j} n_i) dS + \int_{S_e} (p^* p^{n+1}_{,i} n_i) dS \tag{8.126}$$

where p^* is the linear weighting function, and U^n_j represents average value of v^n_i in an element, which is assumed to be constant.

Comparing Eqs. (8.124) and (8.22), we can see that Eq. (8.124) is the same equation as Eq. (8.22) descritized in time. Thus, the mixed interpolation based on the bubble function interpolation for velocity and linear interpolation for pressure is introduced. Following the same procedures described in Sects. 8.5.1–8.5.4, the global forms of the finite element method can be derived. The global equation based on Eq. (8.124) is similar to the equation based on the normal linear interpolation function. The major difference is that the artificial viscosity terms are added on the barycentric nodes. This fact is a substantial advantage for computation. The global equation for pressure is also formulated based on Eq. (8.126). Although we use the bubble function interpolation for velocity in Eq. (8.126), we can derive the same global form as that of linear interpolation. Alternatively, we can solve the global finite element equations based on Eqs. (8.124) and (8.126), and stable time-marching computation can be obtained starting from initial condition imposed. A disadvantage of this method is that the last term in Eq. (8.126) must be considered. Then, the pressure Poisson equation Eq. (8.126) has a non-symmetric coefficient matrix. We must introduce the non-symmetric matrix solver to solve the linear simultaneous equation system.

For verification, pressure coefficient around a circular cylinder is computed using the same boundary conditions in Fig. 9.11 and the finite element mesh in Fig. 9.12 based on the present method. The computed results are shown in Fig. 9.13. The caption in the figure is the incompressible flow method.

8.5.7 Two Phase Flows in Y-Shape Channel

Kurahashi et al. (2012) studied two phase flows in Y-shape channel based on computational methods comparing with experimental results. The computational method employed is the stabilized bubble function method, i.e., a combination of the methods Eqs. (8.124) and (8.126) in Sect. 8.5.6. Figure 8.3 shows the channel model used for the computation. The channel is assumed to be two dimensional and two different liquids, Ethyl acetate, and pure water flow in from the inlets and flow out through outlet. For each fluid, the basic equations are Eqs. (8.1) and (8.11) with different densities $\rho^{(n)}$ $(n = 1, 2)$ and viscosity $\mu^{(n)}$ $(n = 1, 2)$ where $n = 1$ expresses Ethyl

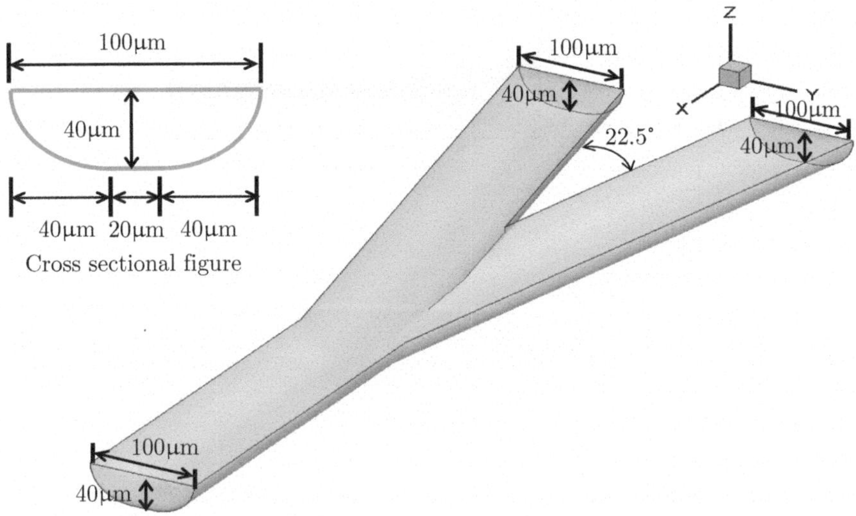

Fig. 8.3 Y-shape channel for computation

acetate, and $n = 2$ is pure water. Two fluids are immiscible and the interface position is given by the following advection equation expressed in Eq. (7.177) as

$$\dot{\mathfrak{f}} + v_i \mathfrak{f}_{,i} = 0$$

where $\mathfrak{f}$ is referred to as the VOF function, which takes the values $\mathfrak{f} = 1$ and $\mathfrak{f} = 0$ for the two types of liquids. The CFS model (Brackbill et al. 1992) is employed for the surface tension s_i as

$$s_i = \sigma \kappa \frac{\rho_{,i} \rho}{[\rho] \langle \rho \rangle}$$

where

$$[\rho] = \rho^{(2)} - \rho^{(1)}$$
$$\langle \rho \rangle = \frac{1}{2}(\rho^{(1)} + \rho^{(2)})$$
$$\rho^{(1)} < \rho^{(2)}$$

The curvature κ is given as

$$\kappa = \frac{1}{|n|} \left[\frac{n_i}{|n|} |n|_{,i} - n_{i,i} \right]$$

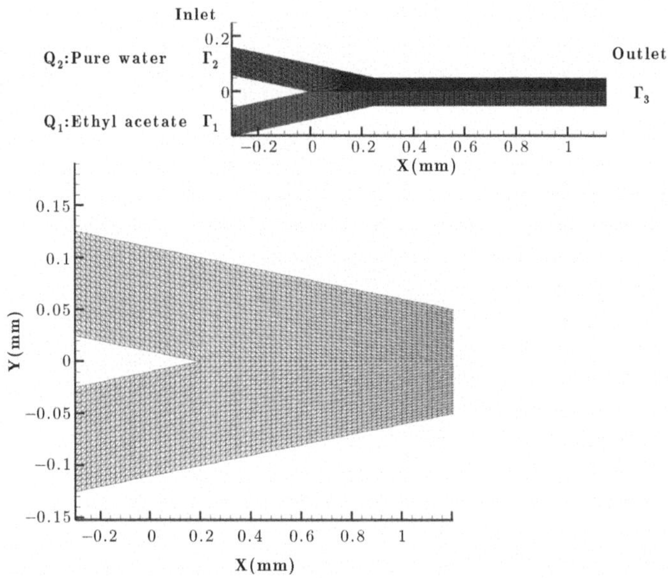

Fig. 8.4 Finite element mesh and boundary conditions

where

$$n_i = \mathfrak{f}_{,i}$$
$$|n| = \sqrt{n_i n_i}$$

Figure 8.4 shows the two-dimensional finite element mesh with boundary conditions. Numbers of elements and nodes are 23,200 and 11,991, respectively. As the boundary conditions, inflow velocities are imposed on Γ_1 and Γ_2, pressure is specified on Γ_3, and interface position is given on Γ_3. Time increment used is 1.0×10^{-7} s. Interface tension coefficient σ is 6.3×10^{-6} N/mm. Densities and viscosities are tabulated in Table 8.1. Discharges of Ethyl acetate are given as $Q^{(1)} = 10$ μl/min., where discharges of water are varied as 10, 20, and 50 μl/min. The comparison of experimental and computational results is shown in Fig. 8.5. The comparisons of $b^{(l)}/H$ are presented in Table 8.2, where $b^{(l)}$ and H are shown in case 1 in Fig. 8.5. Those results show fairly good correlation between the two results. The computational methods developed in this section are adaptable for a wide range of application, such as medicine, cosmetics, electric devices, etc.

Table 8.1 Density and viscosity

Fluid	Density	Viscosity
Ethyl acetate: $Q(1)$	8.984×10^{-7} kg/mm^3	0.557 mm^2/s
Pure water: $Q(2)$	9.982×10^{-7}	1.004

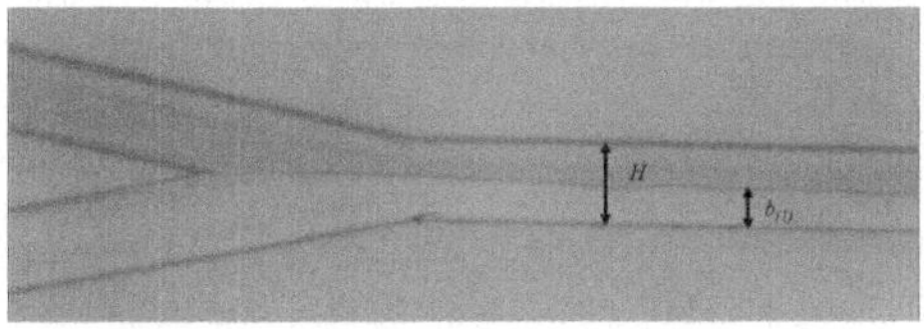

Case 1: Measurement result (Q_1 10 μl/min, Q_2 10 μl/min)

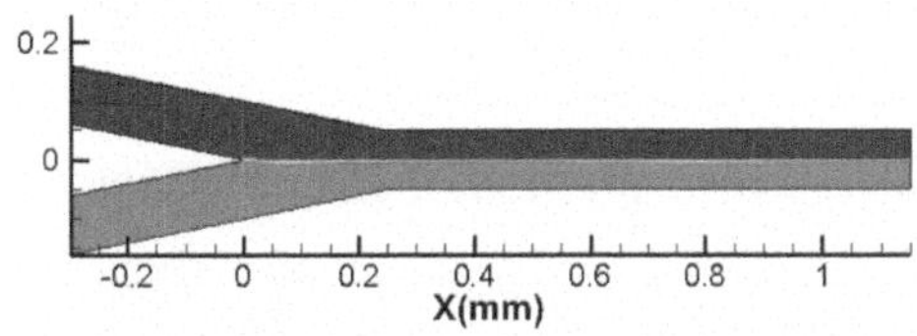

Case 1: Numerical result (Q_1 10 μl/min, Q_2 10 μl/min)

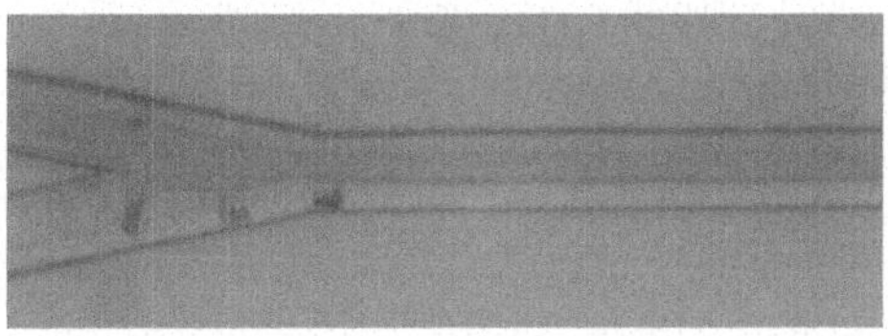

Case 2: Measurement result (Q_1 10 μl/min, Q_2 20 μl/min)

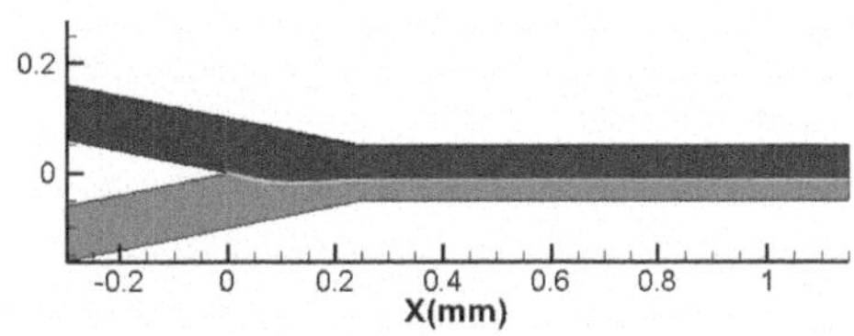

Case 2: Numerical result (Q_1 10 μl/min, Q_2 20 μl/min)

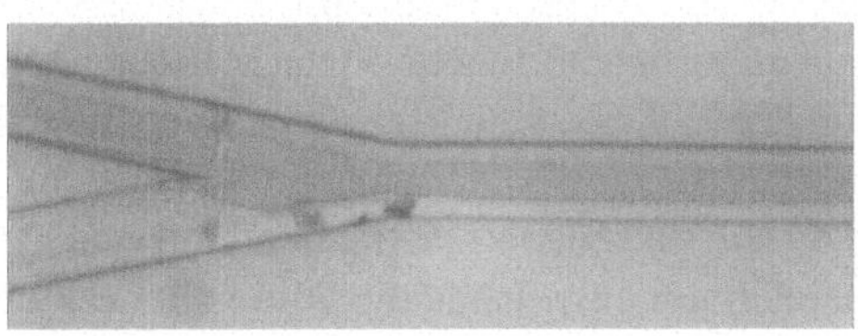

Case 3: Measurement result (Q_1 10 μl/min, Q_2 50 μl/min)

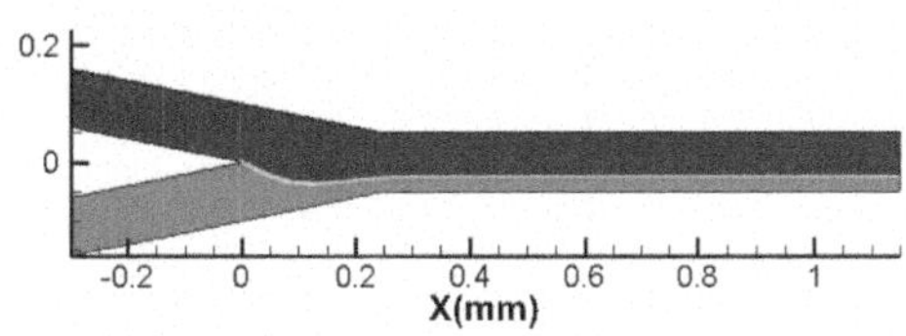

Case 3: Numerical result (Q_1 10 μl/min, Q_2 50 μl/min)

Fig. 8.5 Comparison of experimental and computational results

Table 8.2 Comparisons between computated and experimental results

Case no.	Inflow condition		Computed results	Experimental results
	$Q^{(1)}$	$Q^{(2)}$		
1	10 μl/min	10 μl/min	0.481	0.496
2	10	20	0.347	0.359
3	10	50	0.296	0.243

8.5.8 Coastal Waves over Detached Breakwater

CADMAS-SURF is the general-purpose open program and stands for Computer Aided Design of MAritime Structure. The development of this program for two-dimensional flows started in 1988 by the 'Research Group on Application of Numerical Wave Flume on Maritime Structural Design,' whose chair was Prof. Masahiko Isobe, the University of Tokyo. The two-dimensional code was developed in 2001 by the Coastal Development Institute of Technology. Arikawa et al. (2010) applied this code to three-dimensional model. The Research Group started to investigate the possibility of application of Arikawa et al. (2010) model to three-dimensional flows. They released three-dimensional code, named CADMAS-SURF/3D in 2011 via the Coastal Development Institute of Technology. Okumura and Arikawa (2014) and Okumura (2014) applied GPU technique to CADMAS-SURF/3D to save the computational cost. The purpose of the computer program is to solve velocity, pressure, and surface position of the three-dimensional unsteady incompressible flows for the design of coastal and maritime structures. The program is now a de facto standard in the coastal engineering in Japan. In this section, the computation of the coastal waves over the detached breakwater by CADMAS-SURF/3D is summarized.

The computational method is based on the fractional step finite element method combined with the standard bubble function interpolation applied to the transient incompressible viscous fluid flows, which are in more detail described in Sect. 8.5. To track the position of free surface, VOF function method is employed. The original concept is developed by Hirt and Nicols (1981), Brackbill et al. (1992), etc. The finite element mesh is generated not only inside the fluid volume V_+ but also outside the fluid volume V_-. The VOF function $\mathfrak{f}$ is assigned to distinguish the fluid volume as

$$\mathfrak{f} = \begin{cases} 1 & \text{in} \quad V_+ \\ \frac{1}{2} & \text{on} \quad \Sigma \\ 0 & \text{in} \quad V_- \end{cases}$$

The free surface position can be represented by Eq. (7.177) using the function $\mathfrak{f}$, i.e.,

$$\frac{\partial \mathfrak{f}}{\partial t} + v_i \mathfrak{f}_{,i} = 0 \quad \text{on} \quad \Sigma$$
$$\mathfrak{f} = \mathfrak{f}_0 \quad \text{at} \quad t = t_0$$

where $\mathfrak{f}_0$ is the free surface position at $t = t_0$ and Σ is free surface.

In the computation of the fractional step finite element method, one of the computational bottle necks is to solve the pressure Poisson equation expressed in Eq. (8.126), whose coefficient matrix is sparse but asymmetric. To solve the equation system, ILU–BiCG method is used. ILU means incomplete lower and upper factorization of the coefficient matrix. The asymmetric matrix is once decomposed into lower and upper triangular matrices, and then the inverse matrices of lower and upper matrices are computed for the use of numerical conditioning. At that time, only nonzero elements are used and stored. Hence, this procedure is called the "incomplete factorization." Using the preconditioning, convergence of the computation is accelerated. BiCG stands for the biconjugate gradient iteration method, which is created for the use of solving asymmetric linear simultaneous equation system.

Because iteration solution is based on the multiplication of the row matrix and column matrix, the concept of the parallel computation can be adaptable. To make the computation on a single processor machine more efficient, the implementation of the distributed message passing interface (MPI) is introduced. The adaptability and effectiveness of the CUDA parallelization developed by NVIDA corporation on a graphic processing unit (GPU) accelerator, which is recently used for the scientific

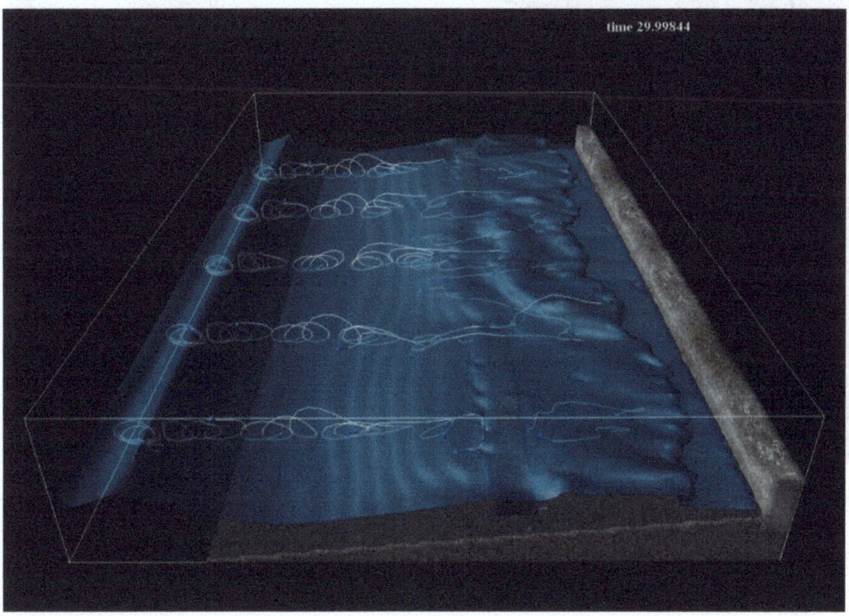

Fig. 8.6 Particle paths

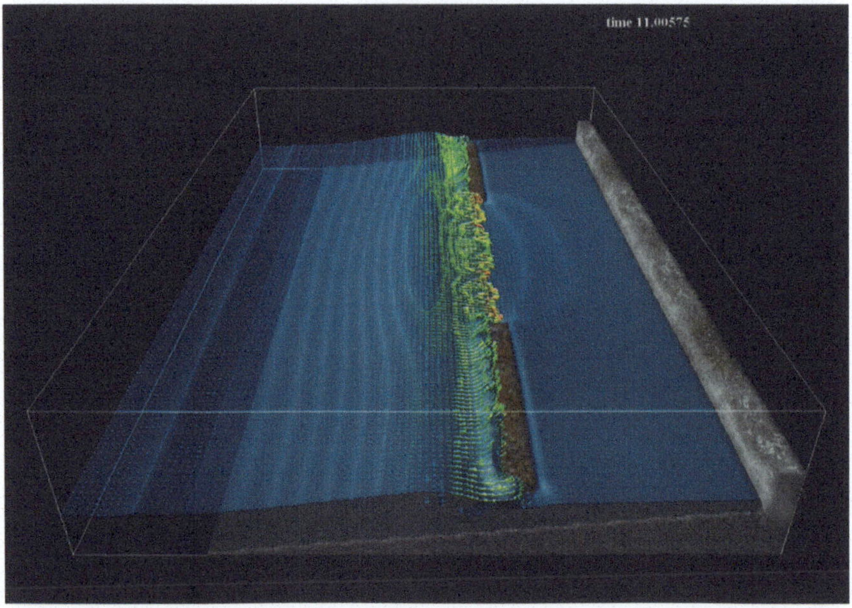

Fig. 8.7 Velocity distributions around detached breakwater

Fig. 8.8 Computed waves over seawall (1)

Fig. 8.9 Computed waves over seawall (2)

computation, is investigated. The performance ratio of the GPU/CPU computational time is reached more than 400 %. The precise information is found in the literatures listed above.

Computations of the waves over the detached breakwater and seawall are performed by the CADMAS-SURF/3D program. Detached breakwater is a coastal structure located inside the surfzone parallel to the coast. Figures 8.6 and 8.7 show the computer graphics of the computational results. Figure 8.6 exhibits the particle paths using white lines. Figure 8.7 shows the distribution of velocity. In Figs. 8.8 and 8.9, the computational results of overflows of the wave over the seawall are shown. The behaviors of the waves flown over the detached breakwater and seawall are clearly represented.

8.6 Improved BTD Finite Element Method

The fractional step finite element method based on the weighted residual equation Eqs. (8.124) and (8.126) using the standard linear weighting and interpolation functions was usually used for the practical computations. However, we noticed that the stability of computations should be improved. One approach is BTD finite element method, and in this section the improved BTD method will be described (Maruoka et al. 2001).

8.6.1 Improved BTD Fractional Step Finite Element Method

Applying the Crank–Nicolson scheme to Eq. (8.17), we can rewrite it as

$$\frac{v_i^{n+1} - v_i^n}{\Delta t} + v_j^n v_{i,j}^{n+1/2} + p_{,i}^{n+1/2} - \mu V_{ij,j}^{n+1/2} = 0 \tag{8.127}$$

where

$$v_i^{n+1/2} = \frac{1}{2}(v_i^{n+1} + v_i^n) \tag{8.128}$$

and V_{ij} is written in Eq. (8.121).

For the equation of continuity, the implicit scheme is

$$v_{i,i}^{n+1} = 0 \tag{8.129}$$

The second term in Eq. (8.127) is rewritten as follows using Eq. (8.128):

$$v_j^n v_{i,j}^{n+1/2} = v_j^n v_{i,j}^n + \frac{1}{2} v_j^n (v_i^{n+1} - v_i^n)_{,j} \tag{8.130}$$

Substituting Eq. (8.127) into Eq. (8.130), we get

$$v_j^n v_{i,j}^{n+1/2} = v_j^n v_{i,j}^n - \frac{\Delta t}{2} v_j^n (v_k^n v_{i,k}^{n+1/2} + p_{,i}^{n+1/2} - \mu V_{ik,k}^{n+1/2})_{,j} \tag{8.131}$$

Introducing Eq. (8.131) into Eq. (8.127) again and rearranging the terms, the equation of motion is expressed as in the following form:

$$\begin{aligned} &\frac{v_i^{n+1} - v_i^n}{\Delta t} + v_j^n v_{i,j}^n + p_{,i}^{n+1/2} - \mu V_{ij,j}^{n+1/2} \\ &- \frac{\Delta t}{2} v_j^n (v_k^n v_{i,k}^{n+1/2} + p_{,i}^{n+1/2} - \mu V_{ik,k}^{n+1/2})_{,j} = 0 \end{aligned} \tag{8.132}$$

Using the weighting function v_i^*, the weighted residual equation can be obtained as follows:

$$\begin{aligned} &\int_V v_i^* \left(\frac{v_i^{n+1} - v_i^n}{\Delta t} + v_j^n v_{i,j}^n \right) dV + \int_V v_{i,j}^* (-p^{n+1/2}\delta_{ij} + \mu V_{ij}^{n+1/2}) dV \\ &- \int_V \left\{ \frac{\Delta t}{2} v_i^* v_j^n (v_k^n v_{i,k}^{n+1/2} + p_{,i}^{n+1/2} - \mu V_{ik,k}^{n+1/2})_{,j} \right\} dV \\ &- \int_S v_i^* (-p^{n+1/2}\delta_{ij} + \mu V_{ij}^{n+1/2}) n_j dS = 0 \end{aligned} \tag{8.133}$$

where V is an arbitrary volume surrounded by a boundary S, and n_j is component of the unit normal of S. We put the third term in Eq. (8.133) as A, and A can be reformulated as

$$
\begin{aligned}
A = &-\frac{\Delta t}{2}\int_S v_i^* v_j^n (v_k^n v_{i,k}^{n+1/2} + p_{,i}^{n+1/2} - \mu V_{ik,k}^{n+1/2}) n_j dS \\
&+\frac{\Delta t}{2}\int_V v_{i,j}^* v_j^n (v_k^n v_{i,k}^{n+1/2} + p_{,i}^{n+1/2} - \mu V_{ik,k}^{n+1/2}) dV \\
&+\frac{\Delta t}{2}\int_V v_i^* v_{j,j}^n (v_k^n v_{i,k}^{n+1/2} + p_{,i}^{n+1/2} - \mu V_{ik,k}^{n+1/2}) dV
\end{aligned}
\tag{8.134}
$$

Introducing Eq. (8.127) into Eq. (8.134) again, A can be reformulated further as follows:

$$
\begin{aligned}
A = &\frac{1}{2}\int_S v_i^* v_j^n (v_i^{n+1} - v_i^n) n_j dS \\
&+\frac{\Delta t}{2}\int_V v_{i,j}^* v_j^n (v_k^n v_{i,k}^{n+1/2} + p_{,i}^{n+1/2}) dV
\end{aligned}
\tag{8.135}
$$

assuming v_j^n is constant and $V_{ik,k} = 0$ because we use the linear interpolation function. The final form of the weighted residual equation of the "Improved BTD method" (IBTD method) can be described, by substituting Eq. (8.135) into Eq. (8.133), and rearranging the terms:

$$
\begin{aligned}
&\int_V v_i^* \left(\frac{v_i^{n+1} - v_i^n}{\Delta t} + v_j^n v_{i,j}^n \right) dV + \int_V v_{i,j}^* (-p^{n+1/2}\delta_{ij} + \mu V_{ij}^{n+1/2}) dV \\
&+ \frac{\Delta t}{2}\int_V v_{i,j}^* v_j^n (v_k^n v_{i,k}^{n+1/2} + p_{,i}^{n+1/2}) dV \\
&= \int_S v_i^* (-p^{n+1/2}\delta_{ij} + \mu V_{ij}^{n+1/2}) n_j dS - \frac{1}{2}\int_S v_i^* v_j^n (v_i^{n+1} - v_i^n) n_j dS
\end{aligned}
\tag{8.136}
$$

The equation of continuity is represented as the following form:

$$
v_{i,i}^{n+1} = v_{i,i}^n + (v_{i,i}^{n+1} - v_{i,i}^n) = 0 \tag{8.137}
$$

Substituting Eq. (8.127) into Eq. (8.137), we get

$$
v_{i,i}^{n+1} = v_{i,i}^n - \Delta t (v_j^n v_{i,j}^{n+1/2} + p_{,i}^{n+1/2} - \mu V_{ij,j}^{n+1/2})_{,i} \tag{8.138}
$$

Equation (8.138) can be rearranged as follows based on Eq. (8.129):

$$
v_{i,i}^n - \Delta t (v_j^n v_{i,j}^{n+1/2} + p_{,i}^{n+1/2} - \mu V_{ij,j}^{n+1/2})_{,i} = 0 \tag{8.139}
$$

The last term in Eq. (8.139) is zero if we use the linear interpolation function. The weighted residual equation can be written as

$$\int_V (p^* v_{i,i}^n) dV + \Delta t \int_V p_{,i}^* (v_j^n v_{i,j}^{n+1/2} + p_{,i}^{n+1/2}) dV$$
$$= \Delta t \int_S p^* (v_j^n v_{i,j}^{n+1/2} + p_{,i}^{n+1/2} - \mu V_{ij,j}^{n+1/2}) n_i dS \tag{8.140}$$

The last term in Eq. (8.140) can be transformed by Eq. (8.127), and we get the final form of the weighted residual equation for pressure:

$$\int_V (p^* v_{i,i}^n) dV + \Delta t \int_V p_{,i}^* (v_j^n v_{i,j}^{n+1/2} + p_{,i}^{n+1/2}) dV$$
$$= -\int_S p^* (v_i^{n+1} - v_i^n) n_i dS \tag{8.141}$$

It is straightforward to obtain the global finite element equation based on Eqs. (8.136) and (8.141). The stable computations are carried out using those equations.

On the other hand, computation of the last term in Eq. (8.136) can be cumbersome. Maruoka et al. (2001) have developed a new means to eliminate the last term in Eq. (8.136). Adding the equality to the right side of Eq. (8.131) expressed in Eq. (8.142)

$$-\frac{1}{2} v_j^n v_{i,j}^n + \frac{1}{2} v_j^n v_{i,j}^{n+1} - \frac{\Delta t}{2} v_j^n \left(\frac{v_i^{n+1} - v_i^n}{\Delta t} \right)_{,j} = 0 \tag{8.142}$$

and rearranging the terms, we get

$$v_j^n v_{i,j}^{n+1/2} = v_j^n v_{i,j}^{n+1/2}$$
$$- \frac{\Delta t}{2} v_j^n \left(\frac{v_i^{n+1} - v_i^n}{\Delta t} + v_k^n v_{i,k}^{n+1/2} + p_{,i}^{n+1/2} - \mu V_{ik,k}^{n+1/2} \right)_{,j} \tag{8.143}$$

Replacing Eq. (8.143) with Eq. (8.131), the weighted residual equation of the IBTD method can be expressed after several transformations:

$$\int_V v_i^* \left(\frac{v_i^{n+1} - v_i^n}{\Delta t} + v_j^n v_{i,j}^{n+1/2} \right) dV + \int_V v_{i,j}^* \left(-p^{n+1/2} \delta_{ij} + \mu V_{ij}^{n+1/2} \right) dV$$
$$+ \frac{\Delta t}{2} \int_V v_{i,j}^* v_j^n \left(\frac{v_i^{n+1} - v_i^n}{\Delta t} + v_k^n v_{i,k}^{n+1/2} + p_{,i}^{n+1/2} - \mu V_{ik,k}^{n+1/2} \right) dV$$
$$= \int_S v_i^* \{ -p^{n+1/2} \delta_{ij} + \mu (v_{i,j}^{n+1/2} + v_{j,i}^{n+1/2}) \} n_j dS \tag{8.144}$$

The third term on the right side of Eq. (8.144) corresponds to the SUPG term, in which the stabilization parameter τ_S is replaced with $\Delta t/2$. Namely, we have shown that IBTD method is equivalent to the SUPG method in the case of the stabilization parameter $\tau_S = \Delta t/2$. We can also carry out stable computations based on Eqs. (8.141) and (8.144). It is also interesting and significant that the first term and the second term in the parenthesis on the third term of Eq. (8.144) are mutually transpose. Thus, the final matrix can be symmetric. The coefficient matrix for pressure in Eq. (8.141) is symmetric. Those are advantages for computer coding.

8.6.2 Wind Flows Around Girder of Long Span Bridge

Computational and experimental studies of long span suspension bridge, whose center span is up to 2800 m, are being carried out to get the information for the future plan. Flows around a girder of suspension bridge are computed and compared with experimental results by Watanabe and Fumoto (2007).

The computational method is the improved balancing tensor diffusivity (IBTD) scheme combined with the fractional step (FS) scheme, which is described in Sect. 8.6.1. For the viscosity, the Smagorinsky SGS model is used, which is added to the normal viscosity:

$$\mu_t = (C_s h)^2 (2 d_{ij} d_{ij})^{1/2}$$

where C_s is the Smagorinsky constant 0.13, and h is the width of filter multiplied by the wall dumping function of Van Driest $f(y)$:

$$f(y) = 1 - \exp(-y/25)$$

in which y is the distance from the wall.

The computational model of the girder is shown in Fig. 8.10. The girder depth is taken to be D. At the girder center, the slots covered by nets are constructed. Both

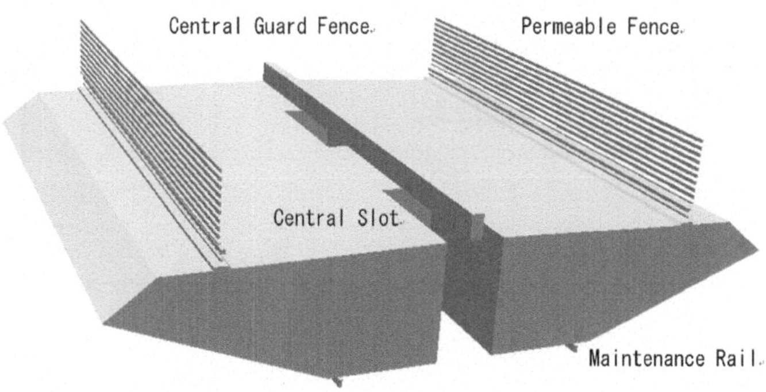

Fig. 8.10 Section model of a girder

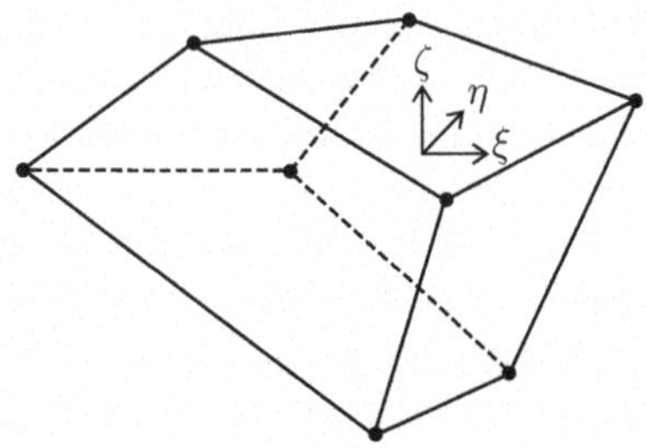

Fig. 8.11 Finite element employed

sides of the girder have fairings to reduce the wind effects. Permeable fences are on the upper face of the girder, which are lumped into nine blocks. On the lower surface of the girder, maintenance rails are equipped.

The finite element used is the iso-parametric linear element as shown in Fig. 8.11. The interpolation function can be expressed as

$$\Phi_\alpha = \frac{1}{8}(1+\xi_0)(1+\eta_0)(1+\zeta_0)$$

where

$$\xi_0 = \xi\xi_\alpha, \qquad \eta_0 = \eta\eta_\alpha, \qquad \zeta_0 = \zeta\zeta_\alpha$$

denoting $(\xi_\alpha, \eta_\alpha, \zeta_\alpha)$ as the coordinates of nodal points. Total numbers of finite elements and nodes are 4,130,014 and 3,997,632, respectively.

The boundary conditions are shown in Fig. 8.12. On the inlet boundary, constant profile free stream, whose magnitude is U, is imposed. Non-slip boundary condition

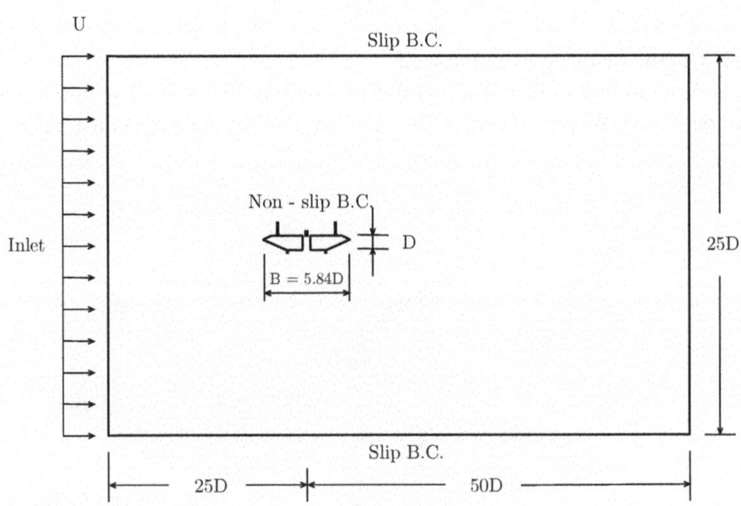

Fig. 8.12 Computational domain and boundary conditions

is imposed on all surfaces of the girder, whereas slip boundary condition at the upper and lower boundaries of the channel. The periodic boundary condition is specified at the end of axial planes.

The Reynolds number of the free stream is 10,000. The time increment used for computation is 0.05 D/U.

Three components of aerodynamic coefficients are defined by

$$C_d = \frac{F_d}{1/2\rho U^2 D}$$

$$C_l = \frac{F_l}{1/2\rho U^2 B}$$

$$C_m = \frac{F_m}{1/2\rho U^2 B^2}$$

where B is the breadth of the girder, which is 5.84D, and F_d, F_l and F_m are the drag force, lift force, and pitching moment, respectively. In those coefficients, ρ is the density of air and taken to be constant. Figure 8.13 shows the comparisons of C_d, C_l, and C_m obtained by computations and experiments (Fumoto et al. 2005). Horizontal axis in Fig. 8.13 is attack angle θ. Both computational and experimental results are in good agreement. We see slight discrepancies in the results of C_l and C_m. The tendency of the aerodynamic forces is understood. In minus attack angle zone ($\theta < 0°$), C_l is also minus. This fact shows that the upward forces to resist the

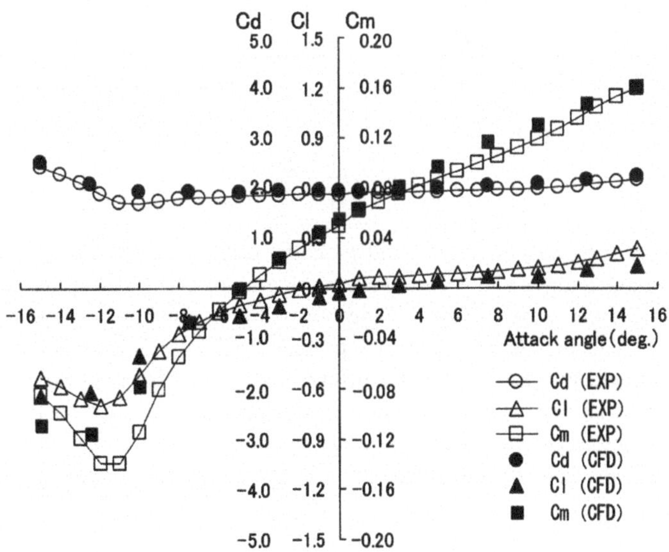

Fig. 8.13 Three components of static aerodynamic wind force

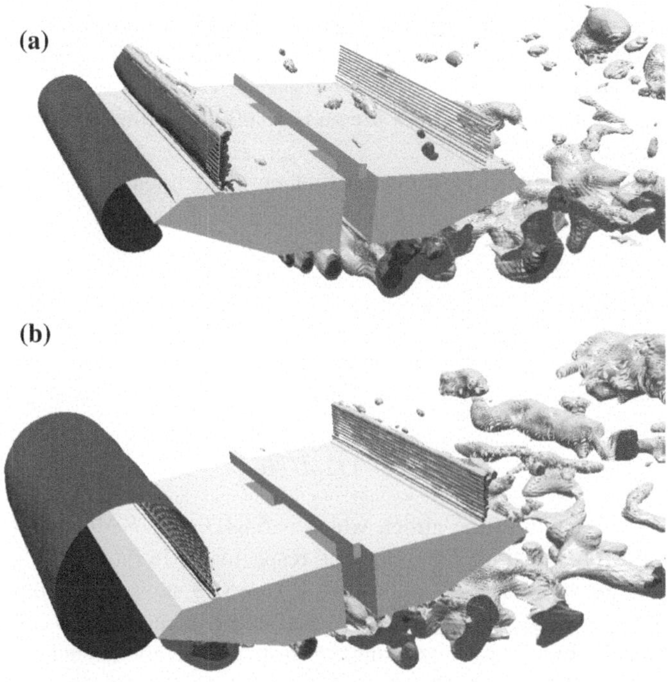

Fig. 8.14 Bird's-eye view of instantaneous pressure contours. **a** $\alpha = 0°$. **b** $\alpha = -10°$

downward movement of the girder will be acted to stabilize the movement of the girder.

Figure 8.14a, b shows bird's-eye view of instantaneous pressure contours at attack angles $\theta = 0°$ and $\theta = -10°$. The pressure is shown by the pressure coefficient defined as

$$C_p = \frac{p - p_0}{1/2\rho U^2}$$

where p is the pressure computed and p_0 is the reference pressure at center point on inflow boundary. We can see high pressure zone of $C_p = 0.4$ at the nose of the girder. The pressure zone at attack angle $\theta = -10°$ is much wider than that attack angle $\theta = 0°$. This fact expresses that the pressure to the girder is higher at the girder which is inclined. We can also see the difference of flow separation zones. At attack angle $\theta = 0°$, the flow is separated at maintenance rail located near the center of the girder, whereas the flow is separated at the point close to the front side of the girder if the girder is inclined. From the above, the overall stability performance was confirmed.

8.7 Summary and Conclusion

This chapter investigated detailed numerical analyses of the finite element method applied to the incompressible fluid flows. We particularly used two methods, i.e., the stabilized bubble function method, simply the bubble function method, and the improved balancing tensor diffusivity method, in short, IBTD method. We limited only two methods combined with the fractional step method in time, which is also referred to the pressure Poisson equation method. The reason is that those methods have superior stability and simplicity for computer coding. We have discussed mutual relationship among those methods and the streamline upwind Petrov Galerkin, SUPG, method. In conclusion, we have found near-equivalence among two methods and the SUPG methods. Those three methods employ in common the stabilization parameters and depend on, more or less, the selection of the parameters to adjust the additional viscosity. We have confirmed that the pressure is not stable in the computation of incompressible flows. Tezduyar et al. (2011) presented "stabilized method" adding additional terms to stabilize the computation. We have discussed several numerical studies applied to practical engineering subjects. The main reason for its instability stems from the fact that the equation of continuity is function of only velocity. In the next chapter, we investigate the possibility that equation of continuity may extend to the function of pressure by physical considerations.

Chapter 9
Analysis of Adiabatic Flows

9.1 Introduction

A large body of studies in the finite element methods have considered that the fluid is incompressible or compressible. In the case of the compressible fluid flow analyses, we must solve the equation of energy in addition to the equations of continuity and momentum. Because this substantially increases computational burdens, we may avoid it if possible. There are a number of actual domains in science and engineering, in which the temperature of the fluid can be assumed to be constant. In those domains, we usually use the incompressible assumption. The incompressibility is a kind of limited ideal state because any continuum has compressibility even if it is a small quantity. For instance, the acoustic velocity in the fluid is not unlimited, but, nevertheless limited even if it can get very large.

As stated in Chap. 8, one of the main causes of the computational instability stems from the fact that the equation of continuity is a function of velocity only and does not include functions of pressure and its derivative with respect to space or time. Therefore, several computational techniques have been presented to weaken the constraint of the incompressibility by adding the penalty terms of a small parameter and a function of pressure to the equation of continuity. These strategies are referred to as the "artificial compressibility method," which was originally proposed by Chorin (1997), and further explored by Temam (1968, 2001), Malan et al. (2002), Madesen and Schäffer (2005), Drikakis (2005), etc. We can consider that those methods such as SUPG, BTD, bubble function, etc., are based on the same concept of the artificial compressibility method. The equation of continuity of the incompressible flow is extended to the equation of continuity disturbed by the penalty terms with reciprocal of the square of acoustic velocity by physical considerations.

Let fluids be in a state in which pressure is a function of density only, and independent of temperature. We can derive the above-mentioned equation of state by positing ideal gas and the constant entropy, as discussed in Sect. 7.13. The same assumption may hold in liquid. We can use the equations of continuity and momentum as the

M. Kawahara, *Finite Element Methods in Incompressible, Adiabatic, and Compressible Flows*, Mathematics for Industry 19,
DOI 10.1007/978-4-431-55450-9_9

governing equation. The pressure can be eliminated from the equation system using the equation of state. Those fluids are referred to as the adiabatic fluids in this book. In this analysis, the primitive field variables are density and velocity. We must pay careful attention to the difference in numeral orders between both variables. We usually assume that gas is compressible and liquid is incompressible, i.e., the density of gas is variable, whereas, that of liquid is (more or less) constant. With these assumptions, we can transform the field variables to pressure and velocity, whose the governing equations include acoustic velocity. This method is referred to as the "acoustic velocity method." To deal with the liquid, we can use the acoustic velocity method with constant acoustic velocity.

Numerical procedures employed in the integration in time are mainly the "characteristic method" and the SUPG method. The characteristic method was discussed by, e.g., Pironneau (1982, 1983), Demcowicz and Rachovicz (1987), Arbogast and Wheeler (1995), Maury (1996), Iannelli (2006), Shi and Wang (2009), and many others. The main concept is described in Sect. 5.9. The pioneering work for the SUPG method is Brooks and Hughes (1982). Hughes and Tezduyar (1984), Tezduyar et al. (1992), Behr and Tezduyar (1994), Vellando et al. (2002), Tezduyar et al. (2011), Nasu et al. (2013), and many others have further explored the method. The basic concept is summarized in Sect. 5.6. Recall that superposed $\dot{}$ represents partial differentiation with respect to time in this chapter.

9.2 Governing Equations of Adiabatic Flows

The field variables for the computation of adiabatic flows are density ρ, velocity v_i and pressure p. The governing equations can be summarized as follows:

(1) Conservation of mass is:

$$\dot{\rho} + v_i \rho_{,i} + \rho v_{i,i} = 0 \tag{9.1}$$

which is Eq. (7.47).

(2) Conservation of momentum is:

$$\rho(\dot{v}_i + v_j v_{i,j}) - \sigma_{ji,j} = \rho f_i \tag{9.2}$$

where σ_{ij} and f_i are total stress and body force, respectively. This equation is referred to as the "equation of motion," which is Eq. (7.65).

(3) Constitutive equation is:

$$\sigma_{ij} = -p\delta_{ij} + \tau_{ij} \tag{9.3}$$

and

$$\tau_{ij} = \lambda d_{kk}\delta_{ij} + 2\mu d_{ij} \tag{9.4}$$

where d_{ij} is the deformation rate defined in Eq. (7.71). In Eq. (9.3), putting $i = j$, we get:

$$\sigma_{ii} = -3p + (3\lambda + 2\mu)d_{ii} \tag{9.5}$$

which means that pressure p is different from the hydrostatic pressure. In practice, we often introduce the relation:

$$\lambda = -\frac{2}{3}\mu \tag{9.6}$$

Then, p is the hydrostatic pressure from Eq. (9.5). In the case of incompressible flows, Eq. (9.6) is valid, and pressure p is indeterminate. Because we are investigating the solution of fluid flows which can be assumed to be incompressible, we sometimes use Eq. (9.6). Equation (9.3) shows that total stress consists of two types of stress, i.e., pressure p and viscous stress τ_{ij}.

(4) Equation of state is:

$$p = p(\rho) \tag{9.7}$$

assuming that the fluid is in the adiabatic state. The equation of state for the ideal gas is:

$$p = \alpha\rho^{\gamma} \tag{9.8}$$

where

$$\alpha = \frac{p_0}{\rho_0^{\gamma}} \tag{9.9}$$

In Eq. (9.8), p_0 and ρ_0 are reference values for pressure and density, respectively, and γ is the ratio of specific heat defined in Eq. (7.142). Equation (9.8) and (9.9) are Eq. (7.160). For the equation of state for liquid, the "Birch–Murnaghan experimental equation" is presented:

$$p = \left(\frac{2}{3}\right)\kappa_0\left[\left(\frac{\rho}{\rho_0}\right)^{\frac{7}{3}} - \left(\frac{\rho_0}{\rho}\right)^{\frac{5}{3}}\right] \tag{9.10}$$

where κ_0 is bulk modulus at zero pressure. We should bear in mind that the pressure p in Eq. (9.10) is the relative pressure. Instead of Eq. (9.7), we sometimes make use of the acoustic velocity $\mathfrak{a}$:

$$\mathfrak{a}^2 = \frac{\partial p}{\partial \rho} \tag{9.11}$$

which could be assumed to be constant.

(5) Boundary conditions are:

$$v_i = \hat{v}_i \quad \text{on} \quad S_1 \tag{9.12}$$

$$t_i = \sigma_{ij}n_j = \hat{t}_i \quad \text{on} \quad S_2 \tag{9.13}$$

$$\rho = \hat{\rho} \quad \text{on} \quad S_3 \tag{9.14}$$

where superscripted carret ˆ stands for the given quantities on the boundary. In Eq. (9.13), t_i is the surface force and n_j is unit normal to the boundary. We assume that:

$$\begin{aligned} S_1 \cup S_2 &= S \\ S_1 \cap S_2 &= \mathbf{0} \end{aligned} \tag{9.15}$$

where S denotes the whole boundary of the flow field V, and $\mathbf{0}$ is a null set. Sometimes, we use the boundary condition that the pressure p is given:

$$p = \hat{p} \quad \text{on} \quad S_4 \tag{9.16}$$

instead of Eq. (9.13).

(6) The initial conditions are:

$$v_i = \hat{v}_i^0 \tag{9.17}$$

$$p = \hat{p}^0 \tag{9.18}$$

$$\rho = \hat{\rho}^0 \tag{9.19}$$

where superscript ^{0} stands for the values at the initial time t_0. Equations (9.17)–(9.19) show that velocity, pressure, and density are given in all flow field as the initial conditions.

9.3 Explicit Euler Finite Element Method

We will show that the adiabatic fluid flow analysis can be achieved using the explicit Euler finite element method, in which we do not use the mixed interpolation. This strategy substantially simplifies the computational procedure.

9.3.1 *Formulation of Explicit Euler Finite Element Method*

Based on the governing equations Eqs. (9.1)–(9.3), the weighted residual equations can be formulated as follows:

$$\int_V (\rho^* \dot{\rho}) dV + \int_V (\rho^* v_i \rho_{,i}) dV + \int_V (\rho^* \rho v_{i,i}) dV = 0 \tag{9.20}$$

$$
\int_V (v_i^* \rho_0 \dot{v}_i) dV + \int_V (v_i^* \rho_0 v_j v_{i,j}) dV
- \int_V (v_{i,i}^* p) dV + \int_V (v_{i,j}^* \tau_{ij}) dV = \int_S (v_i^* t_i) dS \quad (9.21)
$$

where ρ^* and v_i^* are weighting functions corresponding to density ρ and velocity v_i, and ρ_0 represents density, which is assumed to be constant in an element. Linear interpolation and weighting functions are employed for density and velocity as:

$$\rho^* = \Phi_\alpha \rho_\alpha \quad (9.22)$$

$$\rho = \Phi_\beta \rho_\beta \quad (9.23)$$

$$v_i^* = \Phi_\alpha v_{\alpha i}^* \quad (9.24)$$

$$v_i = \Phi_\beta v_{\beta i} \quad (9.25)$$

where Φ_α is the function described in Eq. (4.25), and ρ_α and $v_{\alpha i}$ are density at node α and velocity at node α in the i-direction, and ρ_α^* and $v_{\alpha i}^*$ are corresponding weighting constants. Pressure p is also interpolated as follows:

$$p = \Phi_\beta p_\beta \quad (9.26)$$

Introducing Eqs. (9.22)–(9.26) into Eqs. (9.20) and (9.21), and rearranging the terms, we get the local forms of the finite element method, in which we do not use the mixed interpolation. This process nontrivially simplifies the computational procedure. The resulting equations are as follows:

$$M_{\alpha\beta} \dot{\rho}_\beta + K_{\alpha\beta\gamma i} v_{\beta i} \rho_\gamma + K_{\alpha\beta\gamma i} \rho_\beta v_{\gamma i} = 0 \quad (9.27)$$

$$\rho_0 M_{\alpha\beta} \dot{v}_{\beta i} + \rho_0 K_{\alpha\beta\gamma j} v_{\beta j} v_{\gamma i} + D_{\alpha i\beta} p_\beta + S_{\alpha i\beta j} v_{\beta j} = \Omega_{\alpha i} \quad (9.28)$$

where

$$M_{\alpha\beta} = \int_V (\Phi_\alpha \Phi_\beta) dV, \quad (9.29)$$

$$K_{\alpha\beta\gamma i} = \int_V (\Phi_\alpha \Phi_\beta \Phi_{\gamma,i}) dV \quad (9.30)$$

$$D_{\alpha i\beta} = -\int_V (\Phi_{\alpha,i} \Phi_\beta) dV \quad (9.31)$$

$$S_{\alpha i\beta j} = \lambda \int_V (\Phi_{\alpha,i} \Phi_{\beta,j}) dV + \mu \int_V (\Phi_{\alpha,k} \Phi_{\beta,k}) \delta_{ij} dV + \mu \int_V (\Phi_{\alpha,j} \Phi_{\beta,i}) dV \quad (9.32)$$

$$\Omega_{\alpha i} = \int_S (\Phi_\alpha t_i) dS \tag{9.33}$$

By superimposing Eqs. (9.27) and (9.28) formulated on a finite element over the whole flow field, we get the global forms of the finite element equation in the following form:

$$\mathbf{M}_{\alpha\beta}\mathbf{y}_\beta^{n+1} = \mathbf{M}_{\alpha\beta}\mathbf{y}_\beta^n + \Delta t \mathbf{f}_\alpha \tag{9.34}$$

where $\mathbf{M}_{\alpha\beta}$ denotes the coefficient matrix of the whole flow field corresponding to the first terms of Eqs. (9.27) and (9.28), and $\mathbf{f}_\alpha$ means the second and latter terms in Eqs. (9.27) and (9.28). We use the following numerical integration in time:

$$\dot{\mathbf{y}}_\beta \approx \frac{1}{\Delta t}(\mathbf{y}_\beta^{n+1} - \mathbf{y}_\beta^n) \tag{9.35}$$

where Δt denotes short time increment, and $\mathbf{y}_\beta^n$ means the values on the whole flow field at time point n. As worked out in Sect. 5.5.2, the stable computation may not be obtained using Eq. (9.34). Thus, the simplest stabilization scheme, i.e., lumping matrix technique, is introduced instead:

$$\bar{\mathbf{M}}_{\alpha\beta}\mathbf{y}_\beta^{n+1} = \tilde{\mathbf{M}}_{\alpha\beta}\mathbf{y}_\beta^n + \Delta t \mathbf{f}_\alpha \tag{9.36}$$

where $\bar{\mathbf{M}}_{\alpha\beta}$ means the lumped matrix and:

$$\tilde{\mathbf{M}}_{\alpha\beta} = e\bar{\mathbf{M}}_{\alpha\beta} + (1 - e)\mathbf{M}_{\alpha\beta} \tag{9.37}$$

in which e is the selective lumping parameter to stabilize the computation. Using Eqs. (9.36) and (9.37), computational time can be saved, since the inverse matrix of $\bar{M}_{\alpha\beta}$ is computed by the reciprocals of the diagonal terms.

9.3.2 Flows in a Cavity

For numerical computation, flows in a cavity are carried out to show the adaptability of the computational method presented in Eqs. (9.27) and (9.28) using the lid-driven flows in a cavity. More specifically, the assumption of the adiabatic flow can be available for computation and other purposes. The boundary conditions and the finite element mesh are shown in Figs. 9.1 and 9.2, respectively.

The total numbers of elements and nodes are 28,800 and 14,641, respectively. Reynolds number $Re = 1000$. The lumping parameter $e = 0.99$. The computed velocity at $T = 200$ and with the time increment $\Delta t = 0.00001$ is shown in Fig. 9.3. Because the fluid is almost incompressible, we can compare the results with those obtained by Ghia et al. (1982) (Fig. 9.4). Both results show a fairly good agreement with each other. Therefore, we can see the adiabatic assumption is useful for

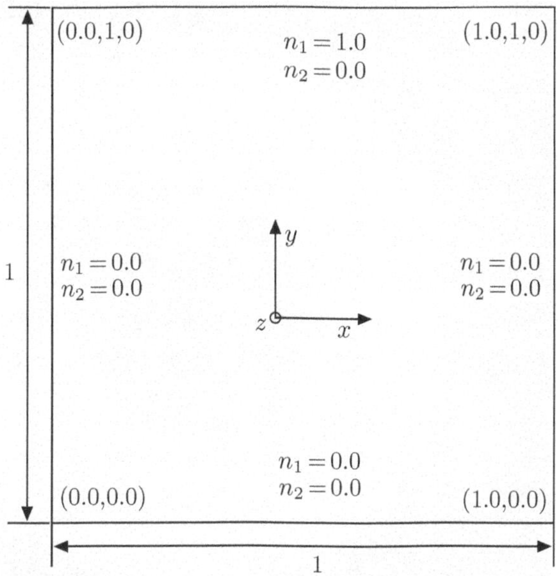

Fig. 9.1 Computational domain and boundary conditions

Fig. 9.2 Finite element mesh

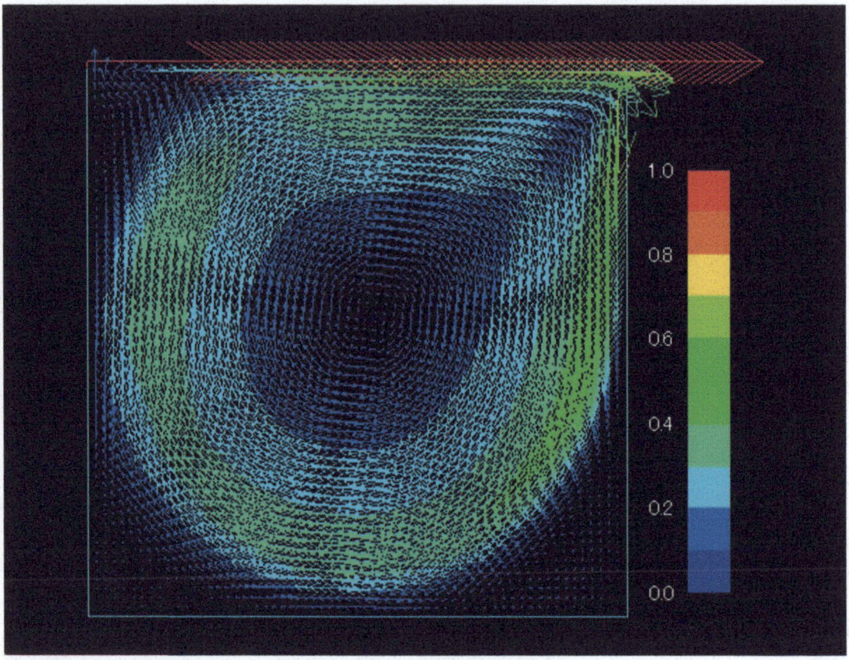

Fig. 9.3 Computed velocity at $Re = 1000$

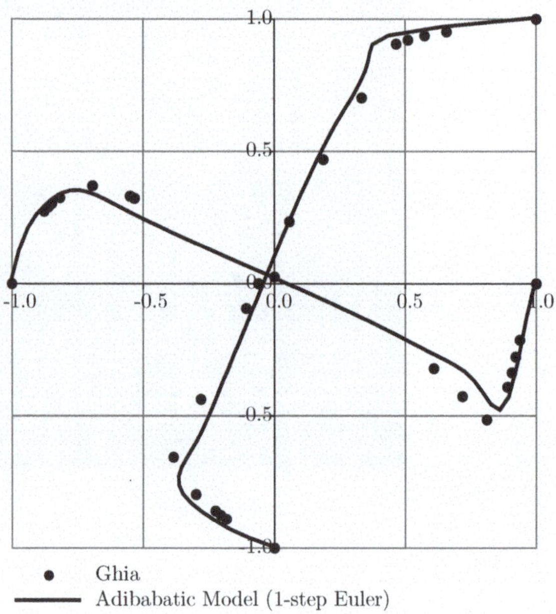

Fig. 9.4 Comparison between the present results and Ghia et al. (1982)

the computation of the incompressible flows. The numerical procedure is based on the normal linear interpolation and the simple explicit method. However, the computational stability is not satisfactory. Thus, we will use the characteristic method combined with the implicit method in Sect. 9.4, and the SUPG method in Sect. 9.6. Moreover, we will present the acoustic velocity method in Sect. 9.7.

9.4 Characteristic Finite Element Method

Consider a fluid particle in the fluid flows to be solved, in which at a short time later the particle will flow to another position. Thus, we can compute the differentiation of the particle position with respect to time as the partial differentiation along the particle path. The path is referred to as the characteristic line.

9.4.1 Characteristic Line Formulation

A position of a fluid particle at time τ, which was at a position x_i at time t, is denoted by $X_i(x_i, t; \tau)$. The trajectory of the particle, which is often referred to as the characteristic line, is given by the following ordinary differential equations:

$$\frac{\mathrm{d}X_i}{\mathrm{d}\tau} = v_i(X_i(x_i, t; \tau), \tau) \tag{9.38}$$

$$X_i(x_i, t; t) = x_i \tag{9.39}$$

The time derivative can be expressed as:

$$\frac{\mathrm{D}\rho}{\mathrm{D}t} = \dot{\rho} + v_i \rho_{,i} = \frac{\mathrm{d}}{\mathrm{d}\tau} \rho(X_i(x_i, t; \tau), \tau)\big|_{\tau=t} \tag{9.40}$$

$$\frac{\mathrm{D}v_i}{\mathrm{D}t} = \dot{v}_i + v_j v_{i,j} = \frac{\mathrm{d}}{\mathrm{d}\tau} v_i(X_i(x_i, t; \tau), \tau)\big|_{\tau=t} \tag{9.41}$$

The governing equations, Eqs. (9.1) and (9.2), consist of two parts: the advection and non-advection terms. For the advection term, the "characteristic method" is applied. The density and velocity at time n are denoted by ρ^n and v_i^n, respectively. After a short time increment Δt, these values will be ρ^{n+1} and v_i^{n+1}. Their corresponding values at the upstream point are denoted by $\bar{\bar{\rho}}$ and $\bar{\bar{v}}_i$. Therefore,

$$\rho^{n+1} = \bar{\bar{\rho}} + \Delta t \left(\frac{\mathrm{D}\rho}{\mathrm{D}t} \right)^{n+1} \tag{9.42}$$

$$v_i^{n+1} = \bar{\bar{v}}_i + \Delta t \left(\frac{\mathrm{D}v_i}{\mathrm{D}t} \right)^{n+1} \tag{9.43}$$

where

$$\bar{\bar{\rho}} = \rho(X_i(x_i, t; t), t) \quad \text{at} \quad t = t^n \tag{9.44}$$

$$\bar{\bar{v}}_i = v_i(X_i(x_i, t; t), t) \quad \text{at} \quad t = t^n \tag{9.45}$$

Using Eqs. (9.40) and (9.42), Eq. (9.1) is transformed as:

$$\frac{1}{\Delta t}\rho^{n+1} + \bar{\rho} v^n_{i,i} = \frac{1}{\Delta t}\bar{\bar{\rho}} \tag{9.46}$$

Similarly, Eq. (9.2) is transformed as:

$$\frac{\bar{\rho}}{\Delta t} v_i^{n+1} + p^n_{,i} - \tau^n_{ij,j} = \frac{\bar{\rho}}{\Delta t}\bar{\bar{v}}_i \tag{9.47}$$

where $\bar{\rho}$ in Eqs. (9.46) and (9.47) is computed as:

$$\bar{\rho} = \frac{1}{2}(3\rho^n - \rho^{n-1}) \tag{9.48}$$

and assumed constant in a finite element and for Δt. In Eq. (9.47), p^n and τ^n_{ij} are obtained by Eqs. (9.7) and (9.4), respectively. Equations (9.46) and (9.47) are the discretized equations of the governing equations Eqs. (9.1) and (9.2) in time by the characteristic formulation. It is straightforward to formulate the global forms of the finite element equation based on Eqs. (9.46) and (9.47).

9.4.2 The Algorithm of Calculation of the Upstream Points

The upstream points corresponding to all nodal points of the finite element mesh in the whole flow field should be obtained. The algorithm is intended to solve Eqs. (9.38) and (9.39), and schematically shown in Fig. 9.5. Because Eqs. (9.38) and (9.39) are nonlinear, iteration calculation should be used. The algorithm is as follows. All calculation is made at $t = t_n$.

(1) Choose a nodal point of the mesh, whose position is denoted by $x_{\alpha i}$.
(2) Assume a position of the upstream point by the following equation:

$$\chi_{\alpha i} = x_{\alpha i} - v_{\alpha i}\Delta t \tag{9.49}$$

where $\chi_{\alpha i}$ represents the initial position of the upstream point.
(3) Search $K(\chi_{\alpha i})$, an element which includes the position $\chi_{\alpha i}$.
(4) Set:

$$X_{\alpha i(0)} = \chi_{\alpha i} \tag{9.50}$$

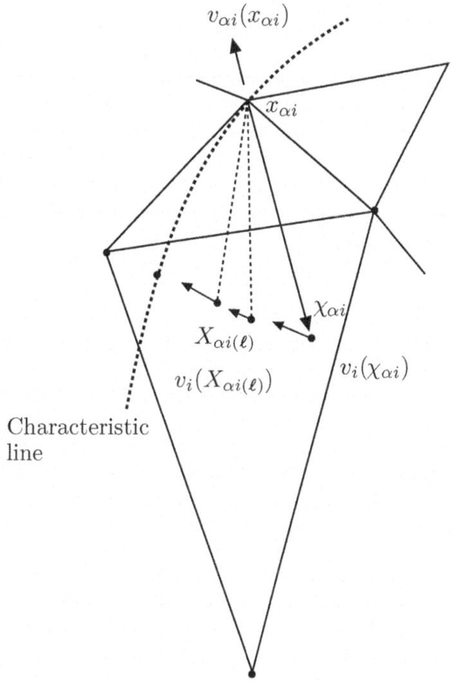

Fig. 9.5 Characteristic line

(5) Interpolate:

$$v_{i(0)} = \Phi_\beta(\chi_{\alpha i})v_{\beta i} \tag{9.51}$$

where $\Phi_\beta(\chi_{\alpha i})$ means the interpolation function based on nodal points of $K(\chi_{\alpha i})$, and $v_{\beta i}$ represents the velocity values at the three nodal points of $K(\chi_{\alpha i})$.

(6) Compute the upstream position $X_{\alpha i(\ell)}$ by iteration for $\ell = 1, 2, \ldots, \ell_{\max}$, where $\ell_{\max}$ is the maximum number of iteration.

6)-1 Compute the upstream position:

$$X_{\alpha i(\ell)} = x_{\alpha i} - \frac{\Delta t}{2}\{v_{\alpha i}(x_{\alpha i}) + v_{i(\ell-1)}(X_{\alpha i(\ell-1)})\} \tag{9.52}$$

6)-2 Search $K(X_{\alpha i(\ell)})$, an element, which includes the upstream position $X_{\alpha i(\ell)}$.

6)-3 Interpolate velocity at the upstream position:

$$v_{i(\ell)} = \Phi_\beta(X_{\alpha i(\ell)})v_{\beta i} \tag{9.53}$$

where $\Phi_\beta(X_{\alpha i(\ell)})$ is the interpolation function based on $K(X_{\alpha i(\ell)})$, and $v_{\beta i}$ is the velocity values at three nodes of $K(X_{\alpha i(\ell)})$.

6)-4 IF $|X_{\alpha i(\ell)} - X_{\alpha i(\ell-1)}| < \varepsilon$ THEN 7) ELSE 6)-1 where ε is a small allowance parameter.

7) Finally, we get

$$\bar{\bar{\rho}} = \rho(X_{\alpha i(\ell)}) = \Phi_\beta(X_{\alpha i(\ell)})\rho_\beta \tag{9.54}$$

$$\bar{\bar{v}}_{\alpha i} = v_{i(\ell)}(X_{\alpha i(\ell)}) = \Phi_\beta(X_{\alpha i(\ell)})v_{\beta i} \tag{9.55}$$

In the above algorithm, we used the simplest integration scheme in Eq. (9.52). It is possible to use the higher order integration scheme as well. However, the simplest one is recommended because it takes longer computational time to search the element in which the candidate of upstream position is included.

9.5 Density Flows in Adiabatic Two-Phase Fluids

In this section, we will illustrate numerical procedures and computational results of adiabatic two-phase immiscible fluids with different densities. The fluids are assumed to be liquids and their equation of state can be characterized by the Birch–Murnaghan equation. The numerical procedure is a combination of the characteristic method and the implicit method. The two fluids have different densities and a surface tension exists at the interface of these fluids. The Hermit interpolation function, which is based on the complete third-order polynomials using triangular element, is employed for the interpolations of both velocity and density. Based on the discontinuous conditions, the "interface translocation method" can be derived. According to the density difference, various flow patterns can be obtained. Such flow patterns are found in a wide range of fields of natural science and industrial engineering.

9.5.1 The Characteristic Finite Element Method of Density Flows in Adiabatic Two-Phase Flows

Consider the two-phase flows schematically shown in Fig. 9.6. We will denote two domains V^+ and V^-, whose densities and velocities are ρ^+ and ρ^-, and v_i^+ and v_i^-. Interface of V^+ and V^- is expressed by S, whose unit normal is represented by n_i. We use velocity v_i and density ρ, which means v_i^+ and ρ^+ in V^+, and v_i^- and ρ^- in V^-. The governing equations are Eqs. (9.1)–(9.4), (9.6) and (9.10) in each domain and Eqs. (7.170) and (7.171) on the interface. Appropriate boundary and initial conditions are imposed.

We use the characteristic finite element method described in Sect. 9.4. Assume that Eqs. (9.46) and (9.47) in V^+ and V^-, and Eqs. (7.170) and (7.171) on the interface are valid. Introducing the weighting functions ρ^* and v_i^*, the weighted residual equations can be expressed as follows:

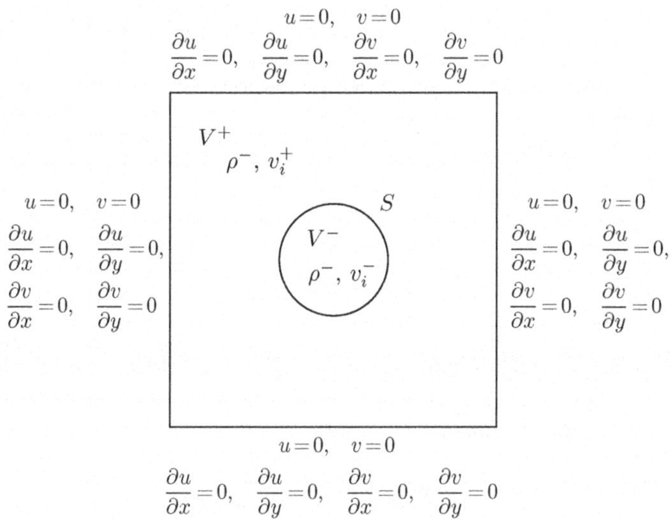

Fig. 9.6 Computational domains, density distributions, and boundary conditions

$$\frac{1}{\Delta t}\int_V (\rho^* \rho^{n+1})dV - \bar{\rho}\int_V (\rho^*_{,i} v_i^n)dV = \frac{1}{\Delta t}\int_V (\rho^* \bar{\bar{\rho}})dV + \int_S \rho^* [\![\rho v_i n_i]\!]\, dS \tag{9.56}$$

$$\frac{\bar{\rho}}{\Delta t}\int_V (v_i^* v_i^{n+1})dV - \int_V (v^*_{i,i} p^n)dV + \int_V (v^*_{i,j}\tau^n_{ij})dV = \frac{\bar{\rho}}{\Delta t}\int_V (v_i^* \bar{\bar{v}}_i)dV + \int_S v_i^* [\![t_i]\!]\, dS \tag{9.57}$$

where $\bar{\bar{\rho}}$ and $\bar{\bar{v}}_i$ represent density and velocity on the characteristic line at time t_n, and the short time increment Δt is $t_{n+1} - t_n$. In Eqs. (9.56) and (9.57), $\bar{\rho}$ is assumed to be constant and computed as in Eq. (9.48). The finite element equation of conservation of mass is expressed as:

$$\frac{1}{\Delta t} M_{\alpha\beta}\rho_\beta^{n+1} - \bar{\rho} G_{\alpha i \beta} v_{\beta i} = \frac{1}{\Delta t} M_{\alpha\beta}\bar{\bar{\rho}}_\beta - \Omega_\alpha \tag{9.58}$$

where

$$M_{\alpha\beta} = \int_V (H_\alpha H_\beta) dV \tag{9.59}$$

$$G_{\alpha i\beta} = \int_V (H_{\alpha,i} H_\beta) dV \tag{9.60}$$

$$\Omega_\alpha = \int_S (H_\alpha [\![\rho v_i n_i]\!] dS) \tag{9.61}$$

where H_α denotes the interpolation function of the third-order polynomials, which will be discussed in Sect. 9.5.2. Equation (9.61) expresses the effect of discontinuity of density at the interface S and:

$$\Omega_\alpha = \int_S H_\alpha (\rho^+ v_n^+ - \rho^- v_n^-) dS \tag{9.62}$$

In Eq. (9.62), we use the average values on the elements to the interface S for v_n^+ and v_n^-, where

$$v_n^+ = v_i^+ n_i \tag{9.63}$$

$$v_n^- = v_i^- n_i \tag{9.64}$$

Similarly, the finite element equation of conservation of momentum is derived as:

$$\begin{aligned} &\frac{\bar{\rho}}{\Delta t} M_{\alpha\beta} v_{\beta i}^{n+1} - \bar{\chi} G_{\alpha i\beta} \rho_\beta^{n+1} + \lambda D_{\alpha i\beta j} v_{\beta j}^{n+1} \\ &+ \mu D_{\alpha j\beta j} v_{\beta i}^{n+1} + \mu D_{\alpha j\beta i} v_{\beta j}^{n+1} = \frac{\bar{\rho}}{\Delta t} M_{\alpha\beta} v_{\beta i} + \Sigma_{\alpha i} \end{aligned} \tag{9.65}$$

where

$$D_{\alpha i\beta j} = \int_V (H_{\alpha,i} H_{\beta,j}) dV \tag{9.66}$$

$$\Sigma_{\alpha i} = \int_S H_\alpha [\![t_i]\!] dS \tag{9.67}$$

Using Eq. (9.10), pressure p is written as

$$\begin{aligned} p &= \frac{2}{3} \kappa_0 \left[\left(\frac{1}{\rho_0} \right)^{\frac{7}{3}} \rho^{\frac{4}{3}} - \left(\frac{1}{\rho_0} \right)^{\frac{5}{3}} \rho^{\frac{2}{3}} \right] \rho \\ &= \bar{\chi} \rho \end{aligned} \tag{9.68}$$

We assume that $\bar{\chi}$ is a constant in Δt and is computed by the density obtained in the previous time. The surface stress is assumed to be given as:

$$t_i = \xi \kappa n_i \tag{9.69}$$

where ξ is a constant and κ is the curvature of the interface S.

9.5.2 Hermit Interpolation Function

To express moving boundary of the two-phase domains, it is necessary to use a higher order interpolation function. For this purpose, we employ the Hermit interpolation function based on the third-order polynomials. In this section, the cartesian coordinate is expressed by (x, y), and the coordinate of a node is denoted by (x_α, y_α) at node α of a triangular element. Fig. 9.7 shows the Hermit interpolation function. The function values, the values of the first derivative at the nodes of element, and the function value at the barycentric node is used as the independent degrees of freedom. Therefore, 10 degrees of freedom are employed for one element in total. The Hermit interpolation functions $H_{0\alpha}$, $H_{x\alpha}$, $H_{y\alpha}$ and H_{0e} are as follows, expressed without summation convention:

$$\begin{aligned} H_{0\alpha} &= \eta_\alpha^2(3 - 2\eta_\alpha) - 7\eta_1\eta_2\eta_3 \\ H_{x\alpha} &= \eta_\alpha^2(x_{\beta\alpha}\eta_\beta - x_{\alpha\gamma}\eta_\gamma) - (x_{\beta\alpha} - x_{\alpha\gamma})\eta_1\eta_2\eta_3 \\ H_{y\alpha} &= \eta_\alpha^2(y_{\beta\alpha}\eta_\beta - y_{\alpha\gamma}\eta_\gamma) - (y_{\beta\alpha} - y_{\alpha\gamma})\eta_1\eta_2\eta_3 \\ H_{0e} &= 27\eta_1\eta_2\eta_3 \end{aligned} \tag{9.70}$$

where

$$x_{\alpha\beta} = x_\alpha - x_\beta, \quad y_{\alpha\beta} = y_\alpha - y_\beta \tag{9.71}$$

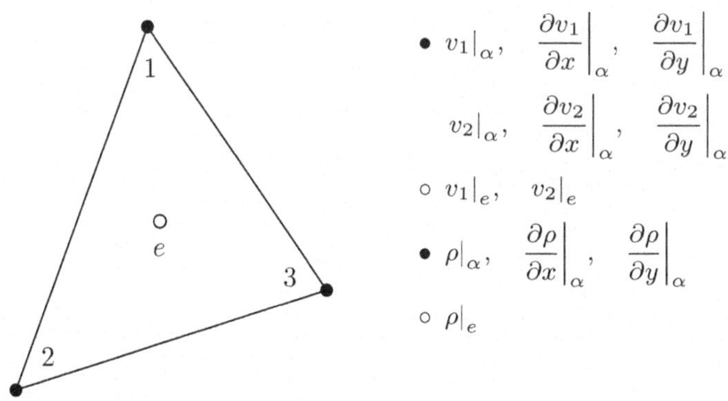

Fig. 9.7 Hermit interpolation function

and (x_α, y_α) represents the nodal coordinate for an element ($\alpha = 1, 2, 3$). Area coordinates η_1, η_2 and η_3 are described in Sect. 6.5, which are functions of coordinate (x, y). In Eqs. (9.70) and (9.71), permutation is used for (α, β, γ).

The finite element interpolations for an element obtained using the Hermit interpolation function Eqs. (9.70) and (9.71) are expressed as follows:

$$v_i = H_\alpha v_{\alpha i} \tag{9.72}$$

(summation convention for α)

$$\rho = H_\alpha \rho_\alpha \tag{9.73}$$

where

$$H_\alpha = \left[H_{0\alpha}\ H_{x\alpha}\ H_{y\alpha}\ H_{0e} \right] \tag{9.74}$$

$$v_{\alpha i} = \left[v_i|_\alpha \ \frac{\partial v_i}{\partial x}\bigg|_\alpha \ \frac{\partial v_i}{\partial y}\bigg|_\alpha \ v_i|_e \right] \tag{9.75}$$

$$\rho_\alpha = \left[\rho|_\alpha \ \frac{\partial \rho}{\partial x}\bigg|_\alpha \ \frac{\partial \rho}{\partial y}\bigg|_\alpha \ \rho|_e \right] \tag{9.76}$$

In Eqs. (9.74)–(9.76), $|_\alpha$ means the values on node α ($\alpha = 1, 2, 3$) and $|_e$ means the value on the barycentric node.

In the computation of the characteristic finite element method, velocities and densities on the characteristic line at $t = t_n$ should be computed. The procedures described in Sect. 9.4.2 are applied to obtain the velocities, their derivatives, densities and their derivatives. Then, we can express the results as follows:

$$\bar{\bar{v}}_{\alpha i} = \left[\bar{\bar{v}}_i|_\alpha \ \frac{\partial \bar{\bar{v}}_i}{\partial x}\bigg|_\alpha \ \frac{\partial \bar{\bar{v}}_i}{\partial y}\bigg|_\alpha \ \bar{\bar{v}}_i|_e \right] \tag{9.77}$$

$$\bar{\bar{\rho}}_\alpha = \left[\bar{\bar{\rho}}|_\alpha \ \frac{\partial \bar{\bar{\rho}}}{\partial x}\bigg|_\alpha \ \frac{\partial \bar{\bar{\rho}}}{\partial y}\bigg|_\alpha \ \bar{\bar{\rho}}|_e \right] \tag{9.78}$$

9.5.3 *Tracking of Interface*

To track the interface position, we have employed the characteristic method. In this case, we can assume $\nu_n = 0$. However, in the practical computation, the accuracy of the position is not satisfactory in this way. Thus, we have corrected the position by the Eulerian type method. To do this, Eq. (7.178) is normally used, as we presented in Sects. 8.5.7 and 8.5.8. On the other hand, in this section, we employed the discontinuous condition to track the interface position.

Consider the interface shown in Fig. 9.6. The position of the interface is moving according to the differences of densities and velocities between in V^+ and in V^-. From Eqs. (7.171) and (7.174), the velocity of the movement of the interface ν_n is expressed as follows:

$$\nu_n = \frac{\rho^+(v_n^+)^2 + \rho^-(v_n^-)^2 - \xi\kappa}{\rho^+ v_n^+ + \rho^- v_n^-} \tag{9.79}$$

Using the velocity ν_n, the position of the interface can be computed:

$$x_i^{n+1} = x_i^n + \Delta t \nu_n n_i \tag{9.80}$$

where x_i^n is the position of the interface at $t = t_n$.

To compute the curvature of the interface, a third-order B-spline function is employed, which is expressed as:

$$\begin{aligned} x(s) &= \sum_{j=-2}^{n+2} N_3(s-j)p_x(j) \quad (-1 \le s \le n+1) \\ y(s) &= \sum_{j=-2}^{n+2} N_3(s-j)p_y(j) \quad (-1 \le s \le n+1) \end{aligned} \tag{9.81}$$

where $x(s)$ and $y(s)$ are the positions of the interface of the coordinate s, $p_x(j)$, $p_y(j)$ are the positions of control points, and n is the total number of control points. The B-spline function $N_3(s)$ is as follows:

$$N_3(s) = \begin{cases} \frac{3|s|^3 - 6|s|^2 + 4}{6} & (-1 < s < 1) \\ -\frac{(|s|-2)^3}{6} & (-2 < s < 2) \\ 0 & \end{cases} \tag{9.82}$$

The curvature κ is computed through the B-spline function and is written as:

$$\kappa = \frac{\dot{x}\ddot{y} - \dot{y}\ddot{x}}{(\dot{x}^2 + \dot{y}^2)^{\frac{3}{2}}} \tag{9.83}$$

where superscripted $\dot{}$ in Eq. (9.83) means differentiation with respect to s.

9.5.4 Two-Phase Flows in a Square Container

The density flows of the two-phase immiscible fluids in a square container are computed to illustrate the numerical aspects of the present method. Two types of flow computation are carried out, i.e., flows with low and high density differences. The computational domain and boundary conditions are shown in Fig. 9.6. There are two

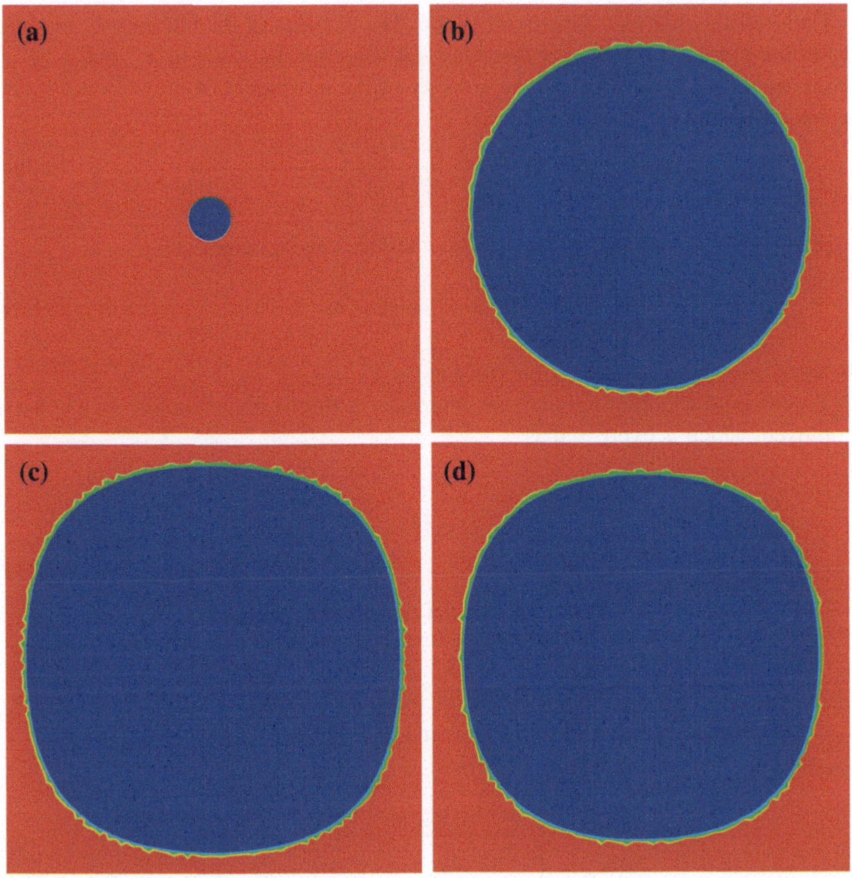

Fig. 9.8 Computed transitions for flows with low density difference. **a** Density distribution ($T = 0.05$). **b** Density distribution ($T = 0.35$). **c** Density distribution ($T = 0.7$). **d** Density distribution ($T = 10.0$)

domains: a high-density domain V^+ and low-density domain V^-. At the initial stage, the circular domain V^- is located at the center of the square domain V^+. The side length of the square container is $10D$, where D is the diameter of the initial inner domain. The boundary conditions of the outer domain are as given in the figure. The total numbers of the nodes and the elements of the finite element mesh are 55,654 and 110,826, respectively. The computations start at the static state, i.e., all velocities are zero. The parameters used are $\mu = 0.1$ and $\chi_0 = 2.19$.

In Fig. 9.8a–d, the computed transitions of flows with low density difference are shown. The computation starts at the static state. The density in V^+ is assumed to be $\rho^+ = 1.0$ and that in V^- is $\rho^- = 0.99$. The surface tension coefficient is $\xi = 0.01$. It is clear that the inner circular domain initially expands with respect to time and reaches maximum in Fig. 9.8c, after which it undergoes a slight contraction

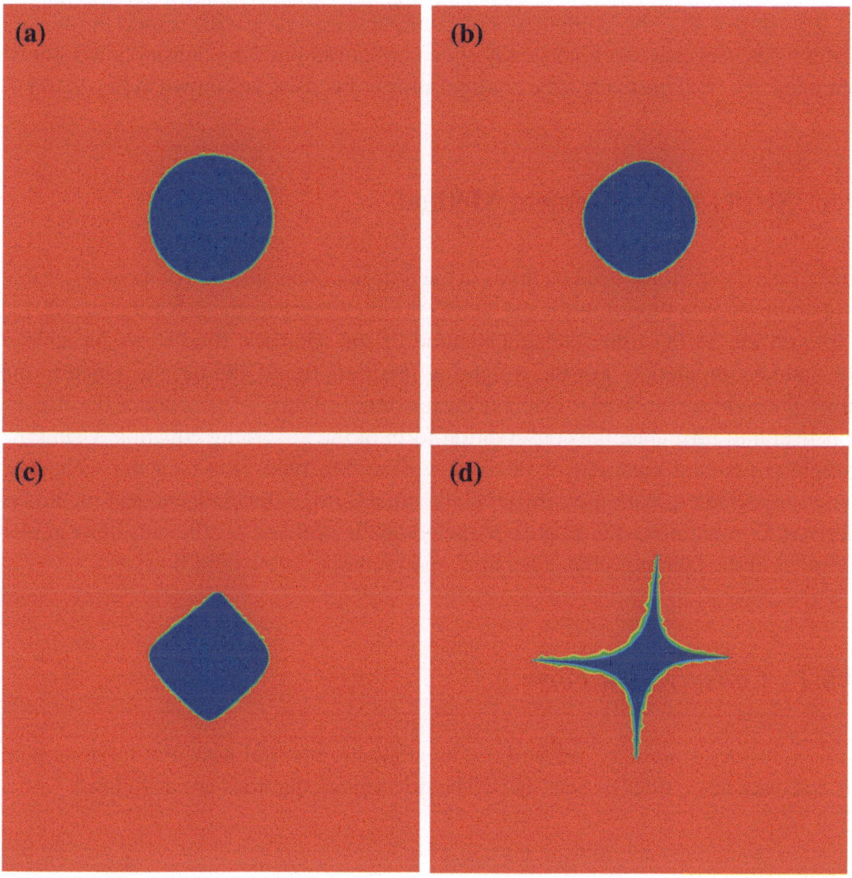

Fig. 9.9 Computed density transitions for flows with high density difference. **a** Density distribution ($T = 0.15$). **b** Density distribution ($T = 0.35$). **c** Density distribution ($T = 0.4$). **d** Density distribution ($T = 0.5$)

as shown in Fig. 9.8d. The computation is terminated, when the velocity distribution approaches zero.

In Fig. 9.9a–d, the computed density transitions for flows with high density differences starting from the same initial shape as that of the computation in Fig. 9.8. For the initial state, the density in V^+ is assumed to be $\rho^+ = 1.0$ and in V^-, $\rho^- = 0.65$. The surface tension coefficient is $\xi = 0.01$. Initially, the inner domain expands circularly with respect to time. Then it takes the shape of thrombus. Finally, as the distortion increases, it takes the shape of a distorted star, whose arms expand along four directions toward the boundary. The computation is terminated at $t = 0.5$, because the mesh cannot represent more complicated patterns. The mesh should be regenerated in order to continue the computation.

It is shown that the computational configurations vary with respect to density differences. Because the final steady state configuration of the density could not be obtained, the final unsteady state configurations of density are shown in Fig. 9.10a–f.

9.6 SUPG Finite Element Method

One of the most widely used methods in the finite element analyses of incompressible and compressible fluid flows is the SUPG finite element method. Nasu et al. (2013) explored the SUPG finite element method of the adiabatic fluid flows to achieve the stable computation. For the weighting function, the SUPG term is added to the normal weighting function to evaluate the upstream effects. To incorporate the effects, it is useful to formulate the conservation form of the governing equation. Using the weighted residual equation of the Petrov–Galerkin type, in which the weighting function and interpolation are different, the final form of the finite element method is derived. Comparative studies are conducted on the pressure coefficients of a circular cylinder in the compressible, adiabatic, and incompressible fluid flows.

9.6.1 Conservation Form

For expository simplicity, we will use an adiabatic inviscid flow. For the treatment of viscosity term, refer to Sect. 10.3. The governing equations are as follows:

$$\mathfrak{A} = \dot{\rho} + m_{j,j} = 0 \tag{9.84}$$

$$\mathfrak{B}_k = \dot{m}_k + \left(\frac{m_k m_j}{\rho}\right)_{,j} + p_{,k} = 0 \tag{9.85}$$

$$p = \alpha \rho^{\gamma} \tag{9.86}$$

where m_k is momentum in the k-direction:

$$m_k = \rho v_k \tag{9.87}$$

and superscripted $\dot{}$ is partial differentiation with respect to time. Eqs. (9.84) and (9.85) can be written as:

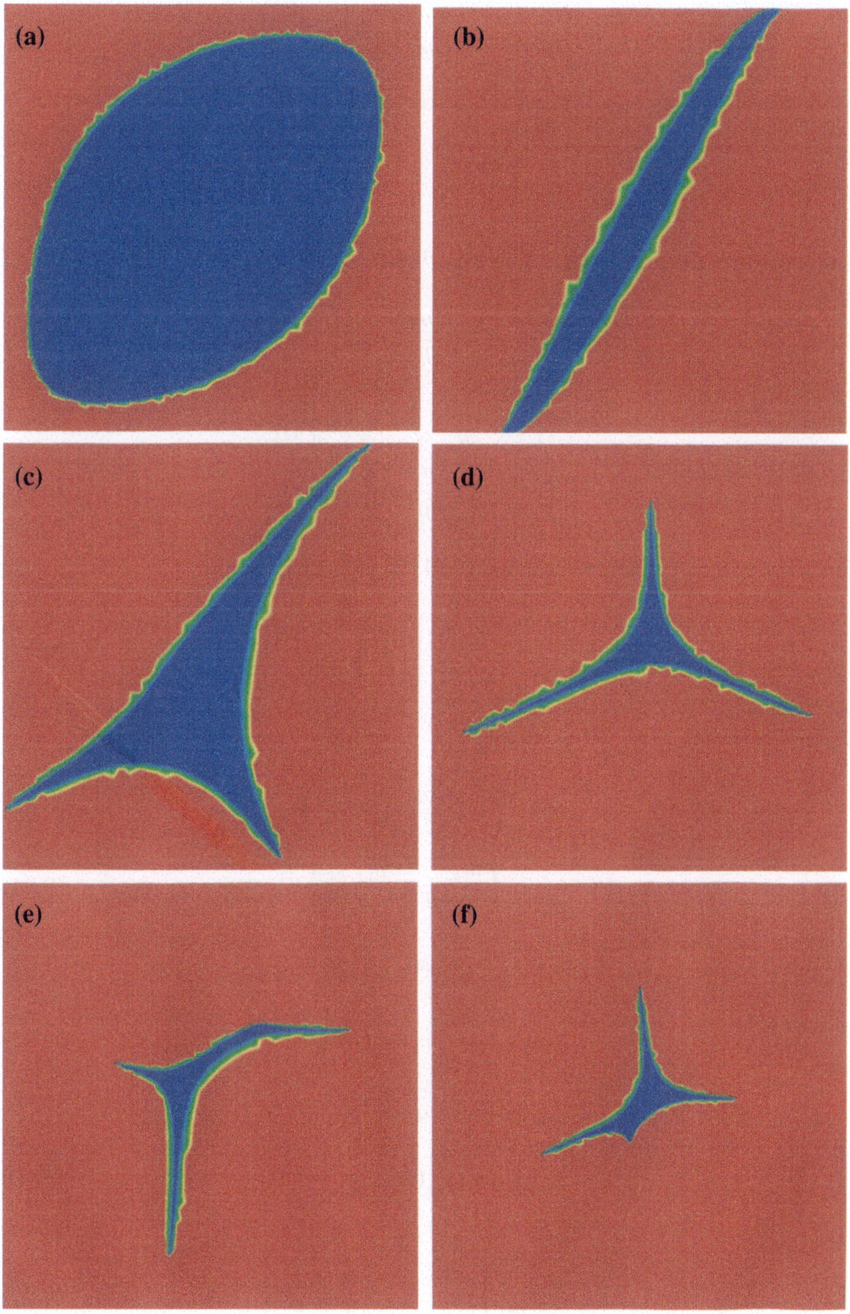

Fig. 9.10 Final configurations of density. **a** Density distribution ($\rho^- = 0.983$ and $T = 10.0$). **b** Density distribution ($\rho^- = 0.95$ and $T = 3.5$). **c** Density distribution ($\rho^- = 0.85$ and $T = 1.25$). **d** Density distribution ($\rho^- = 0.8$ and $T = 1.0$). **e** Density distribution ($\rho^- = 0.7$ and $T = 0.7$). **f** Density distribution ($\rho^- = 0.6$ and $T = 0.45$)

$$\dot{\rho} + \frac{\partial m_i}{\partial \rho}\frac{\partial \rho}{\partial x_j} + \frac{\partial m_j}{\partial m_l}\frac{\partial m_l}{\partial x_j} = 0 \tag{9.88}$$

$$\dot{m}_k + \frac{\partial}{\partial \rho}\left(\frac{m_k m_j}{\rho} + p\delta_{kj}\right)\frac{\partial \rho}{\partial x_j} + \frac{\partial}{\partial m_l}\left(\frac{m_k m_j}{\rho} + p\delta_{kj}\right)\frac{\partial m_l}{\partial x_j} = 0 \tag{9.89}$$

Let us perform the differentiation in Eqs. (9.88) and (9.89), and we get:

$$\dot{\rho} + 0\frac{\partial \rho}{\partial x_j} + \delta_{jl}\frac{\partial m_l}{\partial x_j} = 0 \tag{9.90}$$

$$\dot{m}_k + \left\{\left(-\frac{m_k m_j}{\rho^2}\right) + \mathfrak{a}^2\delta_{kj}\right\}\frac{\partial \rho}{\partial x_j} + \left(\frac{m_k}{\rho}\delta_{lj} + \frac{m_j}{\rho}\delta_{kl}\right)\frac{\partial m_l}{\partial x_j} = 0 \tag{9.91}$$

Further, we can write Eqs. (9.90) and (9.91) in the matrix form:

$$\begin{bmatrix} \dot{\rho} \\ \dot{m}_k \end{bmatrix} + \begin{bmatrix} 0 & \delta_l^j \\ b_k^j & b_{kl}^j \end{bmatrix}\begin{bmatrix} \rho \\ m_l \end{bmatrix}_{,j} = \begin{bmatrix} 0 \\ 0 \end{bmatrix} \tag{9.92}$$

where

$$b_k^j = \left(-\frac{m_k m_j}{\rho^2} + \mathfrak{a}^2\delta_{kj}\right) \tag{9.93}$$

$$b_{kl}^j = \left(\frac{m_k}{\rho}\delta_{lj} + \frac{m_j}{\rho}\delta_{kl}\right) \tag{9.94}$$

and $\mathfrak{a}$ is acoustic velocity, which is expressed as follows:

$$\mathfrak{a}^2 = \frac{\partial p}{\partial \rho} = \alpha\gamma\rho^{\gamma-1} \tag{9.95}$$

using Eq. (9.86). We can also write Eq. (9.92) as in the following form:

$$\dot{\mathbf{U}}_k + \mathbf{A}_{kl}^i \mathbf{U}_{l,i} = 0 \tag{9.96}$$

where

$$\mathbf{U}_k = \begin{bmatrix} \rho \\ m_k \end{bmatrix}, \quad \mathbf{A}_{kl}^i = \begin{bmatrix} 0 & \delta_l^j \\ b_k^i & b_{kl}^i \end{bmatrix} \tag{9.97}$$

Equations (9.92) and (9.96) are referred to as the conservation form of the governing equation of adiabatic fluid flows. In Eqs. (9.92) and (9.97), δ_l^j means Kronecker's delta function.

9.6.2 Formulation of the SUPG Finite Element Method

The SUPG method described in Sect. 5.6 is formulated based on Eq. (9.96). For the interpolation function, a normal linear interpolation function based on a triangular element is used:

$$\mathbf{U}_k = \mathbf{\Phi}_\beta \mathbf{U}_{\beta k} \tag{9.98}$$

where

$$\mathbf{\Phi}_\beta = \begin{bmatrix} \Phi_\beta & \\ & \Phi_\beta \end{bmatrix}, \tag{9.99}$$

$$\mathbf{U}_{\beta k} = \begin{bmatrix} \rho_\beta \\ m_{\beta k} \end{bmatrix} \tag{9.100}$$

In Eq. (9.99), Φ_β denotes the linear interpolation function expressed in Eq. (4.24) in Sect. 4.4. In Eq. (9.100), ρ_β is the density on node β, and $m_{\beta k}$ is the momentum on node β in the k-direction. For the weighting function, the following function is used:

$$\tilde{\mathbf{U}}_k = \mathbf{U}_k^* + \tau_s \mathbf{A}_{lk}^i \mathbf{U}_{l,i}^* \tag{9.101}$$

where $\mathbf{U}_k^*$ is a linear weighting function, such as:

$$\mathbf{U}_k^* = \mathbf{\Phi}_\beta \mathbf{U}_{\beta k}^* \tag{9.102}$$

In Eq. (9.102), $\mathbf{U}_{\beta k}^*$ is:

$$\mathbf{U}_{\beta k}^* = \begin{bmatrix} \rho_\beta^* \\ m_{\beta k}^* \end{bmatrix} \tag{9.103}$$

where ρ_β^* and $m_{\beta k}^*$ are weighting constants corresponding to ρ_β and $m_{\beta k}$. In Eq. (9.101), τ_s is the stabilization parameter and $\mathbf{A}_{lk}^i$ is the transposed matrix of $\mathbf{A}_{kl}^i$ in Eq. (9.97).

Multiplying Eq. (9.101) to both sides of Eq. (9.96) and integrating over the volume V, we get the following equation:

$$\int_V (\mathbf{U}_k^* + \tau_s \mathbf{U}_{m,j}^* \mathbf{A}_{mk}^j)(\dot{\mathbf{U}}_k + \mathbf{A}_{kl}^i \mathbf{U}_{l,i}) dV = 0 \tag{9.104}$$

Expanding Eq. (9.104), the weighted residual equation of the SUPG method can be derived as follows:

$$\int_V (\mathbf{U}_k^* \dot{\mathbf{U}}_k) dV + \int_V (\mathbf{U}_k^* \mathbf{A}_{km}^i \mathbf{U}_{m,i}) dV$$
$$+ \tau_s \int_V (\mathbf{U}_{m,i}^* \mathbf{A}_{mk}^i \dot{\mathbf{U}}_k) dV + \tau_s \int_V (\mathbf{U}_{m,i}^* \mathbf{A}_{mk}^i \mathbf{A}_{kn}^j \mathbf{U}_{n,j}) dV = 0 \qquad (9.105)$$

In Eq. (9.105), the first two terms are those of the normal finite element weighted residual equation, the third term decays shortly in time, and the last term is the stabilization term in the SUPG method. For the integration of the third and fourth terms of Eq. (9.105), $\mathbf{A}_{mk}^i$ is assumed to be constant. Substituting Eqs. (9.98) and (9.102) into Eq. (9.105) and rearranging the terms, we get the following equation:

$$\mathbf{U}_{\alpha k}^* \int_V (\mathbf{\Phi}_\alpha \mathbf{\Phi}_\beta) dV \dot{\mathbf{U}}_{\beta k} + \mathbf{U}_{\alpha k}^* \int_V (\mathbf{\Phi}_\alpha \mathbf{A}_{kl}^i \mathbf{\Phi}_{\beta,i}) dV \mathbf{U}_{\beta l}$$
$$+ \tau_s \mathbf{U}_{\alpha k}^* \int_V (\mathbf{\Phi}_{\alpha,i} \mathbf{A}_{kl}^i \mathbf{\Phi}_\beta) dV \dot{\mathbf{U}}_{\beta l} + \tau_s \mathbf{U}_{\alpha k}^* \int_V (\mathbf{\Phi}_{\alpha,i} \mathbf{A}_{km}^i \mathbf{A}_{ml}^j \mathbf{\Phi}_{\beta,j}) dV \mathbf{U}_{\beta l} = 0 \qquad (9.106)$$

Considering that $\mathbf{U}_{\alpha k}^*$ are arbitrary constants, the local form of the SUPG finite element method can be derived as follows:

$$\mathbf{M}_{\alpha\beta} \dot{\mathbf{U}}_{\beta k} + \mathbf{K}_{\alpha kl\beta} \mathbf{U}_{\beta l} + \tau_s (\mathbf{C}_{\alpha kl\beta} \dot{\mathbf{U}}_{\beta l} + \mathbf{D}_{\alpha kl\beta} \mathbf{U}_{\beta l}) = 0 \qquad (9.107)$$

where

$$\mathbf{M}_{\alpha\beta} = \int_V (\mathbf{\Phi}_\alpha \mathbf{\Phi}_\beta) dV, \qquad (9.108)$$

$$\mathbf{K}_{\alpha kl\beta} = \int_V (\mathbf{\Phi}_\alpha \mathbf{A}_{kl}^i \mathbf{\Phi}_{\beta,i}) dV \qquad (9.109)$$

$$\mathbf{C}_{\alpha kl\beta} = \int_V (\mathbf{\Phi}_{\alpha,i} \mathbf{A}_{kl}^i \mathbf{\Phi}_\beta) dV \qquad (9.110)$$

$$\mathbf{D}_{\alpha kl\beta} = \int_V (\mathbf{\Phi}_{\alpha,i} \mathbf{A}_{km}^i \mathbf{A}_{ml}^j \mathbf{\Phi}_{\beta,j}) dV \qquad (9.111)$$

Superposition of Eq. (9.107) over the whole flow field leads to the global form of the SUPG finite element method. Notice that the last two terms in Eq. (9.107) are not continuous because we assume $\mathbf{A}_{kl}^i$ is constant over an element for integration. We compute these terms including $\mathbf{A}_{kl}^i$ elementwise and add the resulted terms in Eq. (9.107). For the integration in time, simple scheme, such as the explicit scheme, the implicit scheme, or the Crank–Nicolson scheme, etc. is adaptable.

9.6.3 Computational Forms of SUPG Finite Element Method

In this section, we present the SUPG finite element method based on Eq. (9.92) to obtain the computational forms of Eq. (9.107). The weighting function Eq. (9.101) can be written as follows:

$$\begin{bmatrix} \tilde{\rho}^* \\ \tilde{m}_k^* \end{bmatrix} = \begin{bmatrix} \rho^* \\ m_k^* \end{bmatrix} + \tau_s \begin{bmatrix} 0 & b_l^i \\ \delta_k^i & b_{lk}^i \end{bmatrix} \begin{bmatrix} \rho^* \\ m_l^* \end{bmatrix}_{,i} \tag{9.112}$$

where $\tilde{\rho}^*$ and $\tilde{m}_k^*$ are weighting functions of the SUPG method, and ρ^* and m_k^* are the normal weighting functions based on the linear interpolation function. The weighted residual equation is formulated using Eqs. (9.84), (9.85) and (9.112) as:

$$\int_{V_e} [\tilde{\rho}^* \ \tilde{m}_k^*] \begin{bmatrix} \mathfrak{A} \\ \mathfrak{B}_k \end{bmatrix} dV = 0 \tag{9.113}$$

where V_e is the volume of a finite element. Substituting Eq. (9.112) into Eq. (9.113) and arranging the terms, we get:

$$\int_{V_e} \left([\rho^* \ m_k^*] \begin{bmatrix} \mathfrak{A} \\ \mathfrak{B}_k \end{bmatrix} \right) dV + \tau_s \int_{V_e} \left([\rho^*_{,i} \ m^*_{l,i}] \begin{bmatrix} 0 & \delta_k^i \\ b_l^i & b_{lk}^i \end{bmatrix} \begin{bmatrix} \mathfrak{A} \\ \mathfrak{B}_k \end{bmatrix} \right) dV = 0 \tag{9.114}$$

Further, we get the following equations by expanding Eq. (9.114):

$$\int_{V_e} (\rho^* \mathfrak{A}) dV + \tau_s \int_{V_e} (\rho^*_{,i} \delta_l^i \mathfrak{B}_l) dV = 0 \tag{9.115}$$

$$\int_{V_e} (m_k^* \mathfrak{B}_k) dV + \tau_s \int_{V_e} (m^*_{k,i} b_k^i \mathfrak{A}) dV + \tau_s \int_{V_e} (m^*_{k,i} b_{kl}^i \mathfrak{B}_l) dV = 0 \tag{9.116}$$

Substituting Eqs. (9.84) and (9.85) into Eqs. (9.115) and (9.116), and superimposing the resulted equations into the whole flow field, the final form of the weighted residual equations are obtained:

$$\int_V \rho^* \left(\frac{\partial \rho}{\partial t} + m_{j,j} \right) dV + \tau_s \sum_{e=1}^{N_e} \int_{V_e} \rho^*_{,i} \left\{ \frac{\partial m_i}{\partial t} + \left(\frac{m_i m_j}{\rho} \right)_{,j} + p_{,i} \right\} dV = 0 \tag{9.117}$$

$$
\begin{aligned}
&\int_V m_k^* \left\{ \frac{\partial m_k}{\partial t} + \left(\frac{m_k m_j}{\rho} \right)_{,j} \right\} dV + \int_V (-m_{k,k}^* p + m_{k,j}^* \tau_{kj}) dV \\
&+ \tau_s \sum_{e=1}^{N_e} \int_{V_e} m_{k,i}^* b_k^i \left(\frac{\partial \rho}{\partial t} + m_{j,j} \right) dV \\
&+ \tau_s \sum_{e=1}^{N_e} \int_{V_e} m_{k,i}^* b_{kl}^i \left\{ \frac{\partial m_l}{\partial t} + \left(\frac{m_l m_j}{\rho} \right)_{,j} + p_{,l} \right\} dV \\
&= \int_S m_k^* (-p\delta_{kj} + \tau_{kj}) n_j dS
\end{aligned}
\tag{9.118}
$$

The summation symbol in Eqs. (9.117) and (9.118) means that integrated values over each element are summed up from element 1 to element N_e, in which N_e is the total number of elements. This is because the integrated values are discontinuous on the side of the element, assuming that b_k^i and b_{kl}^i are constant in each element. In Eq. (9.118), viscous stress terms are taken into consideration.

Let the weighting and interpolation equations for density ρ and momentum m_k be Eqs. (9.98) and (9.102), and assume the same interpolation equations for pressure p and velocity v_i. Then we get the following SUPG finite element equations:

$$
M_{\alpha\beta} \dot{\rho}_\beta + N_{\alpha\beta j} m_{\beta j} + \mathfrak{A}_\alpha = 0 \tag{9.119}
$$

$$
M_{\alpha\beta} \dot{m}_{\beta k} + (G_{\alpha\beta j\gamma} + G_{\alpha\gamma j\beta}) v_{\beta k} m_{\gamma j} + H_{\alpha k\beta} p_\beta + E_{\alpha k} + \mathfrak{B}_{\alpha k} = \mathfrak{L}_{\alpha k} \tag{9.120}
$$

where $\mathfrak{A}_\alpha$ and $\mathfrak{B}_{\alpha k}$ are the SUPG stabilization terms, and those are expressed in the following forms:

$$
\mathfrak{A}_\alpha = \tau_s \sum_{e=1}^{N_e} \{ A_{\alpha i\beta} \dot{m}_{\beta i} + (B_{\alpha i\beta j\gamma} + B_{\alpha i\gamma j\beta}) v_{\beta i} m_{\gamma j} + C_{\alpha\beta} p_\beta \} \tag{9.121}
$$

$$
\begin{aligned}
\mathfrak{B}_{\alpha k} = \tau_s \sum_{e=1}^{N_e} \{ & D_{\alpha kl\beta} \dot{m}_{\beta l} + (K_{\alpha kl\beta j\gamma} + K_{\alpha kl\gamma j\beta}) v_{\beta l} m_{\gamma j} \\
& + O_{\alpha k\beta} p_\beta + P_{\alpha k\beta} \dot{\rho}_\beta + Q_{\alpha k\beta j} m_{\beta j} \}
\end{aligned}
\tag{9.122}
$$

where

$$M_{\alpha\beta} = \int_V (\Phi_\alpha \Phi_\beta) dV, \tag{9.123}$$

$$N_{\alpha\beta j} = \int_V (\Phi_\alpha \Phi_{\beta,j}) dV \tag{9.124}$$

$$G_{\alpha\beta j\gamma} = \int_V (\Phi_\alpha \Phi_{\beta,j} \Phi_\gamma) dV, \tag{9.125}$$

$$H_{\alpha k\beta} = -\int_V (\Phi_{\alpha,k} \Phi_\beta) dV \tag{9.126}$$

$$E_{\alpha k} = \int_V (\Phi_{\alpha,j} \tau_{kj}) dV, \tag{9.127}$$

$$\mathfrak{L}_{\alpha k} = \int_S (\Phi_\alpha t_k) dS \tag{9.128}$$

$$A_{\alpha i\beta} = \int_{V_e} (\Phi_{\alpha,i} \Phi_\beta) \tag{9.129}$$

$$B_{\alpha i\beta j\gamma} = \int_{V_e} (\Phi_{\alpha,i} \Phi_{\beta,j} \Phi_\gamma) dV \tag{9.130}$$

$$C_{\alpha\beta} = \int_{V_e} (\Phi_{\alpha,i} \Phi_{\beta,i}) dV \tag{9.131}$$

$$D_{\alpha kl\beta} = \int_{V_e} (\Phi_{\alpha,i} b^i_{kl} \Phi_\beta) dV \tag{9.132}$$

$$K_{\alpha kl\beta j\gamma} = \int_{V_e} (\Phi_{\alpha,i} b^i_{kl} \Phi_{\beta,j} \Phi_\gamma) dV \tag{9.133}$$

$$O_{\alpha k\beta} = \int_{V_e} (\Phi_{\alpha,i} b^i_{kl} \Phi_{\beta,l}) dV \tag{9.134}$$

$$P_{\alpha k\beta} = \int_{V_e} (\Phi_{\alpha,i} b^i_k \Phi_\beta) dV, \tag{9.135}$$

$$Q_{\alpha k\beta j} = \int_{V_e} (\Phi_{\alpha,i} b^i_k \Phi_{\beta,j}) dV \tag{9.136}$$

For the discretization in time, we use:

$$\dot{\rho}_\beta \fallingdotseq \frac{1}{\Delta t}(\rho^{n+1}_\beta - \rho^n_\beta) \tag{9.137}$$

$$m_{\beta k} \fallingdotseq \frac{1}{\Delta t}(m^{n+1}_{\beta k} - m^n_{\beta k}) \tag{9.138}$$

where ρ^n_β and $m^n_{\beta k}$ means the values at time n. For the SUPG parameter:

$$\tau_s = \frac{\Delta t}{2} \tag{9.139}$$

is usually employed. The shock capturing concept (Tezduyar and Senga 2006) can also be adapted for the computation of shock waves.

9.6.4 Comparison of Pressure Coefficients

Comparative studies of the pressure coefficients around the surface of a circular cylinder are conducted comparing with experimental results (Okauchi et al. 1977). The experiments are carried out using a two-dimensional cylinder made of wood located in the air tunnel. Two-dimensional computations are carried out assuming that the equation of state is Eq. (9.8) with $\gamma = 1.4$. The Reynolds number is $Re = 1.0 \times 10^5$. The boundary conditions used are in Fig. 9.11. The finite element mesh used is in Fig. 9.12. The pressure coefficient is defined as:

$$C_p = \frac{p - p_0}{\frac{1}{2}\rho_0 \mathcal{V}^2 D}$$

where p is the computed pressure, p_0, ρ_0, $\mathcal{V}$ are fundamental values, and D is the diameter of the cylinder. As the initial condition, a free stream with $\rho_0 = 1.0, v_1 = 1.0$ and $v_2 = 0.0$ are assumed. Four types of computational methods are implemented. One is the flow analysis assuming incompressible fluid based on the standard bubble function interpolation, precisely described in Sect. 8.5. We will refer to the method as the "incompressible flow method." The second is the present method, which is referred to as the "adiabatic flow method." The Mach number $M_c = 0.5$ is used. The third is the acoustic velocity method, which will be described in Sect. 9.7. The acoustic velocity $\mathfrak{a}$ is assumed to be constant and Mach number to be $M_c = 0.5$. For the discritization in time, the SUPG method with the Euler scheme is employed. The last is the compressible flow method based on the SUPG finite element method, which will be described in Sect. 10.5.

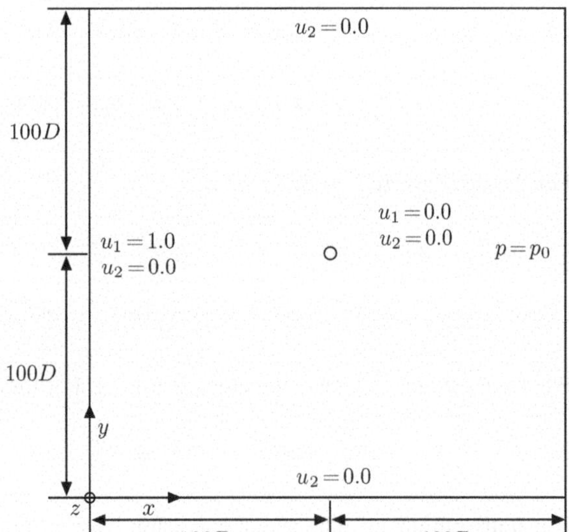

Fig. 9.11 Computational domain and boundary conditions

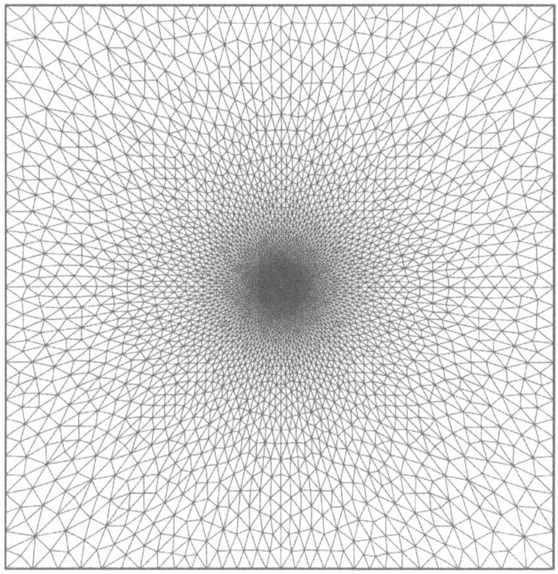

Fig. 9.12 Finite element mesh

The computational results are plotted in Fig. 9.13. The position on the surface is expressed by the angle θ. The tendency of the computational pressure coefficients by four methods are generally comparable to the actual experimental results. Especially, the pressure coefficients obtained both by the compressible flow method and by the adiabatic flow method are similar to each other and also to the experiments. The computational results obtained by the incompressible flow method and those obtained by the acoustic velocity method are also similar to each other. However, these show aspects in which the computational results diverge from the experimental results, which is particularly visible when $80° < \theta < 180°$. This is because $M_c = 0.5$ is not appropriate to compute the incompressible flows. The first two methods employ the variable acoustic velocity, whereas the last two methods utilize the constant acoustic velocity. It seems that the pressure coefficients around the surface of the cylinder should be computed, while taking density variability of the fluid into consideration. The incompressible flow method is not suitable for the computation of the fluid flow, whose density is variable, such as airflow.

9.7 Acoustic Velocity Finite Element Method

In the previous section we have presented the computational method, which employs the density and velocity as field variables. We have shown that the computation can be carried out based on the simple explicit Euler method by the trial computation. However, the computational stability is not satisfactory because the differences of orders of numerals among density, velocity and pressure are so different that the

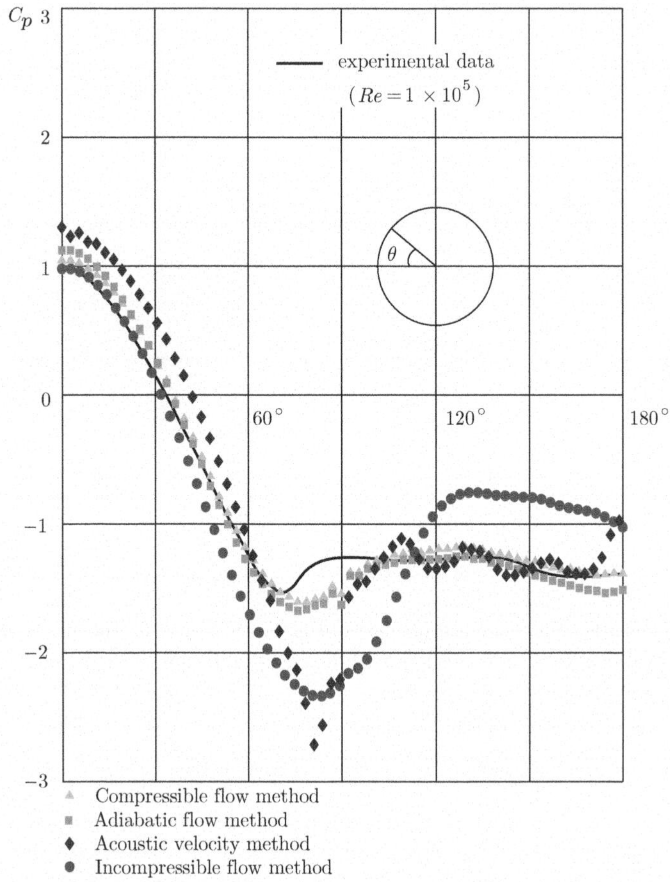

Fig. 9.13 Comparison of C_p

computation tends to be unstable. Thus, in this section, a more stable method is presented. This is originally presented in Kawahara and Hirano (1983a, 1983b), Kawahara and Miwa (1984), Kawahara et al. (1984).

One of the distinct features of adiabatic flows is that pressure is a function of density only. Thus, the conservation equation of mass can be transformed into the form of pressure and velocity including acoustic velocity. The present form corresponds to the equation of continuity of the incompressible flow penalized by pressure. Namely, we can derive the equation of continuity as a function of pressure and velocity by the physical consideration. If we let the acoustic velocity approach infinity, we get the normal equation of continuity of the incompressible fluid flows. In addition, we introduce the nondimensional forms to obtain the stable computation.

9.7.1 Governing Equation of Acoustic Velocity Method

Assuming that pressure is a function of density, i.e. Eq. (9.7), the material differentiation with respect to time leads to the following equation:

$$\frac{Dp}{Dt} = \frac{\partial p}{\partial \rho}\frac{D\rho}{Dt} \tag{9.140}$$

Substituting Eq. (9.1) into Eq. (9.140) and arranging the terms, the following equation can be obtained:

$$\dot{p} + v_i p_{,i} + \rho \mathfrak{a}^2 v_{i,i} = 0 \tag{9.141}$$

where superscripted $\dot{}$ means partial differentiation with respect to time. In Eq. (9.141), $\mathfrak{a}$ is acoustic velocity expressed in Eq. (7.161), and we get the equation of continuity of the incompressible flows, i.e. Eq. (8.1), to let $\mathfrak{a}$ be equal to infinity. Equation of motion can be written in the following form using Eqs. (9.2)–(9.4):

$$\rho(\dot{v}_i + v_j v_{i,j}) + p_{,i} - \lambda v_{k,ki} - \mu(v_{i,j} + v_{j,i})_{,j} = \rho f_i \tag{9.142}$$

Assuming that acoustic velocity $\mathfrak{a}$ and density ρ are constant, during time increment Δt simplifies computation. We will show in the practical computation that the assumption is valid for the solutions close to the incompressible flows. Introducing nondimensional quantities such as:

$$\begin{aligned}
&\tilde{x}_i = \frac{x_i}{\mathfrak{L}}, \quad \tilde{v}_i = \frac{v_i}{\mathcal{V}}, \quad \tilde{t} = \frac{\mathcal{V}}{\mathfrak{L}}t, \\
&\tilde{p} = \frac{p}{\rho \mathfrak{a} \mathcal{V}}, \quad \mathrm{C} = \frac{\mathfrak{a}}{\mathcal{V}} = \frac{1}{M_c}, \\
&\tilde{\lambda} = \frac{\lambda}{\rho \mathcal{V} \mathfrak{L}}, \quad \tilde{\mu} = \frac{\mu}{\rho \mathcal{V} \mathfrak{L}} = \frac{1}{R_e}, \\
&\tilde{f}_i = \frac{\mathfrak{L}}{\mathcal{V}^2} f_i
\end{aligned} \tag{9.143}$$

the nondimensional equations of Eqs. (9.141) and (9.142) can be expressed as follows:

$$\dot{p} + v_i p_{,i} + \mathrm{C} v_{i,i} = 0 \tag{9.144}$$

$$\dot{v}_i + v_j v_{i,j} + \mathrm{C} p_{,i} - \lambda v_{k,ki} - \mu(v_{i,j} + v_{j,i})_{,j} = \rho f_i \tag{9.145}$$

where we omit the superscripted $\tilde{}$. In Eq. (9.143), we denote $\mathfrak{L}$ and $\mathcal{V}$ as fundamental length and velocity, respectively, and M_c and R_e are Mach number and Reynolds number, respectively. We can impose the appropriate boundary and initial conditions for the nondimensional equations Eqs. (9.144) and (9.145).

9.7.2 Acoustic Velocity Finite Element Method

Let p^* and v_i^* be the weighting functions for pressure and velocity, respectively, which are arbitrary functions. Multiplying both sides of Eqs. (9.144) and (9.145), and integrating over the volume V surrounded by the surface S, the following weighted residual equations are obtained:

$$\int_V (p^* \dot{p}) dV + \int_V (p^* v_i p_{,i}) dV + \mathrm{c} \int_V (p^* v_{i,i}) dV = 0 \tag{9.146}$$

$$\begin{aligned} &\int_V (v_i^* \dot{v}_i) dV + \int_V (v_i^* v_j v_{i,j}) dV - \mathrm{c} \int_V (v_{i,i}^* p) dV \\ &\quad + \lambda \int_V (v_{i,i}^* v_{k,k}) dV + \mu \int_V (v_{i,j}^* v_{i,j}) dV + \mu \int_V (v_{i,j}^* v_{j,i}) dV \\ &= \int_S (v_i^* t_i) dS + \int_V (v_i^* f_i) dV \end{aligned} \tag{9.147}$$

For the interpolation functions for velocity and pressure, the following functions are used:

$$v_i = \Phi_\alpha v_{\alpha i} \tag{9.148}$$

$$p = \Phi_\alpha p_\alpha \tag{9.149}$$

where Φ_α is the linear shape function which is in Eq. (4.24), and $v_{\alpha i}$ and p_α are velocity at node α in the ith direction and pressure at node α. For the weighting functions for velocity and pressure, the following functions are used:

$$v_i^* = \Phi_\alpha v_{\alpha i}^* \tag{9.150}$$

$$p^* = \Phi_\alpha p_\alpha^* \tag{9.151}$$

where Φ_α is the linear shape function which is in Eq. (4.25), and $v_{\alpha i}^*$ and p_α^* are nodal values of the weighting functions. We can also use normal interpolation as shown in Eqs. (9.148) and (9.149), i.e., the same shape functions for velocity and pressure.

Substituting Eqs. (9.148)–(9.151) into Eqs. (9.146) and (9.147), and rearranging the terms, we get:

$$M_{\alpha\beta} \dot{p}_\beta + K_{\alpha\beta\gamma j} v_{\beta j} p_\gamma + \mathrm{c} H_{\beta i \alpha} v_{\beta i} = 0 \tag{9.152}$$

$$M_{\alpha i \beta j} \dot{v}_{\beta j} + K_{\alpha\beta\gamma j} v_{\beta j} v_{\gamma i} - \mathrm{c} H_{\alpha i \beta} p_\beta + S_{\alpha i \beta j} v_{\beta j} = \Omega_{\alpha i} \tag{9.153}$$

where

$$M_{\alpha\beta} = \int_V (\Phi_\alpha \Phi_\beta) dV \tag{9.154}$$

$$M_{\alpha i\beta j} = \int_V (\Phi_\alpha \Phi_\beta)\delta_{ij} dV \tag{9.155}$$

$$K_{\alpha\beta\gamma j} = \int_V (\Phi_\alpha \Phi_\beta \Phi_{\gamma,j}) dV \tag{9.156}$$

$$H_{\alpha i\beta} = \int_V (\Phi_{\alpha,i} \Phi_\beta) dV \tag{9.157}$$

$$S_{\alpha i\beta j} = \lambda \int_V (\Phi_{\alpha,i} \Phi_{\beta,j}) dV + \mu \int_V (\Phi_{\alpha,k} \Phi_{\beta,k})\delta_{ij} dV + \mu \int_V (\Phi_{\alpha,j}, \Phi_{\beta,i}) dV \tag{9.158}$$

$$\Omega_{\alpha i} = \int_S (\Phi_\alpha t_i) dS + \int_V (\Phi_\alpha \Phi_\beta) dV f_{\beta i} \tag{9.159}$$

where $f_{\beta i}$ is body force at node β in the ith direction. Coefficient matrices shown in Eqs. (9.154)–(9.159) can easily be integrated using linear shape function in Eq. (4.25) and formula in Eq. (6.36). Equations (9.152) and (9.153) can be written in the matrix form as:

$$\begin{bmatrix} M_{\alpha i\beta j} & \\ & M_{\alpha\beta} \end{bmatrix} \begin{bmatrix} \dot{v}_{\beta j} \\ \dot{p}_\beta \end{bmatrix} + \begin{bmatrix} K_{\alpha\beta\gamma j} v_{\gamma i} + S_{\alpha i\beta j} & -CH_{\alpha i\beta} \\ CH_{\beta j\alpha} & K_{\alpha\gamma\beta k} v_{\beta k} \end{bmatrix} \begin{bmatrix} v_{\beta j} \\ p_\beta \end{bmatrix} = \begin{bmatrix} \Omega_{\alpha i} \\ \end{bmatrix} \tag{9.160}$$

The coefficient matrix of the second term in Eq. (9.160) is full of elements. Namely, equation of continuity of the incompressible flow is the function of velocity only. On the contrary, Eq. (9.160) is the extended equation, which includes the functions of velocity and pressure under the physical consideration. Then, we can execute the stable computation.

9.7.3 Integration in Time

The global form of the acoustic velocity finite element method can be represented as follows based on Eq. (9.160):

$$\mathbf{M}_{\alpha\beta} \dot{\mathbf{v}}_\beta + \mathbf{K}_{\alpha\beta}(\mathbf{v}_\gamma)\mathbf{v}_\beta = \mathbf{\Omega}_\alpha \tag{9.161}$$

where $\mathbf{M}_{\alpha\beta}$ and $\mathbf{K}_{\alpha\beta}(\mathbf{v}_\gamma)$ are coefficient matrices in Eq. (9.160) and:

$$\mathbf{v}_\beta = \begin{bmatrix} v_{\beta j} \\ p_\beta \end{bmatrix}, \quad \mathbf{\Omega}_\alpha = \begin{bmatrix} \Omega_{\alpha i} \\ \end{bmatrix} \tag{9.162}$$

In order to integrate Eq. (9.161) in time, Kawahara and Miwa [1984] used the two-step explicit method, whose algorithm is as follows:

For the first step:

$$\bar{\mathbf{M}}_{\alpha\beta}\mathbf{v}_{\beta}^{n+1/2} = \tilde{\mathbf{M}}_{\alpha\beta}\mathbf{v}_{\beta}^{n} - \frac{\Delta t}{2}\mathbf{K}_{\alpha\beta}(\mathbf{v}_{\gamma}^{n})\mathbf{v}_{\beta}^{n} + \frac{\Delta t}{2}\mathbf{\Omega}_{\alpha}^{n} \tag{9.163}$$

and for the second step:

$$\bar{\mathbf{M}}_{\alpha\beta}\mathbf{v}_{\beta}^{n+1} = \tilde{\mathbf{M}}_{\alpha\beta}\mathbf{v}_{\beta}^{n} - \Delta t\mathbf{K}_{\alpha\beta}(\mathbf{v}^{n+1/2})\mathbf{v}_{\beta}^{n+1/2} + \Delta t\mathbf{\Omega}_{\alpha}^{n} \tag{9.164}$$

where $\bar{\mathbf{M}}_{\alpha\beta}$ means lumped coefficient matrix of $\mathbf{M}_{\alpha\beta}$ and:

$$\tilde{\mathbf{M}}_{\alpha\beta} = e\bar{\mathbf{M}}_{\alpha\beta} + (1-e)\mathbf{M}_{\alpha\beta} \tag{9.165}$$

In Eq. (9.165), e is used as the selective lumping parameter that is used for reducing the numerical damping effect and adjusting the numerical stability. Numerical results of the solitary wave propagation are found in Kawahara and Miwa (1984).

The bubble function finite element method can also be adapted to integrate Eq. (9.161) in time. The mixed interpolation based on the standard bubble function for velocity and linear function for pressure is used. The global form of the finite element equation can be obtained by adding the artificial viscosity on the coefficients on the barycentric nodes as:

$$\mathbf{M}_{\alpha\beta}\dot{\mathbf{v}}_{\beta} + \mathbf{K}_{\alpha\beta}(\mathbf{v}_{\gamma})\mathbf{v}_{\beta} + \sum_{m=1}^{N_e}(\mu+\mu')\mathbf{a}\mathbf{v}_4 = \mathbf{\Omega}_{\alpha} \tag{9.166}$$

where μ' is the artificial viscosity and $\mathbf{a}$ is in Eq. (8.103). Equation (9.166) has the same form as in Eq. (8.100). However, those are fundamentally different, because the coefficient matrices in Eq. (9.160) are full matrices and explicit type numerical integration can be employed.

Uchiyama et al. (2015) solved the solitary wave propagation using the SUPG method based on the acoustic velocity formulation. Equations (9.144) and (9.145) can be written in the following form:

$$\begin{bmatrix}\mathfrak{D}\\ \mathfrak{E}_k\end{bmatrix} = \begin{bmatrix}\dot{p}\\ \dot{v}_k\end{bmatrix} + \begin{bmatrix}v_i & c\delta_l^i\\ c\delta_k^i & v_i\delta_{kl}\end{bmatrix}\begin{bmatrix}p\\ v_l\end{bmatrix}_{,i} + \begin{bmatrix}\\ \tau_{kj,j}+f_k\end{bmatrix} \tag{9.167}$$

Based on Eq (9.167), the weighting function of the SUGP method can be expressed as follows:

$$\begin{bmatrix}\tilde{p}^* & \tilde{v}_k^*\end{bmatrix} = \begin{bmatrix}p^* & v_k^*\end{bmatrix} + \begin{bmatrix}\tau_p p_{,i}^* & \tau_v v_{l,i}^*\end{bmatrix}\begin{bmatrix}v_i & c\delta_k^i\\ c\delta_l^i & v_i\delta_{lk}\end{bmatrix} \tag{9.168}$$

where $\tilde{p}^*$ and $\tilde{v}_k^*$ are the weighting functions of the SUPG method, p^* and v_k^* are the linear weighting functions, and τ_p and τ_v are the stabilization parameters. The weighted residual equations can be described as:

$$\int_V \begin{bmatrix} \tilde{p}^* & \tilde{v}_k^* \end{bmatrix} \begin{bmatrix} \mathfrak{D} \\ \mathfrak{E}_k \end{bmatrix} dV = 0 \tag{9.169}$$

Substituting Eqs. (9.167) and (9.168) into Eq. (9.169), and arranging the terms, we get the following two equations:

$$\int_V (p^*\mathfrak{D})dV + \tau_p \int_V (p^*_{,i} v_i \mathfrak{D})dV + \tau_p \int_V (p^*_{,k} \mathrm{C} \mathfrak{E}_k)dV = 0 \tag{9.170}$$

$$\int_V (v_k^* \mathfrak{E}_k)dV + \tau_v \int_V (v^*_{i,i} \mathrm{C} \mathfrak{D})dV + \tau_v \int_V (v^*_{k,i} v_i \mathfrak{E}_k)dV = 0 \tag{9.171}$$

Based on Eqs. (9.170) and (9.171), we get the finite element equations of the acoustic velocity method based on the SUPG formulation. Using the appropriate values for τ_p and τ_v, we can carry out the stable computation. For the computation of the solitary wave propagation shown in Fig. 9.16, we use:

$$\tau_p = \tau_v \tag{9.172}$$

$$\tau_v = \Delta t/2 \tag{9.173}$$

The present method is characterized by the point by which the SUPG terms can be obtained naturally in Eqs. (9.170) and (9.171). This is because Eqs. (9.144) is the form, which includes the terms of pressure and velocity. Therefore, we can formulate the SUPG method in a straightforward manner. By contrast, the orders of numerals are so different that we must use the different orders of the SUPG parameters τ_p and τ_v in general to secure the computational stability. Eqs. (9.170) and (9.171) can be further developed as follows:

$$\begin{aligned}
&\int_V (p^*\dot{p})dV + \int_V (p^* v_i p_{,i})dV + \int_V (p^* \mathrm{C} v_{i,i})dV \\
&+ \tau_p \int_V (p^*_{,i} v_i \dot{p})dV + \tau_p \int_V (p^*_{,i} v_i v_k p_{,k})dV + \tau_p \int_V (p^*_{,i} v_i \mathrm{C} v_{k,k})dV \\
&+ \tau_p \int_V (p^*_{,i} \mathrm{C} \dot{v}_i)dV + \tau_p \int_V (p^*_{,i} \mathrm{C}^2 p_{,i})dV + \tau_p \int_V (p^*_{,i} \mathrm{C} v_k v_{i,k})dV = 0
\end{aligned} \tag{9.174}$$

$$
\begin{aligned}
&\int_V (v_k^* \dot{v}_k) dV + \int_V (v_k C p_{,k}) dV + \int_V (v_k^* v_i v_{k,i}) dV \\
&+ \tau_v \int_V (v_{i,i}^* C \dot{p}) dV + \tau_v \int_V (v_{i,i}^* C v_k p_{,k}) dV + \tau_v \int_V (v_{i,i}^* C^2 v_{k,k}) dV \\
&+ \tau_v \int_V (v_{k,i}^* v_i \dot{v}_k) dV + \tau_v \int_V (v_{k,i}^* v_i C p_{,k}) dV + \tau_v \int_V (v_{k,i}^* v_i v_j v_{k,j}) dV = 0
\end{aligned}
\tag{9.175}
$$

In Eq. (9.175), the viscosity terms and the body force term are set aside for the sake of simplicity. The fourth and seventh terms in Eq. (9.174) and fourth and seventh terms in Eq. (9.175) are abstracted away from to execute the practical computation.

The computation of pressure coefficient around a circular cylinder is conducted using the same boundary conditions in Fig. 9.11 and finite element mesh in Fig. 9.12. The computed results are shown in Fig. 9.13. The caption in the figure is the acoustic velocity method.

9.7.4 Pressure Around Two-Dimensional Cylinder

Pressure coefficient around a circular cylinder in Sect. 9.6.4 is computed by Terachi and Kawahara (2010) using the bubble function method: Eq. (9.166). The boundary configuration is shown in Fig. 9.11, and the finite element mesh is in Fig. 9.12. The computed pressure coefficient is shown in Fig. 9.13, whose caption is the acoustic velocity method. Fluid flow around the two-dimensional streamline shape cylinder is computed using the mesh shown in Fig. 9.14. The computed pressure at Reynolds

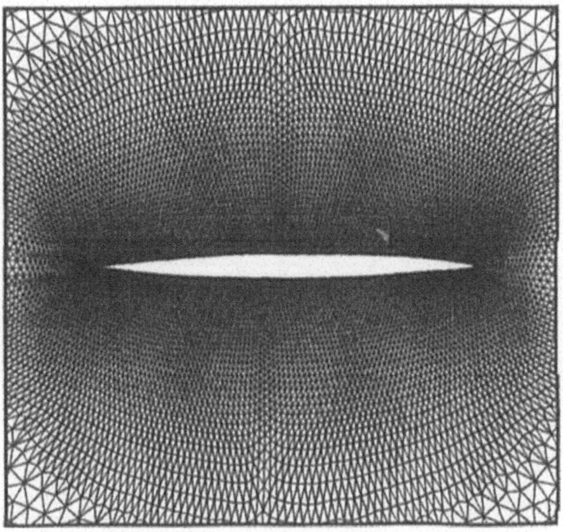

Fig. 9.14 finite element mesh

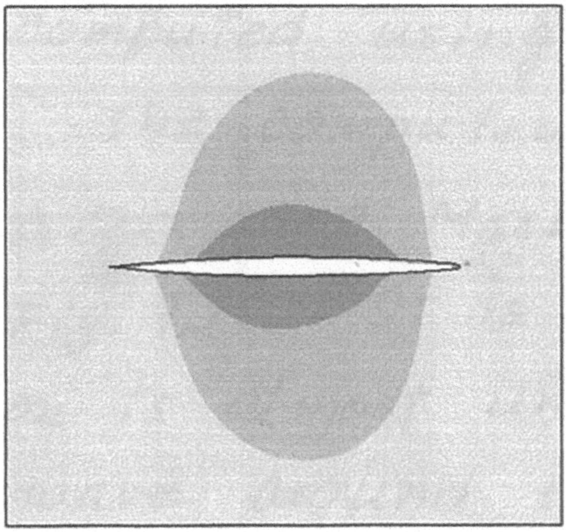

Fig. 9.15 computed pressure

number $Re = 50,000$ and Mach number $M_c = 0.005$ is shown in Fig. 9.15. We can observe that the pressure distribution is almost uniform and there is no shock wave around the cylinder.

9.8 Solitary Wave Propagation

9.8.1 Two-Dimensional Solitary Wave Computation

Uchiyama and Kawahara (2015) carried out the computation of the two-dimensional solitary wave propagation by the acoustic velocity method combined with the SUPG method. The solitary wave is shown in Fig. 9.16, and the configuration is expressed by the Laitone first-order formula as follows: The wave speed c is given as:

$$c = \sqrt{gd\left(1 + \frac{H}{d}\right)} \tag{9.176}$$

where g is the gravity acceleration, d is depth and H is wave height. The surface coordinate y_s is given as the function of x:

$$y_s = d + H\mathrm{sech}^2\left[\sqrt{\frac{3H}{4d^3}}(x - ct)\right] \tag{9.177}$$

and velocity u, v are expressed as follows:

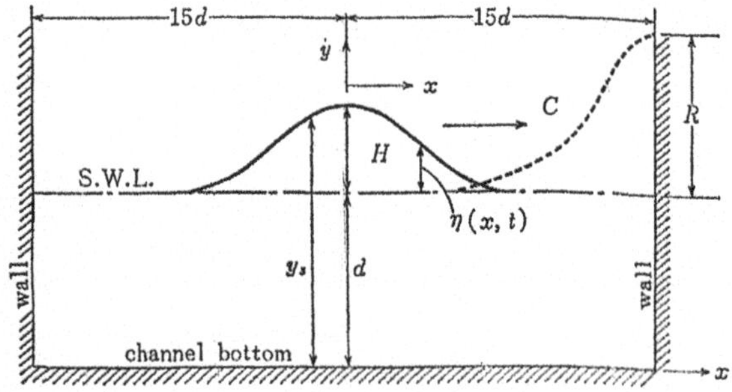

Fig. 9.16 solitary wave

$$u = \sqrt{gd}\frac{H}{d}\text{sech}^2\left[\sqrt{\frac{3}{4}\frac{H}{d^3}}(x - ct)\right]$$
$$v = \sqrt{3gd}\left(\frac{H}{d}\right)^{\frac{3}{2}}\frac{y}{d}\text{sech}^2\left[\sqrt{\frac{3}{4}\frac{H}{d}}(x - ct)\right]\tanh\left[\sqrt{\frac{3}{4}\frac{H}{d}}(x - ct)\right] \tag{9.178}$$

The initial configuration is represented in Fig. 9.17. The computational results are shown in Fig. 9.18a–c. We can see the movement of solitary wave in Fig. 9.19.

Mach number used in the computation is $M_c = 0.88 \times 10^{-3}$. The nonlinear terms, which are the second term in Eq. (9.174) and the third term in Eq. (9.175), are computed using the linearization, i.e., the convective velocity v_i in the equations is approximated as:

$$v_i = \frac{1}{2}(3v_i^n - v_i^{n-1}) \tag{9.179}$$

The movement of free surface is computed as:

$$x_s^{n+1} = x_s^n + \Delta t u_s^n \tag{9.180}$$
$$y_s^{n+1} = y_s^n + \Delta t v_s^n \tag{9.181}$$

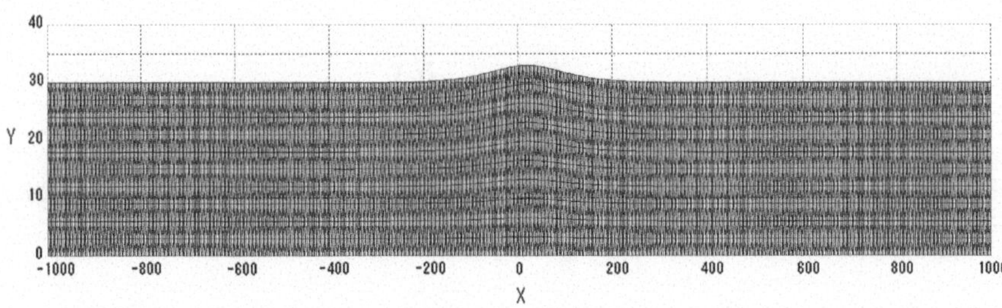

Fig. 9.17 Initial configuration

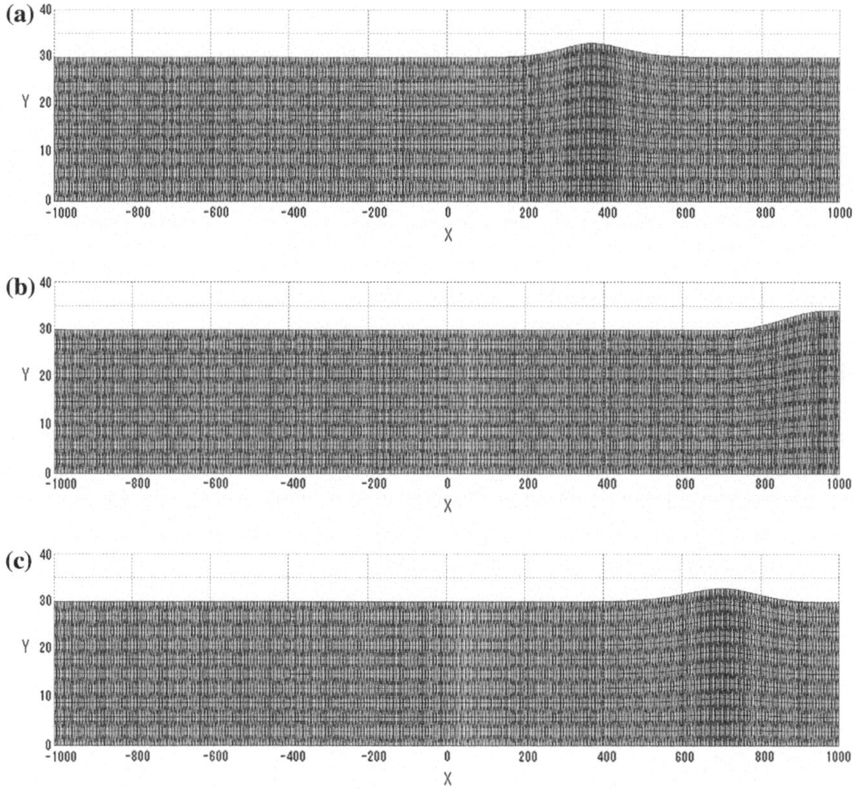

Fig. 9.18 **a** Configuration at $T = 20$. **b** Configuration at $T = 50$. **c** Configuration at $T = 70$

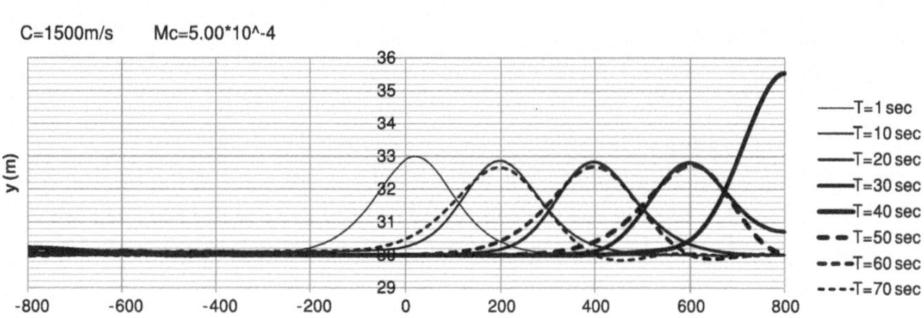

Fig. 9.19 Movement of solitary wave

where x_s and y_s denote the nodal positions and u_s and v_s are velocity on the surface. Fundamentally, the fluid particle positions on the streamline do not necessarily coincide with the positions expressed in Eqs. (9.180) and (9.181). We use Eqs. (9.180) and (9.181) to increase the Computational efficiency. The mesh is regenerated after the computation of time increment Δt based on the free surface points. The computation is implemented by the ALE formulation, which is described in Chap. 11.

9.8.2 Three-Dimensional Solitary Wave Computation

Maruoka et al. (2016) presented the computation of three-dimensional solitary wave propagation using the acoustic velocity method based on the linear interpolation function. The finite element is the trigonal pyramid. For the stabilization method, the SUPG method is used. For the integration in time, nonlinear iteration is introduced. The precise formulation is found in Maruoka et al. (2016). To trace the free surface, ALE formulation described in Chap. 11 is used. On the free surface, the velocity v_i and mesh velocity w_i must satisfy the relation:

$$v_i n_i = w_i n_i$$

where n_i is the unit normal of the free surface. The Newton–Raphson iteration is introduced to get the mesh velocity on the free surface. The mesh is moved only in the z direction.

The velification study is performed by comparing the present results and experiments by Street and Camfield (1966). The boundary conditions and initial configuration are shown in Fig. 9.20. On the sidewalls and bottom bed, the slip condition is used, which means that the normal velocity to the wall is zero. The total numbers of nodes and elements are $251 \times 11 \times 11 = 30,371$ and 125,000, respectively. The water depth is $d = 1\,\text{m}$, and the height of the initial wave $H = 0.12\,\text{m}$. The acoustic velocity $a = 1.48 \times 10^3\,\text{m/sec}$. The total channel length is 50.0 m. The time step is $\Delta t = 0.01$. Propagation of the solitary wave is computed and shown in Fig. 9.21a–d. In those figures, four times amplified wave height is shown. Fig. 9.21a shows the initial wave, Fig. 9.21b is the wave traveling forward, Fig. 9.21c is the wave reflected by the wall, and Fig. 9.21d is the wave traveling backward. The computed results

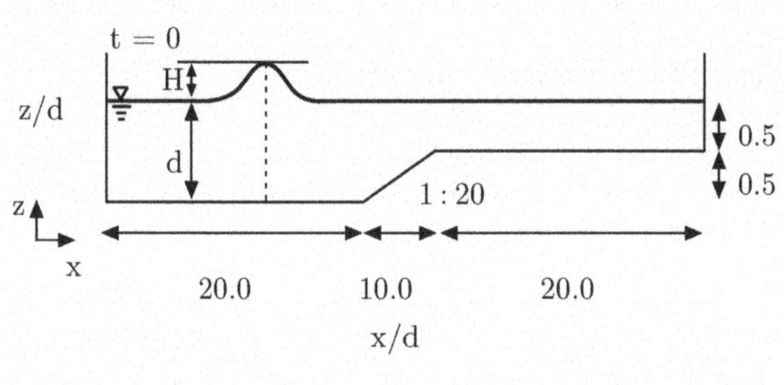

Fig. 9.20 Boundary condition and initial configuration

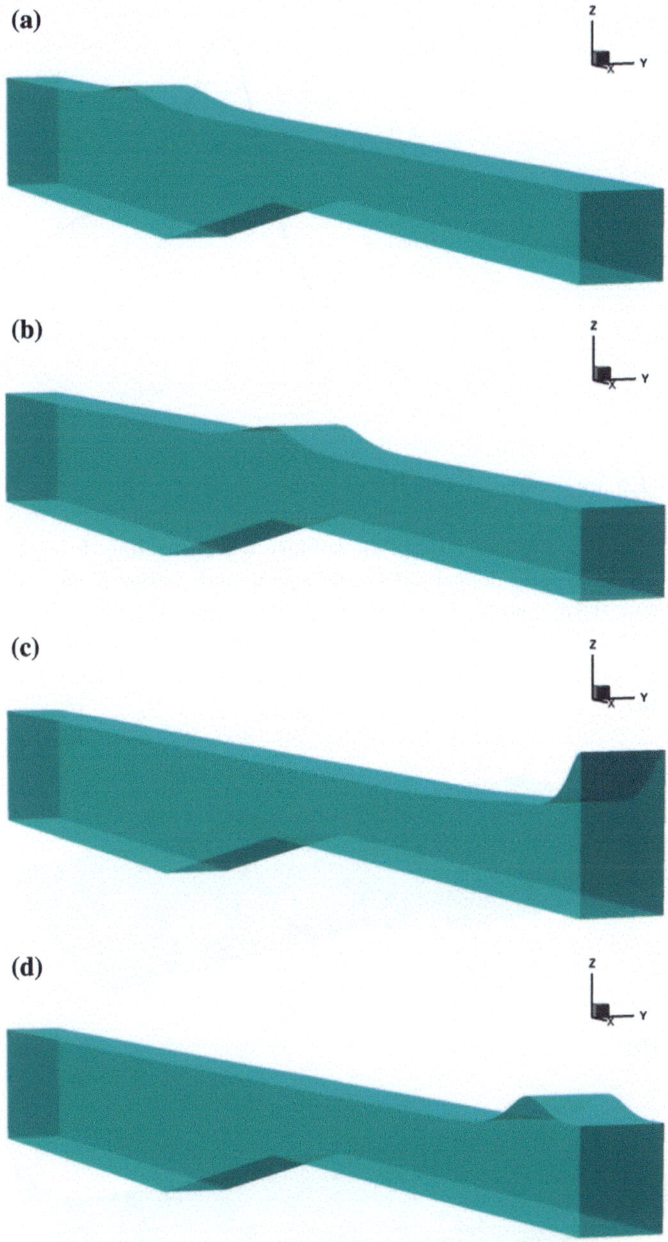

Fig. 9.21 **a** Propagation of solitary wave (initial configuration). **b** Propagation of solitary wave (forward traveling). **c** Propagation of solitary wave (reflection by wall). **d** Propagation of solitary wave (backward traveling)

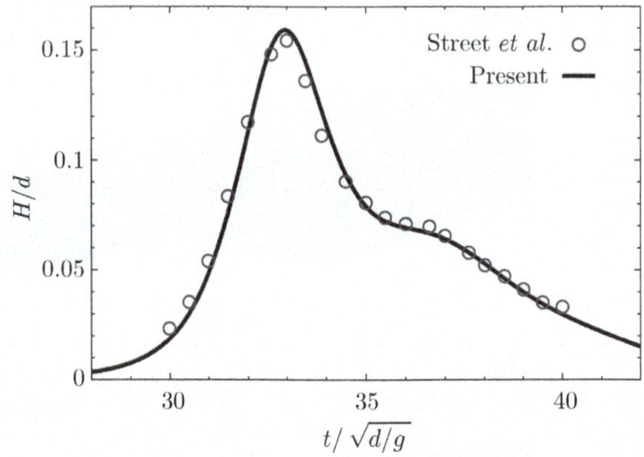

Fig. 9.22 Comparison with the experimental results

at $x/d = 41.6$ are compared with the experimental results obtained by Street and Camfield (1966) in Fig. 9.22. Close agreement between computed and experimental values are obtained. Figure 9.23a, b, the solitary wave traveling through the channel whose width, length, deeper side water depth, and shallower side water depth are 20 m, 50 m, 0.5 m, and 0.1 m, respectively, are shown. The initial wave height is

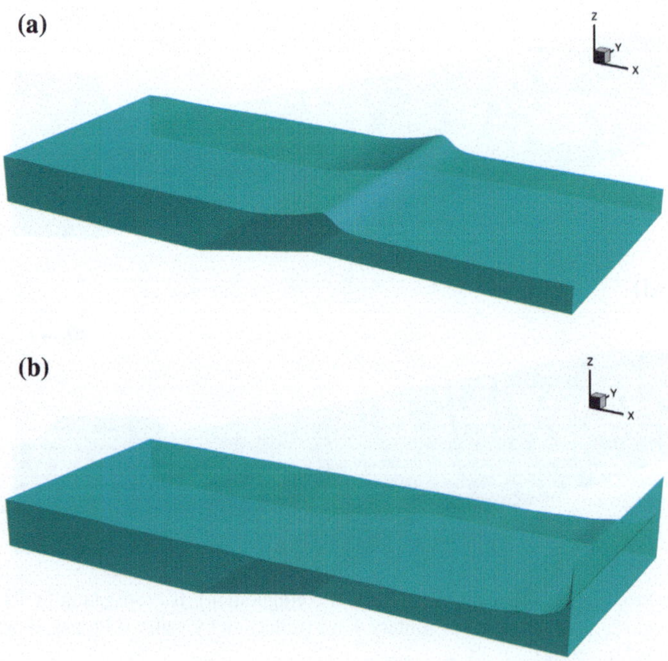

Fig. 9.23 **a** Solitary wave propagation (traveling forward). **b** Solitary wave propagation (reflected by wall)

(a)

(b)

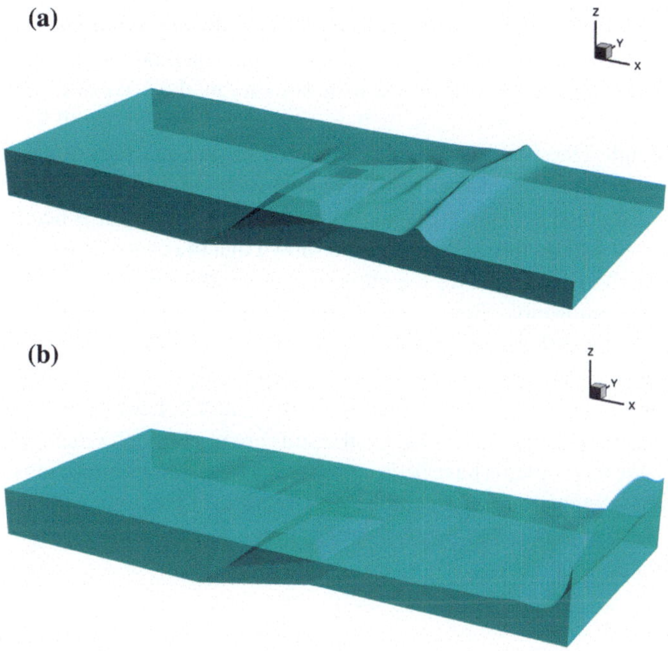

Fig. 9.24 Solitary wave propagation (traveling forward). **b** Solitary wave propagation (reflected by wall)

0.25 m. Fig. 9.24a, b show the solitary wave over the submerged prominent object on the bottom. The top area of the object is 4 m × 8 m. The water depth is 0.3 m. The slant is 1 : 4. It is understood that the wave is distorted through the submerged object. These results are useful for the design of the artificial reef, etc.

9.9 Summary and Conclusion

In this chapter, the finite element methods of the adiabatic flows have been presented. For the last three decades, the assumption about incompressible or compressible flows are commonly used in the computation. We presented the analysis of fluid flows intermediate between incompressible and compressible flows. The main assumption is on the equation of state, in which the pressure is a function of density only, and independent of temperature. Using conservation equation of mass, we can formulate equation of continuity including terms not only of velocity but also of pressure. Considering that the instability of the incompressible flow computation stems mainly from the fact that equation of continuity is a function of velocity only, the adiabatic flow assumption stabilizes the computation based on the physical consideration.

We formulate two types of solution methods with the assumption of variable and constant acoustic velocities. We call the former method as the adiabatic flow method and the latter method as the acoustic velocity method. We have shown by the numerical computations that those methods are useful for almost all kinds of fluid flows. Especially, the former method is available for the analyses of gases with the Poisson type of equation of state and of liquids with Birch–Murnagahan equation, and the latter method is to the analyses of liquids, whose behavior is nearly identical to incompressible fluid flows. We have also shown that the characteristic method and the SUPG method are useful for the numerical integration in time.

Two-phase immiscible flow with discontinuous interface has been computed. According to the difference of densities between two fluids, many kinds of distribution patterns have been computed. To empirically test the present methods, flows through circular cylinder located in the free stream have been computed and compared with experimental results. Both results are in good agreement. The solitary wave propagation problems have been computed. The stable computations have been implemented using the present method. We emphasize that the present methods are useful for the computation of incompressible flows and compressible flows without thermal effects.

Chapter 10
Analysis of Compressible Flows

10.1 Introduction

It is well known that the compressibility of fluids should be considered to solve high Mach number flows. Compressible flow simulation is important not only in the field of airplane engineering, in which high Mach number flows are treated, but also in the field of wind engineering, in which low Mach number flows are considered. There are a plenty of topics to be investigated, such as the flows around the girder of long span bridge, around high-rise building, around racing car, around bullet train, etc. When a fluid is compressible, the fluid density varies as a function of pressure and temperature. To express the behavior of compressible flow, the energy variation should be taken into consideration. The total energy consists of the internal energy and the kinetic energy. The conservation equation of energy should be treated in addition to the conservation equations of mass and momentum for the governing equations. The field variables employed are density, momentum, and total energy. To deal with the nonlinear boundary conditions, various sets of independent variables, such as combinations of enthalpy, temperature, entropy, and some others, are introduced by Elkadri E et al. (2000).

Many research studies in the finite element methods of the compressible flows have been presented: those include Hughes et al. (1986), Hughes and Mallet (1986), Beau et al. (1991), Aliabadi and Tezduyar (1993), Behr and Tezduyar (1994), Elkadri E et al. (2000), Tezduyar and Osawa (2000), Tezduyar and Senga (2006), Hughes et al. (2010), Nakajima and Kawahara (2010a, b), Tezduyar et al. (2011), and many others. Those methods can be classified into two categories: the bubble function method and the SUPG method. In the formulation of the bubble function finite element method, the normal interpolation of the bubble function for density, momentum, and total energy is employed. The final finite element equation is the same form as that of the linear finite element equation including the artificial viscosity terms added to the coefficients on the barycentric nodes. Using the different types of the weighting and

M. Kawahara, *Finite Element Methods in Incompressible, Adiabatic, and Compressible Flows*, Mathematics for Industry 19,
DOI 10.1007/978-4-431-55450-9_10

interpolation functions, the SUPG finite element method is derived. The final finite element equation includes the SUPG terms for the stabilization of the computation. For those formulations, the conservation forms of the governing equations are used. By the numerical computations, computational stability of both methods is compared. We will observe that the stability features of both methods are almost the same. The computation of the flows around a bullet train is carried out as a case study.

10.2 Governing Equation of Compressible Flows

The field variables for the computation of compressible flows are density ρ, velocity v_i, and total energy e. The governing equations can be summarized as follows:

(1) Conservation of mass is

$$\frac{\partial \rho}{\partial t} + (\rho v_i)_{,i} = 0 \tag{10.1}$$

which is Eq. (7.47).

(2) Conservation of momentum is

$$\frac{\partial}{\partial t}(\rho v_i) + (\rho v_i v_j)_{,j} - \sigma_{ji,j} - \rho f_i = 0 \tag{10.2}$$

which is Eq. (7.63).

(3) Constitutive equation is

$$\sigma_{ij} = -p\delta_{ij} + \lambda\delta_{ij}d_{kk} + 2\mu d_{ij} \tag{10.3}$$

which is Eq. (7.74). In Eqs. (10.2) and (10.3), total stress is σ_{ij} and sometimes we will write Eq. (10.3) by separating pressure p and viscous stress τ_{ij}:

$$\sigma_{ij} = -p\delta_{ij} + \tau_{ij} \tag{10.4}$$

$$\tau_{ij} = \lambda\delta_{ij}d_{kk} + 2\mu d_{ij} \tag{10.5}$$

Deformation rate is define by Eq. (7.71):

$$d_{ij} = \frac{1}{2}(v_{i,j} + v_{j,i}) \tag{10.6}$$

Considering that we are dealing with the flows close to the incompressible flows, we assume

$$\lambda = -\frac{2}{3}\mu. \tag{10.7}$$

In Eq. (10.5), μ and λ are viscosity coefficients.

(4) Conservation of energy is

$$\frac{\partial}{\partial t}(\rho e) + (\rho e v_j)_{,j} - (\sigma_{ij} v_i)_{,j} - q_{i,i} = 0 \tag{10.8}$$

which is Eq. (7.94). The total energy e is expressed by the internal energy ε and kinetic energy as follows:

$$e = \varepsilon + \frac{v_i v_i}{2} \tag{10.9}$$

which is Eq. (7.82). In Eq. (10.8), body force and heat source are disregarded. Those equations summarized in (1)–(4) are all derived from the conservation equations and constitutive equation in Chap. 7. Those equations include several variables implicitly, such as pressure p, temperature T, internal energy ε, and thermal flux q_i. To determine those variables, we have to introduce several additional laws. For this purpose, we deploy the notion of ideal gas.

(5) Equation of state:
Assuming the ideal gas, the equation of state can be expressed as the relation between pressure p and internal energy ε as in Eq. (7.150):

$$p = \rho(\gamma - 1)\varepsilon \tag{10.10}$$

where γ is the ratio of specific heat.

(6) Thermal flux is represented in terms of the internal energy ε as

$$q_j = \frac{\kappa}{c_v}\varepsilon_{,j} \tag{10.11}$$

where c_v is the specific heat at a constant volume, and κ is the thermal conduction coefficient. It is useful to use the Prandtl number P_r:

$$P_r = \frac{c_p \mu}{\kappa} \tag{10.12}$$

where c_p is specific heat at constant pressure. Equation (10.11) can be transformed into

$$q_j = \frac{\gamma \mu}{P_r}\varepsilon_{,j} \tag{10.13}$$

(7) Boundary condition:
In addition to the standard boundary condition, we use certain values which are specified on the boundary.

(8) Initial condition:
All the field variables ρ, v_i, e are given as the initial conditions at starting time.

10.3 Conservation Form

Equations (10.1), (10.2), and (10.8) can be reformulated as follows:

$$\frac{\partial}{\partial t}\begin{bmatrix}\rho \\ \rho v_i \\ \rho e\end{bmatrix} + \frac{\partial}{\partial x_j}\begin{bmatrix}\rho v_j \\ \rho v_i v_j + p\delta_{ij} - \tau_{ij} \\ (\rho e + p)v_j - (\tau_{kj} v_k + q_j)\end{bmatrix} = \begin{bmatrix}0 \\ 0 \\ 0\end{bmatrix} \tag{10.14}$$

Here and henceforth, we will restrict expressions to those in two-dimensional cases. However, extending to three-dimensional cases is straightforward. Equation (10.14) is transformed into the following form:

$$\dot{\mathbf{U}}_k + \mathbf{F}_{k_{i,i}} + \mathbf{G}_{k_{i,i}} = 0 \quad \begin{matrix}(k = 1 \text{ to } 4) \\ (i = 1 \text{ to } 4)\end{matrix} \tag{10.15}$$

where

$$\mathbf{U}_k = \begin{bmatrix}\rho \\ \rho v_1 \\ \rho v_2 \\ \rho e\end{bmatrix} = \begin{bmatrix}\rho \\ m \\ n \\ \mathfrak{e}\end{bmatrix} = \begin{bmatrix}U_1 \\ U_2 \\ U_3 \\ U_4\end{bmatrix} \tag{10.16}$$

$$\mathbf{F}_{k_{i,i}} = \begin{bmatrix}\rho v_i \\ v_i \rho v_1 + \delta_{i1} p \\ v_i \rho v_2 + \delta_{i2} p \\ v_i(\rho e + p)\end{bmatrix}_{,i} = \begin{bmatrix}m\delta_{1i} + n\delta_{2i} \\ v_i m + p\delta_{1i} \\ v_i n + p\delta_{2i} \\ v_i(\mathfrak{e} + p)\end{bmatrix}_{,i} \tag{10.17}$$

$$\mathbf{G}_{k_{i,i}} = -\begin{bmatrix}0 \\ \tau_{1i} \\ \tau_{2i} \\ \tau_{ij} v_j + q_i\end{bmatrix}_{,i} \tag{10.18}$$

where m, n, and $\mathfrak{e}$ denote momentum in the one and two directions and total energy per volume, i.e.,

$$\rho v_1 = m \tag{10.19}$$

$$\rho v_2 = n \tag{10.20}$$

$$\rho e = \mathfrak{e} \tag{10.21}$$

Superscripted dot $\dot{}$ represents partial differentiation with respect to time. Equation (10.15) is also rewritten in the following form:

$$\begin{bmatrix} \dot{\rho} \\ \dot{m} \\ \dot{n} \\ \dot{\mathfrak{e}} \end{bmatrix} + \begin{bmatrix} m \\ v_1 m + p \\ v_1 n \\ v_1(\mathfrak{e} + p) \end{bmatrix}_{,1} + \begin{bmatrix} n \\ v_2 m \\ v_2 n + p \\ v_2(\mathfrak{e} + p) \end{bmatrix}_{,2}$$
$$- \begin{bmatrix} 0 \\ \tau_{11} \\ \tau_{21} \\ (\tau_{11} v_1 + \tau_{12} v_2) + q_1 \end{bmatrix}_{,1} - \begin{bmatrix} 0 \\ \tau_{12} \\ \tau_{22} \\ (\tau_{21} v_1 + \tau_{22} v_2) + q_2 \end{bmatrix}_{,2} = \begin{bmatrix} 0 \\ 0 \\ 0 \\ 0 \end{bmatrix} \tag{10.22}$$

It is well known that Eq. (10.15) is reformulated as follows:

$$\dot{\mathbf{U}}_k + \mathbf{A}^i_{kl} \mathbf{U}_{l,i} - \mathbf{K}^{ij}_{kl} (\mathbf{U}_{l,i})_{,j} = 0 \tag{10.23}$$

The second term in Eq. (10.23) is expanded as

$$\mathbf{A}^1_{kl} \mathbf{U}_{l,1} + \mathbf{A}^2_{kl} \mathbf{U}_{l,2}$$
$$= \begin{bmatrix} \frac{\partial m}{\partial \rho} & \frac{\partial m}{\partial m} & \frac{\partial m}{\partial n} & \frac{\partial m}{\partial \mathfrak{e}} \\ \frac{\partial (v_1 m + p)}{\partial \rho} & \frac{\partial (v_1 m + p)}{\partial m} & \frac{\partial (v_1 m + p)}{\partial n} & \frac{\partial (v_1 m + p)}{\partial \mathfrak{e}} \\ \frac{\partial (v_1 n)}{\partial \rho} & \frac{\partial (v_1 n)}{\partial m} & \frac{\partial (v_1 n)}{\partial n} & \frac{\partial (v_1 n)}{\partial \mathfrak{e}} \\ \frac{\partial v_1 (\mathfrak{e} + p)}{\partial \rho} & \frac{\partial v_1 (\mathfrak{e} + p)}{\partial m} & \frac{\partial v_1 (\mathfrak{e} + p)}{\partial n} & \frac{\partial v_1 (\mathfrak{e} + p)}{\partial \mathfrak{e}} \end{bmatrix} \begin{bmatrix} \rho \\ m \\ n \\ \mathfrak{e} \end{bmatrix}_{,1}$$
$$+ \begin{bmatrix} \frac{\partial n}{\partial \rho} & \frac{\partial n}{\partial m} & \frac{\partial n}{\partial n} & \frac{\partial n}{\partial \mathfrak{e}} \\ \frac{\partial (v_2 m)}{\partial \rho} & \frac{\partial (v_2 m)}{\partial m} & \frac{\partial (v_2 m)}{\partial n} & \frac{\partial (v_2 m)}{\partial \mathfrak{e}} \\ \frac{\partial (v_2 n + p)}{\partial \rho} & \frac{\partial (v_2 n + p)}{\partial m} & \frac{\partial (v_2 n + p)}{\partial n} & \frac{\partial (v_2 n + p)}{\partial \mathfrak{e}} \\ \frac{\partial v_2 (\mathfrak{e} + p)}{\partial \rho} & \frac{\partial v_2 (\mathfrak{e} + p)}{\partial m} & \frac{\partial v_2 (\mathfrak{e} + p)}{\partial n} & \frac{\partial v_2 (\mathfrak{e} + p)}{\partial \mathfrak{e}} \end{bmatrix} \begin{bmatrix} \rho \\ m \\ n \\ \mathfrak{e} \end{bmatrix}_{,2} \tag{10.24}$$

$\mathbf{A}^1_{kl}$ and $\mathbf{A}^2_{kl}$ are obtained by differentiating the second and third matrices in Eq. (10.22). In those operations, the terms to be differentiated must be transformed into the functions of independent variables ρ, m, n, and $\mathfrak{e}$. We can eliminate pressure p from the second and third terms in Eq. (10.23) by making use of the equation:

$$p = (\gamma - 1)\rho\varepsilon = \bar{\gamma} \left(\mathfrak{e} - \frac{m^2 + n^2}{2\rho} \right) \tag{10.25}$$

where

$$\bar{\gamma} = \gamma - 1 \tag{10.26}$$

The resulting equations are

$$\begin{bmatrix} m \\ v_1 m + p \\ v_1 m \\ v_1(\mathfrak{e} + p) \end{bmatrix} = \begin{bmatrix} m \\ \bar{\gamma}\mathfrak{e} + \frac{1}{2}(3-\gamma)\frac{m^2}{\rho} - \bar{\gamma}\frac{n^2}{2\rho} \\ \frac{mn}{\rho} \\ \gamma\mathfrak{e}\frac{m}{\rho} - \bar{\gamma}\frac{m^3}{2\rho^2} - \bar{\gamma}\frac{mn^2}{2\rho^2} \end{bmatrix} \tag{10.27}$$

$$\begin{bmatrix} n \\ v_2 m \\ v_2 n + p \\ v_2(\mathfrak{e} + p) \end{bmatrix} = \begin{bmatrix} n \\ \frac{mn}{\rho} \\ \bar{\gamma}\mathfrak{e} - \bar{\gamma}\frac{m^2}{2\rho} + \frac{1}{2}(3-\gamma)\frac{n^2}{\rho} \\ \gamma\mathfrak{e}\frac{n}{\rho} - \bar{\gamma}\frac{nm^2}{2\rho^2} - \bar{\gamma}\frac{n^3}{2\rho^2} \end{bmatrix} \tag{10.28}$$

By performing the differentiation of Eqs. (10.27) and (10.28) with respect to ρ, m, n, and $\mathfrak{e}$, we obtain the coefficient matrices in Eq. (10.24) as follows:

$$\mathbf{A}^1_{kl} = \begin{bmatrix} 0 & 1 & 0 & 0 \\ \frac{1}{2}\bar{\gamma}u^2 - v_1^2 & v_1 - \bar{\gamma}v_1 + v_1 & -\bar{\gamma}v_2 & \bar{\gamma} \\ -v_1 v_2 & v_2 & v_1 & 0 \\ (\bar{\gamma}u^2 - \gamma e)v_1 & \bar{\varepsilon} - \bar{\gamma}v_1^2 & -\bar{\gamma}v_1 v_2 & \gamma v_1 \end{bmatrix} \tag{10.29}$$

$$\mathbf{A}^2_{kl} = \begin{bmatrix} 0 & 0 & 1 & 0 \\ -v_2 v_1 & v_2 & v_1 & 0 \\ \frac{1}{2}\bar{\gamma}u^2 - v_2^2 & -\bar{\gamma}v_1 & v_2 - \bar{\gamma}v_2 + v_2 & \bar{\gamma} \\ (\bar{\gamma}u^2 - \gamma e)v_2 & -\bar{\gamma}v_2 v_1 & \bar{\varepsilon} - \bar{\gamma}v_2^2 & \gamma v_2 \end{bmatrix} \tag{10.30}$$

where

$$u^2 = v_i v_i \tag{10.31}$$

$$\bar{\varepsilon} = \gamma e - \bar{\gamma}v_i v_i/2 \tag{10.32}$$

Both Eqs. (10.29) and (10.30) can be combined, and it is possible to obtain the following equations:

$$\mathbf{A}^i_{kl} = \begin{bmatrix} 0 & \delta^i_1 & \delta^i_2 & 0 \\ b^i_1 & b^i_{11} & b^i_{12} & d^i_1 \\ b^i_2 & b^i_{21} & b^i_{22} & d^i_2 \\ a_i & c^i_1 & c^i_2 & f_i \end{bmatrix} \tag{10.33}$$

where

$$\begin{aligned} b^i_1 &= \frac{1}{2}\delta_{i1}\bar{\gamma}u^2 - v_i v_1 \\ b^i_2 &= \frac{1}{2}\delta_{i2}\bar{\gamma}u^2 - v_i v_2 \\ a_i &= (\bar{\gamma}u^2 - \gamma e)v_i \\ b^i_{11} &= \delta_{i1}v_1 - \delta_{i1}\bar{\gamma}v_1 + v_i \\ b^i_{12} &= \delta_{i2}v_1 - \delta_{i1}\bar{\gamma}v_2 \\ b^i_{21} &= \delta_{i1}v_2 - \delta_{i2}\bar{\gamma}v_1 \\ b^i_{22} &= \delta_{i2}v_2 - \delta_{i2}\bar{\gamma}v_2 + v_i \\ c^i_1 &= \delta_{i1}\bar{\varepsilon} - \bar{\gamma}v_i v_1 \\ c^i_2 &= \delta_{i2}\bar{\varepsilon} - \bar{\gamma}v_i v_2 \\ d^i_1 &= \delta_{i1}\bar{\gamma} \\ d^i_2 &= \delta_{i2}\bar{\gamma} \\ f_i &= \gamma v_i \end{aligned} \tag{10.34}$$

The third term of Eq. (10.23) can be derived from Eq. (10.18). Substituting Eqs. (10.5), (10.6), and (10.13) into Eq. (10.18) and rearranging the terms, the third term of Eq. (10.23) can be expressed as follows:

$$\begin{aligned} &\mathbf{K}^{ij}_{kl}(\mathbf{U}_{l,i})_{,j} \\ &= \left[\begin{bmatrix} 0 & 0 & 0 & 0 \\ K^{11}_{21} & K^{11}_{22} & K^{11}_{23} & 0 \\ K^{11}_{31} & K^{11}_{32} & K^{11}_{33} & 0 \\ K^{11}_{41} & K^{11}_{42} & K^{11}_{43} & K^{11}_{44} \end{bmatrix} \begin{bmatrix} \rho_{,1} \\ (\rho v_1)_{,1} \\ (\rho v_2)_{,1} \\ (\rho e)_{,1} \end{bmatrix}\right]_{,1} + \left[\begin{bmatrix} 0 & 0 & 0 & 0 \\ K^{12}_{21} & K^{12}_{22} & K^{12}_{23} & 0 \\ K^{12}_{31} & K^{12}_{32} & K^{12}_{33} & 0 \\ K^{12}_{41} & K^{12}_{42} & K^{12}_{43} & K^{12}_{44} \end{bmatrix} \begin{bmatrix} \rho_{,1} \\ (\rho v_1)_{,1} \\ (\rho v_2)_{,1} \\ (\rho e)_{,1} \end{bmatrix}\right]_{,2} \\ &+ \left[\begin{bmatrix} 0 & 0 & 0 & 0 \\ K^{21}_{21} & K^{21}_{22} & K^{21}_{23} & 0 \\ K^{21}_{31} & K^{21}_{32} & K^{21}_{33} & 0 \\ K^{21}_{41} & K^{21}_{42} & K^{21}_{43} & K^{21}_{44} \end{bmatrix} \begin{bmatrix} \rho_{,2} \\ (\rho v_1)_{,2} \\ (\rho v_2)_{,2} \\ (\rho e)_{,2} \end{bmatrix}\right]_{,1} + \left[\begin{bmatrix} 0 & 0 & 0 & 0 \\ K^{22}_{21} & K^{22}_{22} & K^{22}_{23} & 0 \\ K^{22}_{31} & K^{22}_{32} & K^{22}_{33} & 0 \\ K^{22}_{41} & K^{22}_{42} & K^{22}_{43} & K^{22}_{44} \end{bmatrix} \begin{bmatrix} \rho_{,2} \\ (\rho v_1)_{,2} \\ (\rho v_2)_{,2} \\ (\rho e)_{,2} \end{bmatrix}\right]_{,2} \end{aligned} \tag{10.35}$$

where

$$K_{21}^{11} = -\frac{(2\mu + \lambda)}{\rho} v_1, \quad K_{22}^{11} = \frac{(2\mu + \lambda)}{\rho}, \quad K_{23}^{11} = 0,$$

$$K_{31}^{11} = -\frac{\mu}{\rho} v_2, \quad K_{32}^{11} = 0, \quad K_{33}^{11} = \frac{\mu}{\rho},$$

$$K_{41}^{11} = -\frac{(2\mu + \lambda)}{\rho} v_1^2 - \frac{\mu}{\rho} v_2^2 + \frac{\gamma\mu}{P_r\rho}(v_1^2 + v_2^2 - e),$$

$$K_{42}^{11} = \frac{(2\mu + \lambda)}{\rho} v_1 - \frac{\gamma\mu}{P_r\rho} v_1,$$

$$K_{43}^{11} = \frac{\mu}{\rho} v_2 - \frac{\gamma\mu}{P_r\rho} v_2, \quad K_{44}^{11} = \frac{\gamma\mu}{P_r\rho},$$

$$K_{21}^{12} = -\frac{\mu}{\rho} v_2, \quad K_{22}^{12} = 0, \quad K_{23}^{12} = \frac{\mu}{\rho},$$

$$K_{31}^{12} = -\frac{\lambda}{\rho} v_1, \quad K_{32}^{12} = \frac{\lambda}{\rho}, \quad K_{33}^{12} = 0,$$

$$K_{41}^{12} = -\frac{\lambda + \mu}{\rho} v_1 v_2, \quad K_{42}^{12} = \frac{\lambda}{\rho} v_2, \quad K_{43}^{12} = \frac{\mu}{\rho} v_1, \quad K_{44}^{12} = 0,$$

$$K_{21}^{21} = -\frac{\lambda}{\rho} v_2, \quad K_{22}^{21} = 0, \quad K_{23}^{21} = \frac{\lambda}{\rho},$$

$$K_{31}^{21} = -\frac{\mu}{\rho} v_1, \quad K_{32}^{21} = \frac{\mu}{\rho}, \quad K_{33}^{21} = 0,$$

$$K_{41}^{21} = -\frac{(\mu + \lambda)}{\rho} v_1 v_2, \quad K_{42}^{21} = \frac{\mu}{\rho} v_2, \quad K_{43}^{21} = \frac{\lambda}{\rho} v_1, \quad K_{44}^{21} = 0,$$

$$K_{21}^{22} = -\frac{\mu}{\rho} v_1, \quad K_{22}^{22} = \frac{\mu}{\rho}, \quad K_{23}^{22} = 0,$$

$$K_{31}^{22} = -\frac{(2\mu + \lambda)}{\rho} v_2, \quad K_{32}^{22} = 0, \quad K_{33}^{22} = \frac{(2\mu + \lambda)}{\rho},$$

$$K_{41}^{22} = -\frac{\mu}{\rho} v_1^2 - \frac{(2\mu + \lambda)}{\rho} v_2^2 + \frac{\gamma\mu}{P_r\rho}(v_1^2 + v_2^2 - e),$$

$$K_{42}^{22} = \frac{\mu}{\rho} v_1 - \frac{\gamma\mu}{P_r\rho} v_1, \quad K_{43}^{22} = \frac{2\mu + \lambda}{\rho} v_2 - \frac{\gamma\mu}{P_r\rho} v_2, \quad K_{44}^{22} = \frac{\gamma\mu}{P_r\rho}$$

10.4 Bubble Function Finite Element Method

We have already treated the bubble function method for advection–diffusion problems in Sect. 5.8 and for incompressible flows in Sect. 8.5. In this section, we will solve compressible flows by the stabilized method based on the bubble function. The mixed interpolation based on the bubble function for velocity and linear function for pressure is used. The linearized advection and constant viscosity terms are assumed. The stabilized scheme can be derived. An artificial viscosity is added on the barycentric nodes. The conservation form is used. Using the standard bubble function, we can achieve the stable computation.

10.4.1 Derivation of Artificial Viscosity Term

Multiplying both sides of Eq. (10.23) by the weighting function $\mathbf{U}_k^*$, and integration by parts, we get

$$\int_{V_e} (\mathbf{U}_k^* \dot{\mathbf{U}}_k) dV + \int_{V_e} (\mathbf{U}_k^* \mathbf{A}_{kl}^i \mathbf{U}_{l,i}) dV + \int_{V_e} (\mathbf{U}_{k,j}^* \mathbf{K}_{kl}^{ij} \mathbf{U}_{l,i}) dV = \int_{S_e} (\mathbf{U}_k^* \Gamma_k) dS \tag{10.36}$$

where

$$\Gamma_k = \mathbf{K}_{kl}^{ij} \mathbf{U}_{l,i} \mathbf{n}_j \tag{10.37}$$

and

$$\mathbf{U}_k^* = \begin{bmatrix} \rho^* \\ m^* \\ n^* \\ \mathfrak{e}^* \end{bmatrix} = \begin{bmatrix} U_1^* \\ U_2^* \\ U_3^* \\ U_4^* \end{bmatrix} \qquad (k = 1 \text{ to } 4) \tag{10.38}$$

In Eq. (10.37), Γ_k means equivalent nodal force, and $\mathbf{n}_i$ is the unit normal. In Eq. (10.38), the weighting functions corresponding to ρ, m, n, and $\mathfrak{e}$ are denoted by ρ^*, m^*, n^*, and $\mathfrak{e}^*$, respectively.

The bubble function interpolation function and weighting function are written as follows:

$$\mathbf{U}_k = \bar{\mathbf{U}}_k + \phi U_{4k} \tag{10.39}$$

$$\mathbf{U}_k^* = \bar{\mathbf{U}}_k^* + \phi U_{4k}^* \tag{10.40}$$

where $\bar{\mathbf{U}}_k$ and $\bar{\mathbf{U}}_k^*$ are linear interpolation and weighting functions, ϕ denotes bubble function, as expressed in Eq. (8.77), and U_{4k} and U_{4k}^* are the values of the interpolation and weighting functions on the barycentric nodes. Eqs. (10.39) and (10.40) can be represented as follows:

$$\begin{bmatrix} \rho \\ m \\ n \\ \mathfrak{e} \end{bmatrix} = \begin{bmatrix} \Phi_\alpha & & & \\ & \Phi_\alpha & & \\ & & \Phi_\alpha & \\ & & & \Phi_\alpha \end{bmatrix} \begin{bmatrix} \rho_\alpha \\ m_\alpha \\ n_\alpha \\ \mathfrak{e}_\alpha \end{bmatrix} + \phi \begin{bmatrix} \rho_4 \\ m_4 \\ n_4 \\ \mathfrak{e}_4 \end{bmatrix} \tag{10.41}$$

$$\begin{bmatrix} \rho^* \\ m^* \\ n^* \\ \mathfrak{e}^* \end{bmatrix} = \begin{bmatrix} \Phi_\alpha & & & \\ & \Phi_\alpha & & \\ & & \Phi_\alpha & \\ & & & \Phi_\alpha \end{bmatrix} \begin{bmatrix} \rho_\alpha^* \\ m_\alpha^* \\ n_\alpha^* \\ \mathfrak{e}_\alpha^* \end{bmatrix} + \phi \begin{bmatrix} \rho_4^* \\ m_4^* \\ n_4^* \\ \mathfrak{e}_4^* \end{bmatrix} \tag{10.42}$$

where α (α = 1 to 3) corresponds to the number of three vertices of a triangular element and 4 is the number of barycentric node. The interpolation function Φ_α (α = 1 to 3) is the standard linear interpolation function given in Eq. (4.26). In Eq. (10.41), $\rho_\alpha, m_\alpha, n_\alpha$, and $\mathfrak{e}_\alpha$ represent nodal values of density, both components of momentum, total energy, respectively, at node α of a triangular element, and ρ_4, m_4, n_4, and $\mathfrak{e}_4$ are nodal values on the barycentric node. In Eq. (10.42), the weighting constants are shown as ρ^*_α, m^*_α, n^*_α, and $\mathfrak{e}^*_\alpha$ at node α and ρ^*_4, m^*_4, n^*_4, and $\mathfrak{e}^*_4$ on the barycentric node. Superscripted over bar $^-$ means linear function in this and next sections.

Replacing $\mathbf{U}^*_k$ in Eq. (10.36) by ϕ, we obtain

$$\int_{V_e}(\phi\dot{\bar{\mathbf{U}}}_k)dV+\int_{V_e}(\phi^2)dV\dot{U}_{4k}+\int_{V_e}(\phi\mathbf{A}^i_{kl}\bar{\mathbf{U}}_{l,i})dV+\int_{V_e}(\phi\mathbf{A}^i_{kl}\phi_{,i})dVU_{4l}$$
$$+\int_{V_e}(\phi_{,j}\mathbf{K}^{ij}_{kl}\bar{\mathbf{U}}_{l,i})dV+\int_{V_e}(\phi_{,j}\mathbf{K}^{ij}_{kl}\phi_{,i})dVU_{4l}=\int_{S_e}(\phi\Gamma_k)dS \tag{10.43}$$

If we assume $\mathbf{A}^i_{kl}$ and $\mathbf{K}^{ij}_{kl}$ are constants, the following equalities hold:

$$\int_{V_e}(\phi\mathbf{A}^i_{kl}\phi_{,i})dV=0 \tag{10.44}$$

$$\int_{V_e}(\phi_{,j}\mathbf{K}^{ij}_{kl}\bar{\mathbf{U}}_{l,i})dV=0 \tag{10.45}$$

Using Eqs. (10.44) and (10.45) and setting aside the second term of Eq. (10.43), the following equation is derived:

$$\int_{V_e}(\phi\dot{\bar{\mathbf{U}}}_k)dV+\int_{V_e}(\phi\mathbf{A}^i_{kl}\bar{\mathbf{U}}_{l,i})dV+\int_{V_e}(\phi_{,j}\mathbf{K}^{ij}_{kl}\phi_{,i})dVU_{4l}=\int_{S_e}(\phi\Gamma_k)dS \tag{10.46}$$

Disregarding the first and the last terms in Eq. (10.46), we get

$$U_{4k}=-[\mathbf{R}_{kl}]\left(\int_{V_e}\phi\mathbf{A}^i_{ln}\bar{\mathbf{U}}_{n,i}dV\right) \tag{10.47}$$

where

$$\mathbf{R}_{kl}=\left[\int_{V_e}(\phi_{,j}\mathbf{K}^{ij}_{kl}\phi_{,i})dV\right]^{-1} \tag{10.48}$$

This is because significant terms for stability are the second and third terms.

To formulate the stabilization parameters τ_{CB}, we will perform the following procedures. Replace $\mathbf{U}^*_k$ in Eq. (10.36) by the linear weighting function $\bar{\mathbf{U}}^*_k$:

$$\int_{V_e}(\bar{\mathbf{U}}_k^*\dot{\bar{\mathbf{U}}}_k)dV+\int_{V_e}(\bar{\mathbf{U}}_k^*\phi)dV\dot{U}_{4k}+\int_{V_e}(\bar{\mathbf{U}}_k^*\mathbf{A}_{kl}^i\bar{\mathbf{U}}_{l,i})dV+\int_{V_e}(\bar{\mathbf{U}}_k^*\mathbf{A}_{kl}^i\phi_{,i})dVU_{4l}$$
$$+\int_{V_e}(\bar{\mathbf{U}}_{k,j}^*\mathbf{K}_{kl}^{ij}\bar{\mathbf{U}}_{l,i})dV+\int_{V_e}(\bar{\mathbf{U}}_{k,j}^*\mathbf{K}_{kl}^{ij}\phi_{,i})dVU_{4l}=\int_{S_e}(\bar{\mathbf{U}}_k^*\Gamma_k)dS \tag{10.49}$$

Using the equality

$$\int_{V_e}(\bar{\mathbf{U}}_{k,j}^*\mathbf{K}_{kl}^{ij}\phi_{,i})dVU_{4l}=0 \tag{10.50}$$

and disregarding the second term, which has no significant effects on the stability, and the fifth term, which itself is a viscosity term, Eq. (10.49) is transformed into

$$\int_{V_e}(\bar{\mathbf{U}}_k^*\dot{\bar{\mathbf{U}}}_k)dV+\int_{V_e}(\bar{\mathbf{U}}_k^*\mathbf{A}_{kl}^i\bar{\mathbf{U}}_{l,i})dV+\int_{V_e}(\bar{\mathbf{U}}_k^*\mathbf{A}_{kl}^i\phi_{,i})dVU_{4l}=\int_{S_e}(\bar{\mathbf{U}}_k^*\Gamma_k)dS \tag{10.51}$$

Consider the third term in Eq. (10.51):

$$\begin{aligned}&\int_{V_e}(\bar{\mathbf{U}}_k^*\mathbf{A}_{kl}^i\phi_{,i})dVU_{4l}\\&=\int_{S_e}(\bar{\mathbf{U}}_k^*\mathbf{A}_{kl}^i\phi n_i)dSU_{4l}-\int_{V_e}(\mathbf{U}_{k,i}\mathbf{A}_{kl}^i\phi)dVU_{4l}\\&=-\int_{V_e}(\mathbf{U}_{k,i}\mathbf{A}_{kl}^i\phi)dVU_{4l}\end{aligned} \tag{10.52}$$

because the first term on the right side in Eq. (10.52) is zero. Substitute Eq. (10.47) into Eq. (10.52) and eliminate U_{4l}; then, we can write the third term in Eq. (10.51) as follows:

$$\text{the third term}=\int_{V_e}(\bar{\mathbf{U}}_{k,i}^*\mathbf{A}_{kl}^i)\frac{1}{\mu}[\mathbf{R}_{lm}]\frac{\left(\int_{V_e}\phi dV\right)^2}{\Delta}(\mathbf{A}_{mn}^j\mathbf{U}_{n,j})dV \tag{10.53}$$

In Eq. (10.53), μ is written outside of $[\mathbf{R}_{lm}]$, because μ is multiplied for all the terms in $[\mathbf{K}_{kl}^{ij}]$ as shown in Eq. (10.35). We will consider $[\mathbf{R}_{lm}]$ in Eq. (10.53) is the $[\mathbf{R}_{lm}]$ in Eq. (10.48) divided by $1/\mu$. If we put

$$\tau_{CB}[\mathbf{R}_{lm}]=\frac{1}{\mu+\mu'}[\mathbf{R}_{lm}]\frac{\left(\int_{V_e}\phi dV\right)^2}{\Delta} \tag{10.54}$$

where τ_{CB} is the stabilization parameter and μ' is the artificial viscosity. Then, we obtain μ' as follows:

$$
\mu + \mu' = \frac{1}{\tau_{CB}} \frac{\left(\int_{V_e} \phi dV \right)^2}{\Delta} \tag{10.55}
$$

To determine concrete form of τ_{CB}, Eq. (8.97) is useful. If we use

$$
\tau_{CB} = \left\{ \frac{2|U_j|}{h_e} + \left(\frac{4\nu}{h_e^2} \right)^2 + \left(\frac{2}{\Delta t} \right)^2 \right\}^{-1/2} \tag{10.56}
$$

where $|U_j|$ is the magnitude of velocity, h_e is the characteristic length of an element, ν is dynamic viscosity, and Δt is time increment, then we can obtain the artificial viscosity μ' in Eq. (10.55).

10.4.2 Finite Element Equation

To derive the final form of the weighted residual equation and the finite element equation, substitute Eqs. (10.39) and (10.40) into Eq. (10.36):

$$
\begin{aligned}
&\int_{V_e} (\bar{\mathbf{U}}_k^* + \phi U_{4k}^*)(\dot{\bar{\mathbf{U}}}_k + \phi \dot{U}_{4k}) dV + \int_{V_e} (\bar{\mathbf{U}}_k^* + \phi U_{4k}^*) \mathbf{A}_{kl}^i (\bar{\mathbf{U}}_{l,i} + \phi_{,i} U_{4l}) dV \\
&+ \mu \int_{V_e} (\bar{\mathbf{U}}_{k,j}^* + \phi_{,j} U_{4k}^*) \mathbf{K}_{kl}^{ij} (\bar{\mathbf{U}}_{l,i} + \phi_{,i} U_{4l}) dV = \int_{S_e} (\bar{\mathbf{U}}_k^* + \phi U_{4k}) \Gamma_k^* dS
\end{aligned} \tag{10.57}
$$

In Eq. (10.57), μ is written outside of integral symbol because all terms of $\mathbf{K}_{kl}^{ij}$ are multiplied by μ. Equation (10.57) can be reformulated as follows:

$$
\begin{aligned}
&\int_{V_e} (\bar{\mathbf{U}}_k^* \dot{\bar{\mathbf{U}}}_k) dV + \int_{V_e} (\bar{\mathbf{U}}_k^* \phi) dV \dot{U}_{4k} + U_{4k}^* \int_{V_e} (\phi \dot{\bar{\mathbf{U}}}_k) dV + U_{4k}^* \int_{V} (\phi^2) dV \dot{U}_{4k} \\
&+ \int_{V_e} (\bar{\mathbf{U}}_k^* \mathbf{A}_{kl}^i \bar{\mathbf{U}}_{l,i}) dV + \int_{V_e} (\bar{\mathbf{U}}_k^* \mathbf{A}_{kl}^i \phi_{,i}) dV U_{4l} + U_{4k}^* \int_{V_e} (\phi \mathbf{A}_{kl}^i \bar{\mathbf{U}}_{l,i}) dV \\
&+ U_{4k}^* \int_{V_e} (\phi \mathbf{A}_{kl}^i \phi_{,i}) dV U_{4l} + \mu \int_{V_e} (\bar{\mathbf{U}}_{k,j}^* \mathbf{K}_{kl}^{ij} \bar{\mathbf{U}}_{l,i}) dV \\
&+ \mu \int_{V_e} (\bar{\mathbf{U}}_{k,j}^* \mathbf{K}_{kl}^{ij} \phi_{,i}) dV U_{4l} + U_{4k}^* \mu \int_{V_e} (\phi_{,j} \mathbf{K}_{kl}^{ij} \bar{\mathbf{U}}_{l,i}) dV \\
&+ U_{4k}^* \mu \int_{V} (\phi_{,j} \mathbf{K}_{kl}^{ij} \phi_{,i}) dV U_{4l} \\
&= \int_{S_e} (\bar{\mathbf{U}}_k^* \Gamma_k) dS + U_{4k}^* \int_{S_e} (\phi \Gamma_k) dS
\end{aligned} \tag{10.58}
$$

We can separate Eq. (10.58) into the following two equations:

$$\int_{V_e} (\bar{\mathbf{U}}_k^* \dot{\bar{\mathbf{U}}}_k) dV + \int_{V_e} (\bar{\mathbf{U}}_k^* \phi) dV \dot{U}_{k4}$$
$$+ \int_{V_e} (\bar{\mathbf{U}}_k^* \mathbf{A}_{kl}^i \bar{\mathbf{U}}_{l,i}) dV + \int_{V_e} (\bar{\mathbf{U}}_k^* \mathbf{A}_{kl}^i \phi_{,i}) dV U_{4l} + \mu \int_{V_e} (\bar{\mathbf{U}}_{k,j}^* \mathbf{K}_{kl}^{ij} \bar{\mathbf{U}}_{l,i}) dV$$
$$= \int_{S_e} (\bar{\mathbf{U}}_k^* \Gamma_k) dS \tag{10.59}$$

$$U_{4k}^* \int_{V_e} (\phi \dot{\bar{\mathbf{U}}}_k) dV + U_{4k}^* \int_{V_e} (\phi^2) dV \dot{U}_{4k}$$
$$+ U_{4k}^* \int_{V_e} (\phi \mathbf{A}_{kl}^i \bar{\mathbf{U}}_{l,i}) dV + U_{4k}^* (\mu + \mu') \int_{V_e} (\phi_{,j} \mathbf{K}_{kl}^{ij} \phi_{,i}) dV U_{4l} = 0 \tag{10.60}$$

Equations (10.59) and (10.60) are the final form of the local weighted residual equation. The linear interpolation and weighting functions are the first terms in Eqs. (10.41) and (10.42), respectively:

$$\bar{\mathbf{U}}_k = \Phi_\beta \bar{\mathbf{U}}_{\beta k} \tag{10.61}$$
$$\bar{\mathbf{U}}_k^* = \Phi_\beta \bar{\mathbf{U}}_{\beta k}^* \tag{10.62}$$

where Φ_α, $\bar{\mathbf{U}}_{\beta k}$, and $\bar{\mathbf{U}}_{\beta k}^*$ are interpolation function matrix, nodal values, and corresponding weighting constants, respectively. Substituting Eqs. (10.61) and (10.62) into Eqs. (10.59) and (10.60), and rearranging the terms, we get

$$\bar{\mathbf{U}}_{\alpha k}^* \int_{V_e} (\Phi_\alpha \Phi_\beta) dV \dot{\bar{\mathbf{U}}}_{\beta k} + \bar{\mathbf{U}}_{\alpha k}^* \int_{V_e} (\Phi_\alpha \phi) dV \dot{U}_{4k}$$
$$+ \bar{\mathbf{U}}_{\alpha k}^* \int_{V_e} (\Phi_\alpha \mathbf{A}_{kl}^i \Phi_{\beta,i}) dV \bar{\mathbf{U}}_{\beta l} + \bar{\mathbf{U}}_{\alpha k}^* \int_{V_e} (\Phi_\alpha \mathbf{A}_{kl}^i \phi_{,i}) dV U_{4l}$$
$$+ \bar{\mathbf{U}}_{\alpha k}^* \mu \int_{V_e} (\Phi_{\alpha,j} \mathbf{K}_{kl}^{ij} \Phi_{\beta,i}) dV \bar{\mathbf{U}}_{\beta l}$$
$$= \bar{\mathbf{U}}_{\alpha k}^* \int_{S_e} (\Phi_\alpha \Gamma_k) dS \tag{10.63}$$

$$U_{4k}^* \int_{V_e} (\phi \Phi_\beta) dV \dot{\bar{\mathbf{U}}}_{\beta k} + U_{4k}^* \int_{V_e} (\phi^2) dV U_{4k}$$
$$+ U_{4k}^* \int_{V_e} (\phi \mathbf{A}_{kl}^i \Phi_{\beta,i}) dV \bar{\mathbf{U}}_{\beta l} + U_{4k}^* (\mu + \mu') \int_{V_e} (\phi_{,j} \mathbf{K}_{kl}^{ij} \phi_{,i}) dV U_{4l} = 0$$
$$\tag{10.64}$$

Consider $\bar{\mathbf{U}}^*_{\alpha k}$ and U^*_{4k} to be arbitrary constants. Then, the finite element equations can be obtained:

$$\mathbf{M}_{\alpha\beta}\dot{\bar{\mathbf{U}}}_{\beta k} + \mathbf{Q}_{\alpha}\dot{U}_{4k} + \mathbf{A}_{\alpha kl\beta}\bar{\mathbf{U}}_{\beta l} + \mathbf{B}_{\alpha kl}U_{4l} + \mu S_{\alpha kl\beta}\bar{\mathbf{U}}_{\beta l} = \Omega_{\alpha k} \tag{10.65}$$

$$\mathbf{a}_{\beta}\dot{\bar{\mathbf{U}}}_{\beta k} + b\dot{U}_{4k} + \mathbf{C}_{kl\beta}\bar{\mathbf{U}}_{\beta l} + (\mu + \mu')\mathbf{D}_{kl}U_{4l} = 0 \tag{10.66}$$

where

$$\mathbf{M}_{\alpha\beta} = \int_{V_e}(\Phi_\alpha\Phi_\beta)dV, \quad \mathbf{Q}_\alpha = \int_{V_e}(\Phi_\alpha\phi)dV \tag{10.67}$$

$$\mathbf{A}_{\alpha kl\beta} = \int_{V_e}(\Phi_\alpha\mathbf{A}^i_{kl}\Phi_{\beta,i})dV$$

$$\mathbf{B}_{\alpha kl} = \int_{V_e}(\Phi_\alpha\mathbf{A}^i_{kl}\phi_{,i})dV$$

$$\mathbf{S}_{\alpha kl\beta} = \int_{V_e}(\Phi_{\alpha,j}\mathbf{K}^{ij}_{kl}\Phi_{\beta,i})dV$$

$$\mathbf{a}_\beta = \int_{V_e}(\phi\Phi_\beta)dV, \quad b = \int_{V_e}\phi^2 dV$$

$$\mathbf{C}_{kl\beta} = \int_{V_e}(\phi\mathbf{A}^i_{kl}\Phi_{\beta,i})dV, \quad \mathbf{D}_{kl} = \int_{V_e}(\phi_{,j}\mathbf{K}^{ij}_{kl}\phi_{,i})dV$$

$$\Omega_{\alpha k} = \int_{S_e}(\Phi_\alpha\Gamma_k)dS$$

Superpositions of Eqs. (10.65) and (10.66) lead to the global form of the finite element method. We assume that $\mathbf{A}^i_{kl}$ and $\mathbf{K}^{ij}_{kl}$ are constant at the integrations in Eq. (10.67). If necessary, several iterations are performed in a time cycle. One or two iterations are sufficient in the author's experiences. Superimposing Eqs. (10.65) and (10.66) over the whole flow field, we get the global finite element equation. As is shown in Eq. (8.100), we may distinguish the discontinuous coefficients in Eq. (10.67).

Imposing the boundary and initial conditions, and discretizing the time-dependent terms, we can solve the global finite element equations and carry out computations of density, momentum, and total energy. In these equations, we can see that an artificial viscosity appears on the term only on the barycentric node. This fact is one of advantageous features of the bubble function finite element method. We need to solve the global finite element equation based on Eqs. (10.65) and (10.66) simultaneously because these two equation systems are coupled with variables $\dot{\bar{\mathbf{U}}}_{\beta k}$, $\dot{U}_{4k}$, $\bar{U}_{\beta k}$, and U_{4k}. If we disregard the second term in Eq. (10.65), we can eliminate U_{4k} from Eq. (10.65) using Eq. (10.66). This procedure is convenient to accelerate the computation, because we need not to store U_{4l} during time advancing procedure.

10.5 SUPG Finite Element Method of Compressible Flow

A number of the SUPG finite element methods of compressible flow have been presented by Hughes et al. (2010), Tezduyar et al. (2011), etc. The methods are of great benefit for practical computations. The SUPG methods employ the Petrov–Glerkin method, i.e., we use a different type of weighting function from interpolation function. The weighting function includes the up-stream effects, which are adjusted by the stabilization parameter. This parameter is significantly related to that of the bubble function method. In this section, we deal with the SUPG method for comparative studies.

10.5.1 Formulation of the SUPG Finite Element Method

The basic equation that we will use in this section is Eq. (10.23), repeated as Eq. (10.68):

$$\dot{\mathbf{U}}_k + \mathbf{A}^i_{kl}\mathbf{U}_{l,i} + \mathbf{G}_{ki,i} = 0 \tag{10.68}$$

where we use Eq. (10.18). For the interpolation function, normal linear interpolation function based on a triangular element is used:

$$\mathbf{U}_k = \Phi_\beta \mathbf{U}_{\beta k} \tag{10.69}$$

which is the same function shown in Eq. (10.61). The superscripted over bar $^-$ is not shown for the sake of simplicity. Equation (10.69) can be written as follows:

$$\begin{bmatrix} \rho \\ m \\ n \\ \mathfrak{e} \end{bmatrix} = \begin{bmatrix} \Phi_\alpha & & & \\ & \Phi_\alpha & & \\ & & \Phi_\alpha & \\ & & & \Phi_\alpha \end{bmatrix} \begin{bmatrix} \rho_\alpha \\ m_\alpha \\ n_\alpha \\ \mathfrak{e}_\alpha \end{bmatrix} \tag{10.70}$$

where ρ_α, m_α, n_α, and $\mathfrak{e}_\alpha$ are density, both components of momentum, and total energy, respectively, at node α of the triangular element and Φ_α is the linear interpolation function.

For the weighting function, the following function is used:

$$\tilde{\mathbf{U}}^*_k = \mathbf{U}^*_k + \tau_s \mathbf{A}^i_{lk}\mathbf{U}^*_{l,i} \tag{10.71}$$

where $\mathbf{U}^*_k$ is the linear weighting function expressed in Eq. (10.62). The superscripted over bar $^-$ is deleted for the sake of simplicity, and τ_s is the stabilization parameter, $\mathbf{A}^i_{lk}$ is the transpose matrix of $\mathbf{A}^i_{kl}$ in Eq. (10.33), and $\mathbf{U}^*_{l,i}$ is the derivative of the linear weighting function. The weighting function can be rewritten as follows:

$$\tilde{\mathbf{U}}_k^* = \Phi_\alpha \mathbf{U}_{\alpha k}^* + \tau_s \mathbf{A}_{lk}^i \Phi_{\alpha,i} \mathbf{U}_{\alpha l}^* \tag{10.72}$$

where $\mathbf{U}_{\alpha k}^*$ means weighting constants of weighting functions at node α of a triangular element. To clarify the expression, Eq. (10.72) is expanded as

$$\begin{bmatrix} \tilde{\rho}^* \\ \tilde{m}^* \\ \tilde{n}^* \\ \tilde{\mathfrak{e}}^* \end{bmatrix} = \begin{bmatrix} \Phi_\alpha & & & \\ & \Phi_\alpha & & \\ & & \Phi_\alpha & \\ & & & \Phi_\alpha \end{bmatrix} \begin{bmatrix} \rho_\alpha^* \\ m_\alpha^* \\ n_\alpha^* \\ \mathfrak{e}_\alpha^* \end{bmatrix} + \tau_s \left(\begin{bmatrix} 0 & b_1^1 & b_2^1 & a_1 \\ 1 & b_{11}^1 & b_{21}^1 & c_1^1 \\ 0 & b_{12}^1 & b_{22}^1 & c_2^1 \\ 0 & d_1^1 & d_2^1 & f_1 \end{bmatrix} \begin{bmatrix} \Phi_{\alpha,1} & & & \\ & \Phi_{\alpha,1} & & \\ & & \Phi_{\alpha,1} & \\ & & & \Phi_{\alpha,1} \end{bmatrix} \begin{bmatrix} \rho_\alpha^* \\ m_\alpha^* \\ n_\alpha^* \\ \mathfrak{e}_\alpha^* \end{bmatrix} + \begin{bmatrix} 0 & b_1^2 & b_2^2 & a_2 \\ 0 & b_{11}^2 & b_{21}^2 & c_1^2 \\ 1 & b_{12}^2 & b_{22}^2 & c_2^2 \\ 0 & d_1^2 & d_2^2 & f_2 \end{bmatrix} \begin{bmatrix} \Phi_{\alpha,2} & & & \\ & \Phi_{\alpha,2} & & \\ & & \Phi_{\alpha,2} & \\ & & & \Phi_{\alpha,2} \end{bmatrix} \begin{bmatrix} \rho_\alpha^* \\ m_\alpha^* \\ n_\alpha^* \\ \mathfrak{e}_\alpha^* \end{bmatrix} \right) \tag{10.73}$$

where ρ_α^*, m_α^*, n_α^*, and $\mathfrak{e}_\alpha^*$ are nodal values of the weighting functions corresponding to ρ_α, m_α, n_α, and $\mathfrak{e}_\alpha$, respectively.

Multiplying both sides of Eq. (10.68) by Eq. (10.71), the following equation is obtained:

$$(\mathbf{U}_k^* + \tau_s \mathbf{A}_{mk}^i \mathbf{U}_{m,i}^*)(\dot{\mathbf{U}}_k + \mathbf{A}_{kn}^j \mathbf{U}_{n,j} + \mathbf{G}_{kj,j}) = 0 \tag{10.74}$$

Equation (10.74) can be expanded as follows:

$$\mathbf{U}_k^*(\dot{\mathbf{U}}_k + \mathbf{A}_{kn}^i \mathbf{U}_{n,i} + \mathbf{G}_{kj,j}) + \tau_s \mathbf{U}_{m,i}^* \mathbf{A}_{mk}^i \dot{\mathbf{U}}_k + \tau_s \mathbf{U}_{m,i}^* \mathbf{A}_{mk}^i \mathbf{A}_{kn}^j \mathbf{U}_{n,j} + \tau_s \mathbf{U}_{m,i}^* \mathbf{A}_{mk}^i \mathbf{G}_{kj,j} = 0 \tag{10.75}$$

Integrating Eq. (10.75) over a finite element V_e surrounded by S_e, the weighted residual equation can be obtained:

$$\int_{V_e} (\mathbf{U}_k^* \dot{\mathbf{U}}_k) dV + \int_{V_e} (\mathbf{U}_k^* \mathbf{A}_{km}^i \mathbf{U}_{m,i}) dV - \int_{V_e} (\mathbf{U}_{k,j}^* \mathbf{G}_{kj}) dV + \tau_s \int_{V_e} (\mathbf{U}_{m,i}^* \mathbf{A}_{mk}^i \dot{\mathbf{U}}_k) dV + \tau_s \int_{V_e} (\mathbf{U}_{m,i}^* \mathbf{A}_{mk}^i \mathbf{A}_{kn}^j \mathbf{U}_{n,j}) dV + \tau_s \int_{V_e} (\mathbf{U}_{m,i}^* \mathbf{A}_{mk}^i \mathbf{G}_{kj,j}) dV = \int_{S_e} (\mathbf{U}_k^* \Gamma_k) dS \tag{10.76}$$

where Γ_k is defined in Eq. (10.37). In Eq. (10.76), the first three terms are the weighted residual equation of the normal linear finite element method, and the last three terms on the left side are the stabilization terms. As worked out in Sect. 5.6, the fifth term is important for stabilization, and τ_s is the stabilization parameter, for which we use the same value as in Eq. (10.56). Using the interpolation function Eq. (10.69) and the weighting function Eq. (10.72), we get the following equation:

$$\begin{aligned}
&\mathbf{U}^*_{\alpha k}\int_{V_e}(\Phi_\alpha\Phi_\beta)dV\dot{\mathbf{U}}_{\beta k}+\mathbf{U}^*_{\alpha k}\int_{V_e}(\Phi_\alpha\mathbf{A}^i_{kl}\Phi_{\beta,i})dV\mathbf{U}_{\beta l}\\
&\quad+\mathbf{U}^*_{\alpha k}\int_{V_e}(\Phi_{\alpha,j}\mathbf{K}^{ij}_{kl}\Phi_{\beta,i})dV\mathbf{U}_{\beta l}+\tau_s\mathbf{U}^*_{\alpha k}\int_{V_e}(\Phi_{\alpha,i}\mathbf{A}^i_{kl}\Phi_\beta)dV\dot{\mathbf{U}}_{\beta l}\\
&\quad+\tau_s\mathbf{U}^*_{\alpha k}\int_{V_e}(\Phi_{\alpha,i}\mathbf{A}^i_{km}\mathbf{A}^j_{ml}\Phi_{\beta,j})dV\mathbf{U}_{\beta l}\\
&=\mathbf{U}^*_{\alpha k}\int_{S_e}(\Phi_\alpha\Gamma_k)dS
\end{aligned}\tag{10.77}$$

The last term on the left side in Eq. (10.76) is zero because we use the linear interpolation function. The first three terms on the left side in Eq. (10.77) express the normal linear finite element equation, and those are easily integrated. Considering the fact that $\mathbf{U}^*_{\alpha k}$ are arbitrary constants, we obtain the local forms of the SUPG finite element equation as follows:

$$\mathbf{M}_{\alpha\beta}\dot{\mathbf{U}}_{\beta k}+\mathbf{K}_{\alpha kl\beta}\mathbf{U}_{\beta l}+\mathbf{S}_{\alpha kl\beta}\mathbf{U}_{\beta l}+\tau_s(\mathbf{C}_{\alpha kl\beta}\dot{\mathbf{U}}_{\beta l}+\mathbf{D}_{\alpha kl\beta}\mathbf{U}_{\beta l})=\Omega_{\alpha k}\tag{10.78}$$

where

$$\begin{aligned}
\mathbf{M}_{\alpha\beta}&=\int_{V_e}(\Phi_\alpha\Phi_\beta)dV\\
\mathbf{K}_{\alpha kl\beta}&=\int_{V_e}(\Phi_\alpha\mathbf{A}^i_{kl}\Phi_{\beta,i})dV\\
\mathbf{S}_{\alpha kl\beta}&=\int_{V_e}(\Phi_{\alpha,j}\mathbf{K}^{ij}_{kl}\Phi_{\beta,i})dV\\
\mathbf{C}_{\alpha kl\beta}&=\int_{V_e}(\Phi_{\alpha,i}\mathbf{A}^i_{kl}\Phi_\beta)dV\\
\mathbf{D}_{\alpha kl\beta}&=\int_{V_e}(\Phi_{\alpha,i}\mathbf{A}^i_{km}\mathbf{A}^j_{ml}\Phi_{\beta,j}dV\\
\Omega_{\alpha i}&=\int_{S_e}(\Phi_\alpha\Gamma_k)dS
\end{aligned}\tag{10.79}$$

Superposition of Eq. (10.78) into the whole flow field leads to the global forms of the finite element method. Notice that global equations due to the first three terms on the left side in Eq. (10.78) are continuous. However, the last two terms may not be continuous because $\mathbf{A}^i_{kl}$ are assumed constants over an element. For the time discretization, a simple scheme, such as the Crank–Nicolson scheme, is adaptable.

10.5.2 Computational Forms of SUPG Finite Element Method

In this section, we will present computational forms of the SUPG method, which is useful for practical usages and for understanding of the concept. The basic equations in this section are Eq. (10.14):

$$\mathfrak{A} = \dot{\rho} + m_{j,j} = 0 \tag{10.80}$$

$$\mathfrak{B}_k = \dot{m}_k + \left(\frac{m_k m_j}{\rho}\right)_{,j} + p_{,k} - \tau_{kj,j} = 0 \tag{10.81}$$

$$\mathfrak{C} = \dot{\mathfrak{e}} + \left\{(\mathfrak{e} + p)\frac{m_j}{\rho}\right\}_{,j} - \left(\tau_{ij}\frac{m_i}{\rho} + q_j\right)_{,j} = 0 \tag{10.82}$$

Advection parts of Eqs. (10.80)–(10.82) are rewritten as follows:

$$\begin{bmatrix} \dot{\rho} \\ \dot{m}_k \\ \dot{\mathfrak{e}} \end{bmatrix} + \begin{bmatrix} 0 & \delta^i_l & 0 \\ b^i_k & b^i_{kl} & d^i_k \\ a_i & c^i_l & f_i \end{bmatrix} \begin{bmatrix} \rho \\ m_l \\ \mathfrak{e} \end{bmatrix}_{,i} = \begin{bmatrix} 0 \\ 0 \\ 0 \end{bmatrix} \tag{10.83}$$

where we use Eq. (10.33). We employ the linear interpolation functions:

$$\begin{bmatrix} \rho \\ m_l \\ \mathfrak{e} \end{bmatrix} = \begin{bmatrix} \Phi_\beta & & \\ & \Phi_\beta & \\ & & \Phi_\beta \end{bmatrix} \begin{bmatrix} \rho_\beta \\ m_{\beta l} \\ \mathfrak{e}_\beta \end{bmatrix} \tag{10.84}$$

where ρ_β, $m_{\beta l}$, and $\mathfrak{e}_\beta$ are nodal values of density, both components of momentum, and total energy, respectively, at node β of a triangular element. We also employ the following weighting functions, which are expressed by $\tilde{\rho}^*$, $\tilde{m}^*_k$, and $\tilde{\mathfrak{e}}^*$ corresponding to density, both components of momentum, and total energy, respectively:

$$\begin{bmatrix} \tilde{\rho}^* \\ \tilde{m}^*_k \\ \tilde{\mathfrak{e}}^* \end{bmatrix} = \begin{bmatrix} \rho^* \\ m^*_k \\ \mathfrak{e}^* \end{bmatrix} + \tau_s \begin{bmatrix} 0 & b^i_l & a_i \\ \delta^i_k & b^i_{lk} & c^i_k \\ 0 & d^i_l & f_i \end{bmatrix} \begin{bmatrix} \rho^* \\ m^*_l \\ \mathfrak{e}^* \end{bmatrix}_{,i} \tag{10.85}$$

where ρ^*, m_k^*, and $\mathfrak{e}^*$ are linear weighting functions, τ_s is the SUPG stabilization parameter, and the coefficient matrix of the second term is the transposed matrix in Eq. (10.83). The linear weighting functions are described as follows:

$$\begin{bmatrix} \rho^* \\ m_k^* \\ \mathfrak{e}^* \end{bmatrix} = \begin{bmatrix} \Phi_\alpha & & \\ & \Phi_\alpha & \\ & & \Phi_\alpha \end{bmatrix} \begin{bmatrix} \rho_\alpha^* \\ m_{\alpha k} \\ \mathfrak{e}_\alpha^* \end{bmatrix} \tag{10.86}$$

where ρ_α^*, $m_{\alpha k}^*$, and $\mathfrak{e}_\alpha^*$ are nodal values of the weighting functions at node α.

The weighted residual equation of the SUPG finite element method is derived by multiplying the weighting functions to the basic equations:

$$\int_V \begin{bmatrix} \tilde{\rho}^* & \tilde{m}_k^* & \tilde{\mathfrak{e}}^* \end{bmatrix} \begin{bmatrix} \mathfrak{A} \\ \mathfrak{B}_k \\ \mathfrak{C} \end{bmatrix} dV = 0 \tag{10.87}$$

Substituting Eq. (10.85) into Eq. (10.87) and rearranging the terms, the weighted residual equation can be transformed into

$$\int_{V_e} \begin{bmatrix} \rho^* & m_k^* & \mathfrak{e}^* \end{bmatrix} \begin{bmatrix} \mathfrak{A} \\ \mathfrak{B}_k \\ \mathfrak{C} \end{bmatrix} dV + \tau_s \int_{V_e} \begin{bmatrix} \rho_{,i}^* & m_{k,i}^* & \mathfrak{e}_{,i}^* \end{bmatrix} \begin{bmatrix} 0 & \delta_l^i & 0 \\ b_k^i & b_{kl}^i & d_k^i \\ a_i & c_l^i & f_i \end{bmatrix} \begin{bmatrix} \mathfrak{A} \\ \mathfrak{B}_l \\ \mathfrak{C} \end{bmatrix} dV$$
$$= 0 \tag{10.88}$$

we get the following equations by expanding Eq. (10.88):

$$\int_{V_e} (\rho^* \mathfrak{A}) dV + \tau_s \int_{V_e} (\rho_{,i}^* \delta_{il} \mathfrak{B}_l) dV = 0 \tag{10.89}$$

$$\int_{V_e} (m_k^* \mathfrak{B}_k) dV + \tau_s \int_{V_e} (m_{k,i}^* b_k^i \mathfrak{A} + m_{k,i}^* b_{kl}^i \mathfrak{B}_l + m_{k,i}^* d_k^i \mathfrak{C}) dV = 0 \tag{10.90}$$

$$\int_{V_e} (\mathfrak{e}^* \mathfrak{C}) dV + \tau_s \int_{V_e} (\mathfrak{e}_{,i}^* a_i \mathfrak{A} + \mathfrak{e}_{,i}^* c_l^i \mathfrak{B}_l + \mathfrak{e}_{,i}^* f_i \mathfrak{C}) dV = 0 \tag{10.91}$$

Substituting Eqs. (10.80)–(10.82) into Eqs. (10.89)–(10.91), and superposing the resulting equations into the whole flow field, the final forms of the weighted residual equations are obtained:

$$\int_V \rho^* \left(\frac{\partial \rho}{\partial t} + m_{j,j} \right) dV + \tau_s \sum_{e=1}^{N_e} \int_{V_e} \rho_{,i}^* \left\{ \frac{\partial m_i}{\partial t} + \left(\frac{m_j m_i}{\rho} \right)_{,j} + p_{,i} \right\} dV = 0 \tag{10.92}$$

$$
\begin{aligned}
&\int_V m_k^* \left\{ \frac{\partial m_k}{\partial t} + \left(\frac{m_k m_j}{\rho} \right)_{,j} \right\} dV + \int_V (-m_{k,k}^* p + m_{k,j}^* \tau_{kj}) dV \\
&\quad + \tau_s \sum_{e=1}^{N_e} \int_{V_e} m_{k,i}^* b_k^i \left(\frac{\partial \rho}{\partial t} + m_{j,j} \right) dV \\
&\quad + \tau_s \sum_{e=1}^{N_e} \int_{V_e} m_{k,i}^* b_{kl}^i \left\{ \frac{\partial m_l}{\partial t} + \left(\frac{m_l m_j}{\rho} \right)_{,j} + p_{,l} \right\} dV \\
&\quad + \tau_s \sum_{e=1}^{N_e} \int_{V_e} m_{k,i}^* d_k^i \left[\frac{\partial \mathfrak{e}}{\partial t} + \left\{ (\mathfrak{e} + p) \frac{m_j}{\rho} \right\}_{,j} \right] dV \\
&\quad = \int_S m_k^* (-p\delta_{kj} + \tau_{kj}) n_j dS
\end{aligned}
\tag{10.93}
$$

$$
\begin{aligned}
&\int_V \mathfrak{e}^* \left[\frac{\partial \mathfrak{e}}{\partial t} + \left\{ (\mathfrak{e} + p) \frac{m_j}{\rho} \right\}_{,j} \right] dV + \int_V \mathfrak{e}_{,j}^* \left(\tau_{kj} \frac{m_k}{\rho} + q_j \right) dV \\
&\quad + \tau_s \sum_{e=1}^{N_e} \int_{V_e} \mathfrak{e}_{,i}^* a_i \left(\frac{\partial \rho}{\partial t} + m_{j,j} \right) dV + \tau_s \sum_{e=1}^{N_e} \int_{V_e} \mathfrak{e}_{,i}^* c_l^i \left\{ \frac{\partial m_l}{\partial t} + \left(\frac{m_l m_j}{\rho} \right)_{,j} + p_{,l} \right\} dV \\
&\quad + \tau_s \sum_{e=1}^{N_e} \int_{V_e} \mathfrak{e}_{,i}^* f_i \left[\frac{\partial \mathfrak{e}}{\partial t} + \left\{ (\mathfrak{e} + p) \frac{m_j}{\rho} \right\}_{,j} \right] dV \\
&\quad = \int_S \mathfrak{e}^* \left(\tau_{kj} \frac{m_k}{\rho} + q_j \right) n_j dS
\end{aligned}
\tag{10.94}
$$

where the summation symbol means that integrated values over each element are summed up from element 1 to element N_e, in which N_e is the total number of elements. This is because the integrated values are discontinuous on the side of the elements. Based on Eqs. (10.92)–(10.94), the global forms of the finite element equations can be derived using the ordinary procedures. Introducing the discretization in time, we can obtain the solution of density, momentum, and total energy. For the discretization in time, simple schemes, such as Euler scheme, two-step scheme, Crank–Nicolson Scheme, etc. can be adaptable. The computation of pressure coefficient around the circular cylinder is implemented using the same boundary conditions in Fig. 9.11 and finite element mesh in Fig. 9.12. The computed results are shown in Fig. 9.13. The caption in the figure is compressible flow method. Those are in good agreement with the experimental results (Okauchi et al. 1977).

10.6 Flows Around Bullet Train

Nojima and Kawahara (2011) have presented the computation of compressible flows around bullet train based on the standard bubble function finite element method described in Sect. 10.4. The computational domain employed is schematically shown in Fig. 10.1. A solid body is located in the external flows.

The following boundary conditions are imposed.

(1) On the inlet boundary S_{I}:

$$\left.\begin{array}{l} u_t = 0 \\ R = R_\infty \\ \eta = \eta_\infty \end{array}\right. \quad \text{on} \quad S_{\mathrm{I}}$$

where u_t denotes the tangential velocity on S_{I}, and to compute the free stream, Rieman invariant R and entropy η are imposed, those are

$$R = u_n - \frac{2\mathfrak{a}}{\gamma - 1}$$

$$\mathfrak{a}^2 = \gamma(\gamma - 1)\left(e - \frac{1}{2}u_i u_i\right)$$

$$\eta = \ln(p\rho^{-\gamma})$$

R_∞ and η_∞ stand for the values of the free stream. The acoustic velocity is $\mathfrak{a}$. The ratio of the specific heat γ is 1.4. The Reynolds number Re is 10,000. The Prandtle number P_r is 0.72. The Mach number M_∞ is 0.8. The free stream parameters are taken to be $\rho = 1.0$, $u_1 = 1.0$, $u_2 = 0.0$, $\theta = 1.0$, and $\theta_\infty = 216.7$. To get free stream, density ρ and total energy e are specified on S_{I} so as to Rieman invariant R and entropy η constant during computation.

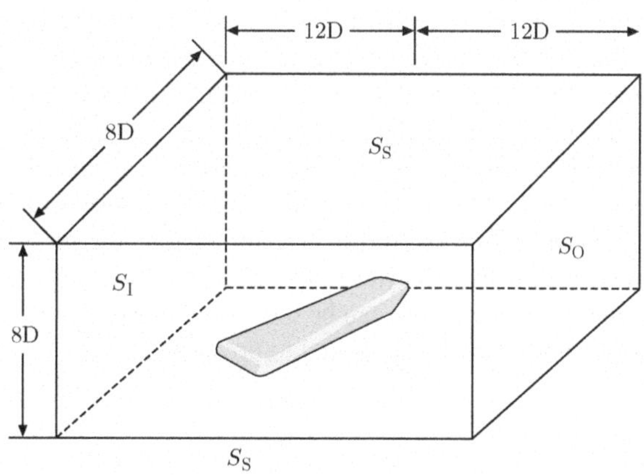

Fig. 10.1 Computation domain

(2) On the outlet boundary S_O:

$$\begin{aligned} p &= p_\infty \\ \tau_{1j} &= 0 \qquad \text{on} \quad S_O \\ q_n &= 0 \end{aligned}$$

where p is the pressure, τ_{1j} is the viscous stress, and q_n is the thermal flux. The pressure of the free stream is denoted by P_∞.

(3) On the side boundary S_S:

$$\begin{aligned} u_2 &= 0 \\ u_3 &= 0 \\ \tau_{2j} &= 0 \\ \tau_{3j} &= 0 \end{aligned} \qquad \text{on} \quad S_S$$

For the computation of the specific heat at constant volume is computed as

$$c_v = [\gamma(\gamma - 1)M_\infty^2]^{-1}$$

The viscosity is given by the Sutherland's law as

$$\mu = \frac{1}{Re} \cdot \frac{\theta_\infty + C}{\theta_\infty \theta + C} \cdot \theta^{3/2}$$

where C is a constant and $C = 110$.

The finite element mesh used is shown in Fig. 10.2. Total numbers of nodes are 1,565,937 and elements are 9,088,753. The streamlines are computed and shown in Figs. 10.3 and 10.4. Solid lines exhibit the streamlines and the colored balls show the magnitude of velocity. Figure 10.4a–d are amplified figures around the nose of the body during a certain period of time. In Fig. 10.4a, the color of balls at the nose of the body is green, which means that its velocity is relatively slow. After several times passed, the green balls are conveyed over the body as shown in Fig. 10.4b. Then, color of balls changes to orange as shown in Fig. 10.4c, which means relatively high magnitude. This is due to the fact that fluids intend to compensate for the space to keep continuity. Finally, the velocity slows down again as shown in Fig. 10.4d. Namely, oscillatory velocity changes can be seen in the fluids around the nose of the body. This fact may be related to the oscillation of the body. Figures 10.5 and 10.6 represent the pressure distributions around the body at a certain time. Around the nose, the pressure barrier can be clearly seen in Fig. 10.5. At the rear of the body, the separation of the flow can be computed in Fig. 10.6. This fact indicates that the rear of the body may be subjected to upward force.

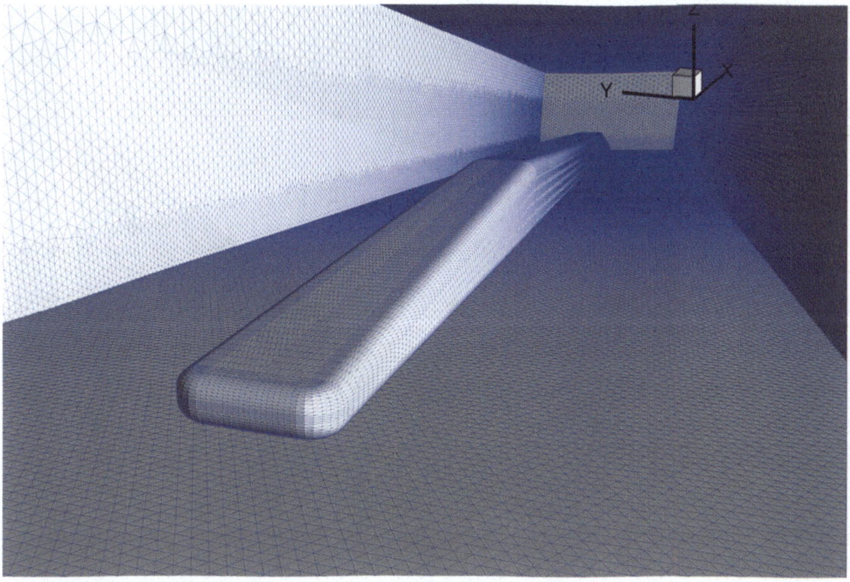

Fig. 10.2 Finite element mesh

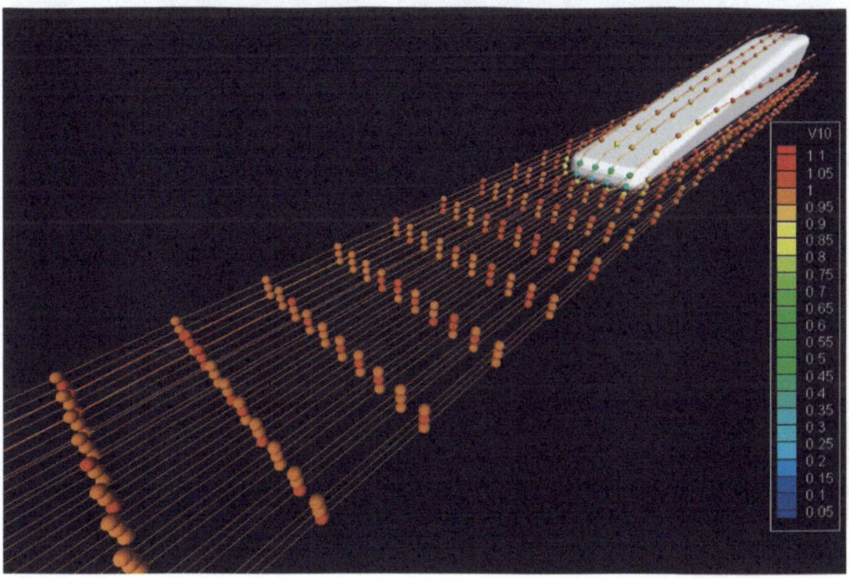

Fig. 10.3 Streamline computed

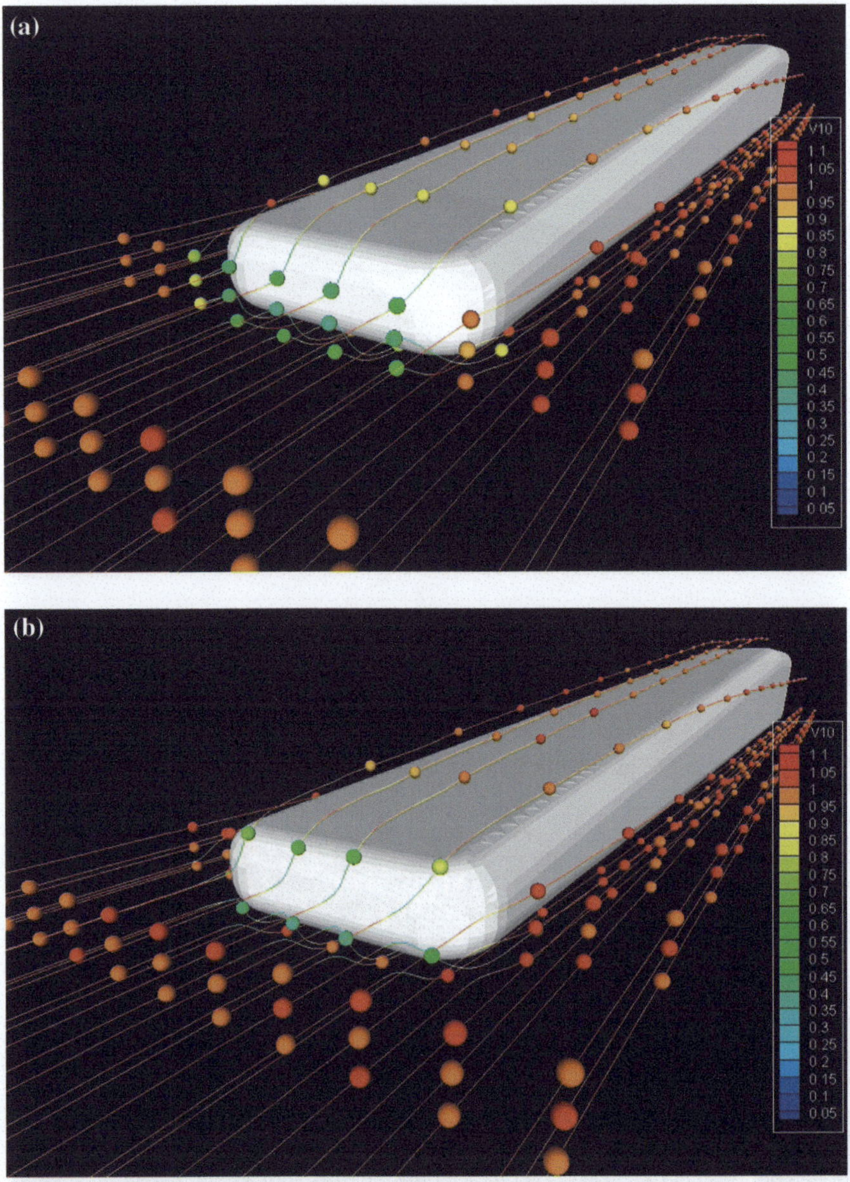

Fig. 10.4 Computed streamline

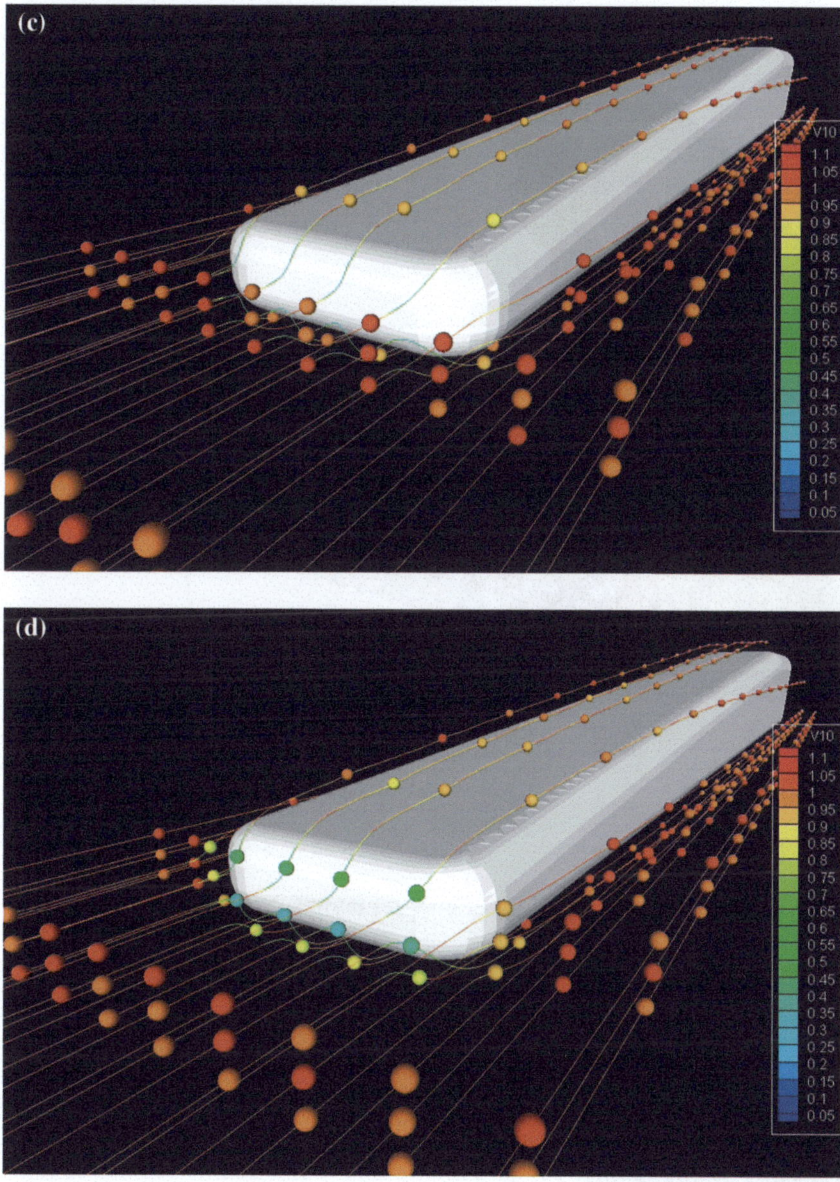

Fig. 10.4 (continued)

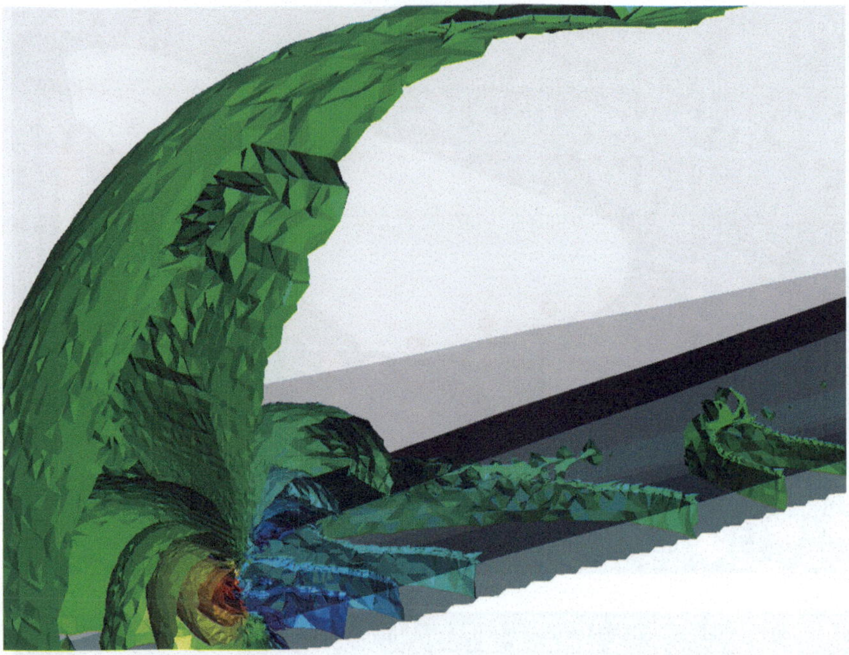

Fig. 10.5 Pressure distribution around nose of the body

Fig. 10.6 Pressure distribution around rear of the body

10.7 Summary and Conclusion

In this chapter, the finite element methods of compressible fluid flows have been presented. The field variables employed are density, momentum, and total energy. The governing equations are conservations of mass, momentum, and energy. The conservative forms are used. For the equation of state, the ideal gas assumption is introduced. We have shown that the SUPG method and the bubble function method are useful. For the bubble function method, we employ the mixed interpolation of bubble function for velocity and linear functions for pressure and total energy. The nonlinear boundary conditions are imposed. However, those are one of the future research topics to be explored. The flows around bullet train have been analyzed. At any rate we get useful qualitative information for the practical use.

Chapter 11
ALE Formulation

11.1 Introduction

Finite element methods to trace the moving boundary problems may be classified into three categories: (a) Lagrangian, (b) Eulerian, and (c) arbitrary Lagrangian–Eulerian methods, which are schematically shown in Fig. 11.1. In the Lagrangian method, the mesh moves along with fluid particles. The equation and computation become simple. However, mesh is distorted, and as a consequence, sometimes, computation is unstable. In the Eulerian method, the mesh is fixed and does not move until the end of computation of free boundary. However, inventive approaches to trace the free boundary are necessary. The mesh movement does not need to coincide with the movement of fluids; we can assign the mesh independent of the fluid flows. The method is referred to as the "arbitrary Lagrangian–Eulerian method," i.e., ALE method. Hughes et al. (1981), Okamoto and Kawahara (1992), Nomura (1992), Nomura and Hughes (1992), Anju et al. (1997), Kawahara and Anju (1988), Soulaimani and Saad (1998), Donea et al. (2004), Duarte et al. (2004), Sawanobori and Kawahara (2013), and many others presented original research work using this method. As discussed in detail in Chap. 7, there are two description methods, Lagrangian and Eulerian descriptions. We also introduced the material and spatial domains. The Lagrangian method corresponds to the method which employs the Lagrangian description based on the material domain. The Eulerian method corresponds to the method which employs the Eulerian description based on the spatial domain. In the arbitrary Lagrangian–Eulerian method, we introduce the referential domain for origins of coordinate of position. Actually, the referential domain is the mesh itself, because nodes of the mesh correspond to the computed fluid particles.

M. Kawahara, *Finite Element Methods in Incompressible, Adiabatic, and Compressible Flows*, Mathematics for Industry 19,
DOI 10.1007/978-4-431-55450-9_11

Fig. 11.1 Lagrangian, Eulerian, and ALE mesh

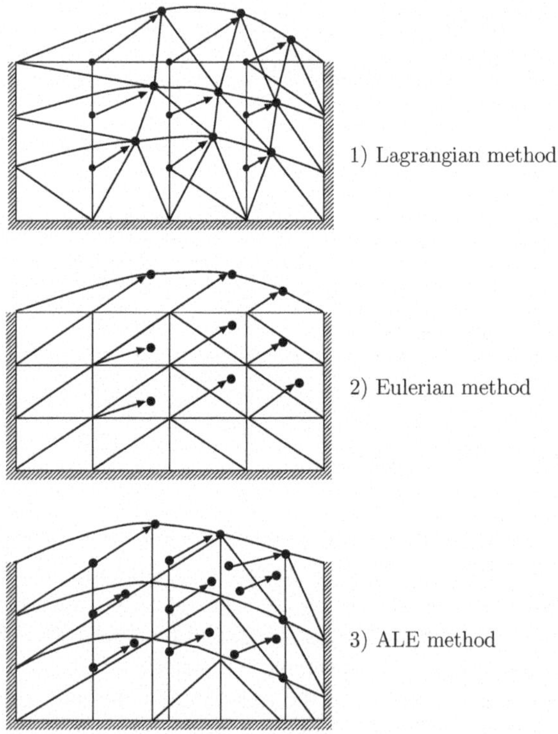

Fundamentally, the ALE formulation must be implemented in the referential domain. In that case, the governing equations become nonlinear. To overcome the nonliniality, a second-order approximate variational method is derived by Soulaimani and Saad (1998). We usually use the formulation in the spatial domain because of its simplicity. Actually, the deformation gradient between the spatial and referential domains is approximated to be a unit matrix, the governing equations in both domains coincide with each other which means that the approximation is valid at least for a short-time increment.

For the integration in time, it is necessary to introduce the stable computational schemes. Soulaimani and Saad (1998) used the mixed interpolation method based on the bubble and linear functions. Uchiyama and Kawahara (2015) employed the SUPG method. Sawanobori and Kawahara (2013) used the method based on the predictor–corrector iteration method. In the computation of ALE formulation, the mesh refinement is important, although we do not address this issue in this book. Interested readers refer to, e.g., Shewchuck (2012), Cheng et al. (2012), etc.

11.2 ALE Formulation

11.2.1 Material, Spatial, and Referential Domains

Based on Fig. 11.2, we will define the material, spatial, and referential domains. The material domain consists of material point $\boldsymbol{X}$ before deformation, and spatial domain consists of spatial point $\boldsymbol{x}$ after deformation. To measure the position $\boldsymbol{X}$ and $\boldsymbol{x}$, we introduce the position of origin $\mathfrak{x}$ in the referential domain. The positions in the domains can be defined as follows:

$$\boldsymbol{X} = X_i \boldsymbol{G}_i \tag{11.1}$$

$$\boldsymbol{x} = x_k \boldsymbol{g}_k \tag{11.2}$$

$$\mathfrak{x} = \mathfrak{x}_\alpha \mathfrak{g}_\alpha \tag{11.3}$$

where $\boldsymbol{G}_i$, $\boldsymbol{g}_k$, and $\mathfrak{g}_\alpha$ are base vectors and X_i, x_k, and $\mathfrak{x}_\alpha$ are coordinates.

Suppose that a fluid particle whose position is $\boldsymbol{X}$ in the material domain moves to the position $\boldsymbol{x}$ in the spatial domain. Motion of a body can be expressed by the positions $\boldsymbol{X}$ and $\boldsymbol{x}$ as

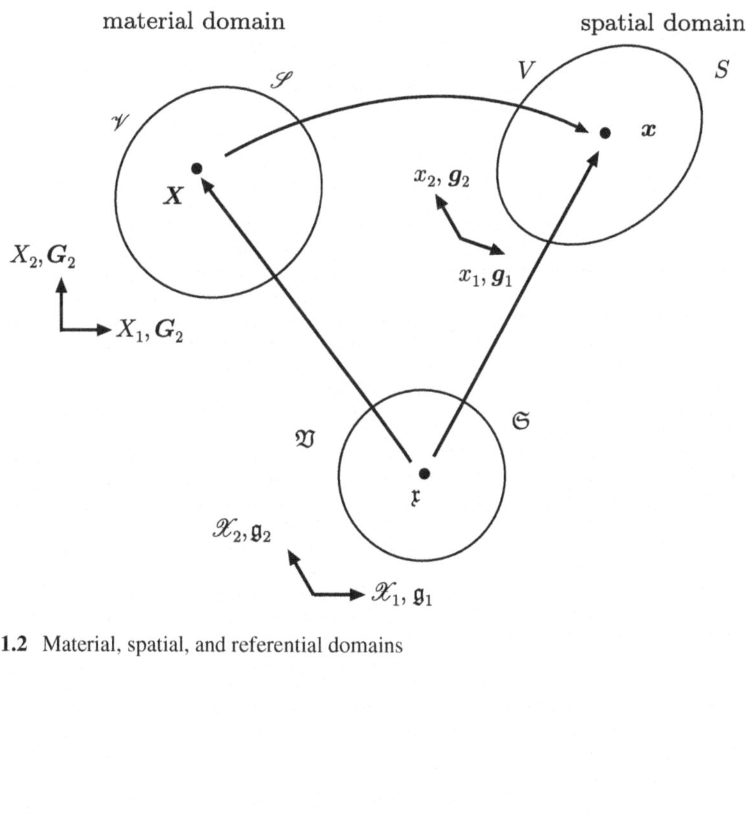

Fig. 11.2 Material, spatial, and referential domains

$$\boldsymbol{x} = \boldsymbol{x}(\boldsymbol{X}, t) \tag{11.4}$$

$$\boldsymbol{X} = \boldsymbol{X}(\boldsymbol{x}, t) \tag{11.5}$$

where t denotes time. The material and spatial positions can be described by the referential position $\boldsymbol{\mathfrak{x}}$ as follows:

$$\boldsymbol{X} = \boldsymbol{X}(\boldsymbol{\mathfrak{x}}, t) \tag{11.6}$$

$$\boldsymbol{\mathfrak{x}} = \boldsymbol{\mathfrak{x}}(\boldsymbol{X}, t) \tag{11.7}$$

and

$$\boldsymbol{x} = \boldsymbol{x}(\boldsymbol{\mathfrak{x}}, t) \tag{11.8}$$

$$\boldsymbol{\mathfrak{x}} = \boldsymbol{\mathfrak{x}}(\boldsymbol{x}, t) \tag{11.9}$$

We can introduce the deformation gradient based on the relation between $\boldsymbol{x}$ and $\boldsymbol{\mathfrak{x}}$ as

$$\boldsymbol{\mathfrak{g}}_\alpha = \frac{\partial x_k}{\partial \mathfrak{x}_\alpha} \boldsymbol{g}_k = \hat{F}_{\alpha k} \boldsymbol{g}_k \tag{11.10}$$

$$\boldsymbol{g}_k = \frac{\partial \mathfrak{x}_\alpha}{\partial x_k} \boldsymbol{\mathfrak{g}}_\alpha = \hat{\mathfrak{F}}_{\alpha k} \boldsymbol{\mathfrak{g}}_\alpha \tag{11.11}$$

where

$$\hat{F}_{\alpha k} = \frac{\partial x_k}{\partial \mathfrak{x}_\alpha} \tag{11.12}$$

$$\hat{\mathfrak{F}}_{\alpha k} = \frac{\partial \mathfrak{x}_\alpha}{\partial x_k} \tag{11.13}$$

The deformation gradients $\hat{\boldsymbol{F}}$ and $\hat{\boldsymbol{\mathfrak{F}}}$ can be derived as follows:

$$\hat{\boldsymbol{F}} = \hat{\boldsymbol{F}}_{\alpha k} \boldsymbol{g}_k \boldsymbol{\mathfrak{g}}_\alpha \tag{11.14}$$

$$\hat{\boldsymbol{\mathfrak{F}}} = \hat{\mathfrak{F}}_{\alpha k} \boldsymbol{g}_k \boldsymbol{\mathfrak{g}}_\alpha \tag{11.15}$$

Then, we get the following relations:

$$\hat{\boldsymbol{F}} \cdot \boldsymbol{\mathfrak{g}}_\beta = \frac{\partial x_k}{\partial \mathfrak{x}_\alpha} \boldsymbol{g}_k \boldsymbol{\mathfrak{g}}_\alpha \boldsymbol{\mathfrak{g}}_\beta = \frac{\partial x_k}{\partial \mathfrak{x}_\beta} \boldsymbol{g}_k = \hat{F}_{\beta k} \boldsymbol{g}_k \tag{11.16}$$

$$\hat{\boldsymbol{\mathfrak{F}}} \cdot \boldsymbol{g}_k = \frac{\partial \mathfrak{x}_\alpha}{\partial x_j} \boldsymbol{g}_j \boldsymbol{\mathfrak{g}}_\alpha \boldsymbol{g}_k = \frac{\partial \mathfrak{x}_\alpha}{\partial x_k} \boldsymbol{\mathfrak{g}}_\alpha = \hat{\mathfrak{F}}_{\alpha k} \boldsymbol{\mathfrak{g}}_\alpha \tag{11.17}$$

Since the referential domain is actually the finite element mesh, we consider the referential domain and the mesh to be the same domain. From Eqs. (11.12) and (11.13), the following relations hold:

$$\hat{F}_{\alpha k}\hat{\mathfrak{F}}_{\beta k} = \frac{\partial x_k}{\partial \mathfrak{x}_\alpha}\frac{\partial \mathfrak{x}_\beta}{\partial x_k} = \delta_{\alpha\beta} \tag{11.18}$$

$$\hat{F}_{\alpha k}\hat{\mathfrak{F}}_{\alpha l} = \frac{\partial x_k}{\partial \mathfrak{x}_\alpha}\frac{\partial \mathfrak{x}_\alpha}{\partial x_l} = \delta_{kl} \tag{11.19}$$

Namely, $\hat{F}_{\alpha k}$ and $\hat{\mathfrak{F}}_{\beta k}$ are mutually inverse relations. Thus, sometimes, we write as follows:

$$\hat{\mathfrak{F}}_{\alpha k} = \hat{F}_{\alpha k}^{-1} \tag{11.20}$$

11.2.2 Velocity and Convective Velocity

In Chap. 7, we define velocity as

$$\boldsymbol{v}(\boldsymbol{x}, t) = \left.\frac{\partial \boldsymbol{x}}{\partial t}\right|_{\boldsymbol{X}} (\boldsymbol{X}, t) \tag{11.21}$$

based on the motion expressed in Eq. (11.4). The mesh velocity $\boldsymbol{w}(\boldsymbol{x}, t)$ is the time variation of the spatial position $\boldsymbol{x}$, holding the referential position $\mathfrak{x}$ constant based on the motion expressed in Eq. (11.8). The mesh velocity $\boldsymbol{w}(\boldsymbol{x}, t)$ can be described as follows:

$$\boldsymbol{w}(\boldsymbol{x}, t) = \left.\frac{\partial \boldsymbol{x}}{\partial t}\right|_{\mathfrak{x}} = \left.\frac{\partial x_k}{\partial t}\right|_{\mathfrak{x}} \boldsymbol{g}_k = \left.\frac{\partial x_k}{\partial t}\right|_{\mathfrak{x}} \hat{\mathfrak{F}}_{\beta k}\mathfrak{g}_\beta \tag{11.22}$$

If we put

$$w_k = \left.\frac{\partial x_k}{\partial t}\right|_{\mathfrak{x}} \tag{11.23}$$

and

$$\hat{w}_\beta = \left.\frac{\partial x_k}{\partial t}\right|_{\mathfrak{x}} \hat{\mathfrak{F}}_{\beta k} = w_k\hat{\mathfrak{F}}_{\beta k} \tag{11.24}$$

then, we get

$$\boldsymbol{w}(\boldsymbol{x}, t) = w_k\boldsymbol{g}_k = \hat{w}_\alpha\mathfrak{g}_\alpha = \hat{\boldsymbol{w}}(\boldsymbol{x}, t) \tag{11.25}$$

where

$$\hat{\boldsymbol{w}} = \hat{w}_\alpha\mathfrak{g}_\alpha \tag{11.26}$$

Namely, $\boldsymbol{w}(\boldsymbol{x}, t)$ is the mesh velocity in the spatial domain and $\hat{\boldsymbol{w}}(\boldsymbol{x}, t)$ is the one in the referential, i.e., mesh, domain.

The moving rate of mesh $\mathfrak{v}$ is defined as the time variation of the referential position $\mathfrak{x}$, holding the material position constant based on the motion expressed in Eq. (11.7):

$$\mathfrak{v}(\boldsymbol{X}, t) = \left.\frac{\partial \mathfrak{x}}{\partial t}\right|_{\boldsymbol{X}} = \left.\frac{\partial \mathfrak{x}_\alpha}{\partial t}\right|_{\boldsymbol{X}} \mathfrak{g}_\alpha \tag{11.27}$$

If we posit

$$\mathfrak{v}(\boldsymbol{X}, t) = \mathfrak{v}_\alpha \mathfrak{g}_\alpha \tag{11.28}$$

then, we get

$$\mathfrak{v}_\alpha = \left.\frac{\partial \mathfrak{x}_\alpha}{\partial t}\right|_{\boldsymbol{X}} (\boldsymbol{X}, t) \tag{11.29}$$

Equation (11.29) is the componential form of the moving rate of mesh.

To express the material time rate of change of a function, we introduce the convective velocity. From Eqs. (11.2), (11.7), and (11.8), we get

$$\boldsymbol{x} = x_k(\mathfrak{x}(\boldsymbol{X}, t), t)\boldsymbol{g}_k \tag{11.30}$$

The following equation holds:

$$\begin{aligned}
\boldsymbol{v} = \left.\frac{\partial \boldsymbol{x}}{\partial t}\right|_{\boldsymbol{X}} &= \left(\left.\frac{\partial x_k}{\partial t}\right|_{\mathfrak{x}} + \left.\frac{\partial x_k}{\partial \mathfrak{x}_\alpha} \frac{\partial \mathfrak{x}_\alpha}{\partial t}\right|_{\boldsymbol{X}} \right) \boldsymbol{g}_k \\
&= \left(\left.\frac{\partial x_k}{\partial t}\right|_{\mathfrak{x}} \right) \boldsymbol{g}_k + \left(\mathfrak{v}_a \frac{\partial x_k}{\partial \mathfrak{x}_\alpha} \right) \boldsymbol{g}_k \\
&= \left(\left.\frac{\partial x_k}{\partial t}\right|_{\mathfrak{x}} \right) \boldsymbol{g}_k + \left(\mathfrak{v}_a \mathfrak{g}_\alpha \frac{\partial x_k}{\partial \mathfrak{x}_\beta} \mathfrak{g}_\beta \right) \boldsymbol{g}_k \\
&= \boldsymbol{w} + \mathfrak{v} \cdot \hat{\boldsymbol{F}}
\end{aligned} \tag{11.31}$$

Using Eq. (11.31), we can define convective velocity $\boldsymbol{c}$ as follows:

$$\boldsymbol{c} = \boldsymbol{v} - \boldsymbol{w} = \mathfrak{v} \cdot \hat{\boldsymbol{F}} \tag{11.32}$$

In the componential form

$$c_k \boldsymbol{g}_k = v_k \boldsymbol{g}_k - w_k \boldsymbol{g}_k = \mathfrak{v}_\alpha \frac{\partial x_k}{\partial \mathfrak{x}_\alpha} \boldsymbol{g}_k \tag{11.33}$$

Finally, we get

$$c_k = v_k - w_k = \mathfrak{v}_\alpha \frac{\partial x_k}{\partial \mathfrak{x}_\alpha} \tag{11.34}$$

In Eq. (11.32), $\hat{\boldsymbol{F}}$ is defined in Eq. (11.14), and

$$\hat{\boldsymbol{F}} = \frac{\partial x_k}{\partial \mathfrak{x}_\beta} \boldsymbol{\mathfrak{g}}_\beta \boldsymbol{g}_k = \frac{\partial}{\partial \mathfrak{x}_\beta} \boldsymbol{\mathfrak{g}}_\beta x_k \boldsymbol{g}_k = \hat{\nabla} \cdot \boldsymbol{x} \tag{11.35}$$

where

$$\hat{\nabla} = \frac{\partial}{\partial \mathfrak{x}_\beta} \boldsymbol{\mathfrak{g}}_\beta \tag{11.36}$$

11.2.3 Governing Equations in the ALE Formulation

Let us consider the material rate of change of a vector $\boldsymbol{f}(\mathfrak{x}(\boldsymbol{X}, t), t)$, which can be written as follows:

$$\boldsymbol{f}(\mathfrak{x}(\boldsymbol{X}, t), t) = f_k(\mathfrak{x}(\boldsymbol{X}, t), t)\boldsymbol{g}_k \tag{11.37}$$

The time change of the vector can be expressed as

$$\begin{aligned} \frac{d\boldsymbol{f}}{dt} &= \left(\left.\frac{\partial f_k}{\partial t}\right|_{\mathfrak{x}} + \left.\frac{\partial f_k}{\partial \mathfrak{x}_\alpha} \frac{\partial \mathfrak{x}_\alpha}{\partial t}\right|_X \right) \boldsymbol{g}_k \\ &= \left(\left.\frac{\partial f_k}{\partial t}\right|_{\mathfrak{x}} + \frac{\partial f_k}{\partial x_l} \frac{\partial x_l}{\partial \mathfrak{x}_\alpha} \mathfrak{v}_\alpha \right) \boldsymbol{g}_k \\ &= \left(\left.\frac{\partial f_k}{\partial t}\right|_{\mathfrak{x}} \right) \boldsymbol{g}_k + \left(\frac{\partial f_k}{\partial x_l} c_l \right) \boldsymbol{g}_k \\ &= \left(\left.\frac{\partial f_k}{\partial t}\right|_{\mathfrak{x}} \right) \boldsymbol{g}_k + \left(c_l \boldsymbol{g}_l \frac{\partial f_k}{\partial x_m} \boldsymbol{g}_m \right) \boldsymbol{g}_k \\ &= \left.\frac{\partial \boldsymbol{f}}{\partial t}\right|_{\mathfrak{x}} + \boldsymbol{c} \cdot \nabla \boldsymbol{f} \end{aligned} \tag{11.38}$$

where

$$\nabla = \frac{\partial}{\partial x_m} \boldsymbol{\mathfrak{g}}_m \tag{11.39}$$

Based on Eq. (11.38), the acceleration in the ALE formulation can be expressed as follows:

$$\begin{aligned}\frac{d\boldsymbol{v}}{dt} &= \left(\left.\frac{\partial v_k}{\partial t}\right|_{\mathfrak{x}} + \left.\frac{\partial v_k}{\partial \mathfrak{x}_\alpha}\frac{\partial \mathfrak{x}_\alpha}{\partial t}\right|_X\right)\boldsymbol{g}_k \\ &= \left.\frac{d\boldsymbol{v}}{dt}\right|_{\mathfrak{x}} + \boldsymbol{c}\cdot\nabla\boldsymbol{v}\end{aligned} \tag{11.40}$$

From Eq. (11.40), we get

$$\frac{d\boldsymbol{v}}{dt} = \left(\frac{Dv_k}{Dt}\right)\boldsymbol{g}_k = \left(\left.\frac{\partial v_k}{\partial t}\right|_{\mathfrak{x}} + \left.\frac{\partial v_k}{\partial \mathfrak{x}_\alpha}\frac{\partial \mathfrak{x}_\alpha}{\partial t}\right|_X\right)\boldsymbol{g}_k \tag{11.41}$$

Therefore, the componential form of the material derivative can be expressed as follows:

$$\frac{Dv_k}{Dt} = \left.\frac{\partial v_k}{\partial t}\right|_{\mathfrak{x}} + c_l\frac{\partial v_k}{\partial x_l} \tag{11.42}$$

Applying Eq. (11.4) to Eqs. (10.1), (10.2), and (10.8), the conservation equations can be obtained in the following forms:

$$\left.\frac{\partial \rho}{\partial t}\right|_{\mathfrak{x}} + c_l\rho_{,l} = -\rho v_{l,k} \tag{11.43}$$

$$\rho\left(\left.\frac{\partial v_k}{\partial t}\right|_{\mathfrak{x}} + c_l v_{k,l}\right) = \sigma_{lk,l} + \rho b_k \tag{11.44}$$

$$\rho\left(\left.\frac{\partial \mathfrak{e}}{\partial t}\right|_{\mathfrak{x}} + c_l \mathfrak{e}_{,l}\right) = (\sigma_{kl}v_l)_{,k} + q_{k,k} + \rho v_k b_k \tag{11.45}$$

where ρ, v_k, $\mathfrak{e}$, σ_{kl}, q_k, and b_k represent density, velocity, total energy, total stress, thermal flux, and body force, respectively. Introducing the constitutive laws for σ_{kl} and q_k, and appropriate boundary and initial conditions, we get the governing equations in the ALE formulation. Conservation equations of mass, momentum, and total energy and constitutive laws are defined by the spatial functions. Therefore, strictly speaking, we must transform those equation systems into the forms of referential functions (Soulaimani and Saad 1998). However, because we apply the ALE formulation for the period during a short-time increment, we can assume that the movements are small and thus negligible. Thus, we can use the deformation gradient as

$$\hat{F}_{\alpha i} \fallingdotseq \delta_{\alpha i} \tag{11.46}$$

Finally, the equation system expressed in Eqs. (11.43)–(11.45) and associate constitutive laws, which are Eqs. (10.4) and (10.11) for the compressible fluid flows, are the governing equations in the ALE formulation. For the incompressible fluid flows, the governing equations are as follows:

$$\left.\frac{\partial v_k}{\partial t}\right|_{\chi} + c_l v_{k,l} + p_{,k} - \mu(v_{k,l} + v_{l,k})_{,l} = 0 \tag{11.47}$$

$$v_{l,l} = 0 \tag{11.48}$$

$$c_l = v_l - w_l \tag{11.34}$$

where w_l is the mesh velocity. For the acoustic velocity formulation of the adiabatic fluid flows, the governing equation can be expressed as

$$\left.\frac{\partial v_k}{\partial t}\right|_{\chi} + c_l v_{k,l} + \mathrm{C} p_{,k} - \lambda v_{k,kl} - \mu(v_{k,l} + v_{l,k})_{,l} = 0 \tag{11.49}$$

$$\frac{\partial p}{\partial t} + c_l p_{,l} + \mathrm{C} v_{k,k} = 0 \tag{11.50}$$

where C is the acoustic velocity.

11.3 Flows Around a Girder Bridge

Let us now turn to a case study. Recently, the oscillation of the steel girder bridges is sometimes caused by low-velocity wind. These phenomena are due to the fact that the weight and rigidity of the girder are reduced because of high quality materials which became available in recent years. The oscillation sometimes influences traffic on the bridge. To prevent the oscillation, some bridges are equipped with wind-resistant wings. Sawanobori and Kawahara (2013) explored the optimal shape determination of the wing adjusting the attached angle to the girder as shown in Fig. 11.3. Schematic view of the computational domain is shown in Fig. 11.4. To express the fluid flows, two-dimensional incompressible flows in the ALE form are assumed. The governing equations are Eqs. (11.47) and (11.48) with Eq. (11.34) and appropriate boundary and initial conditions. For the finite element discretization, the mixed interpolation of the bubble function for velocity and linear function for pressure described in Sect. 8.5 is used.

The finite element mesh, which is generated by the Delaunay triangularization method, is shown in Fig. 11.5. A part of the mesh around the girder is shown in Fig. 11.6 in its enlarged form. We introduce the mesh regeneration to express the smooth movements of fluid flows in response to the movement of the girder.

Fig. 11.3 Wind-resistant wing

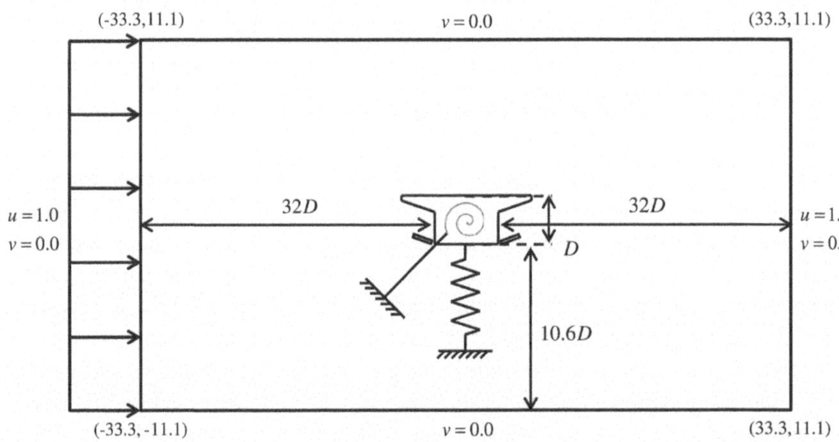

Fig. 11.4 Computational domain

In Fig. 11.6, we can see two zones bordered by the quarter circular arc. One is outside zone, in which all meshes are fixed, and the other is inside zone, in which meshes are regenerated to compensate for the distortion according to the movement of the girder. The mesh movement is independent of the fluid flows. Thus, we can apply ALE formulation for the fluid flows. The girder is assumed to be a rigid body with three degrees of freedom: horizontal displacement, vertical displacement, and

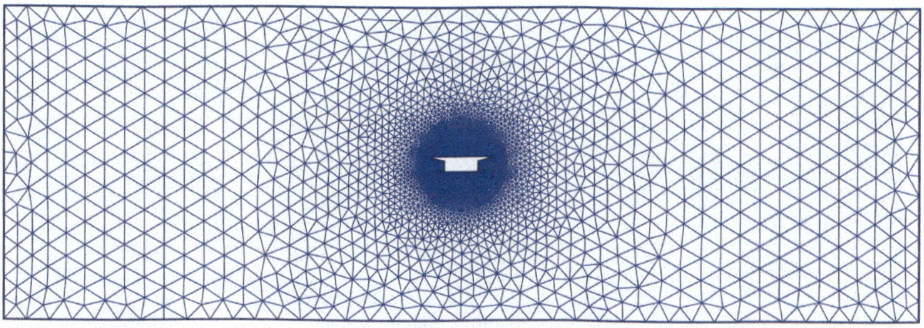

Fig. 11.5 Finite element mesh

Fig. 11.6 Remesh and non-remeshing zones

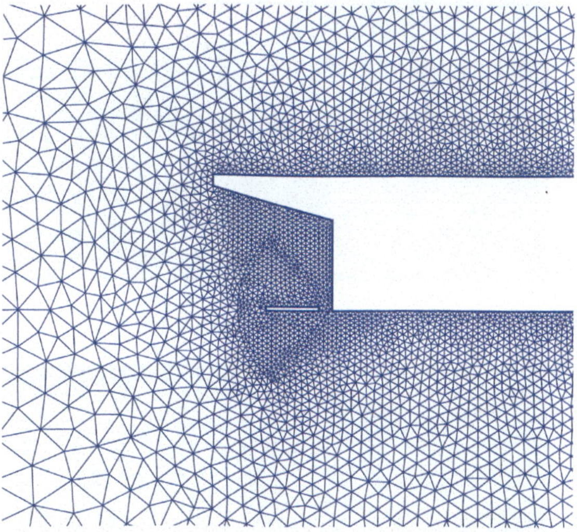

rotation around the barycenter of the girder. The fluid movement on the surface of the girder is assumed to be the same as that of the fluids.

The computed vorticities are shown in Figs. 11.7 and 11.8. Figure 11.7 shows the computed vorticity around the girder without wind-resistant wing, and Fig. 11.8 shows the vorticity around the girder with the wing. Comparing those results, we can see that the wind-resistant wing plays an important role in controlling the flows around the girder. The optimal determination of the attached angle of the girder is determined by the adjoint equation method. The precise formulation is found in Sawanobori and Kawahara (2013).

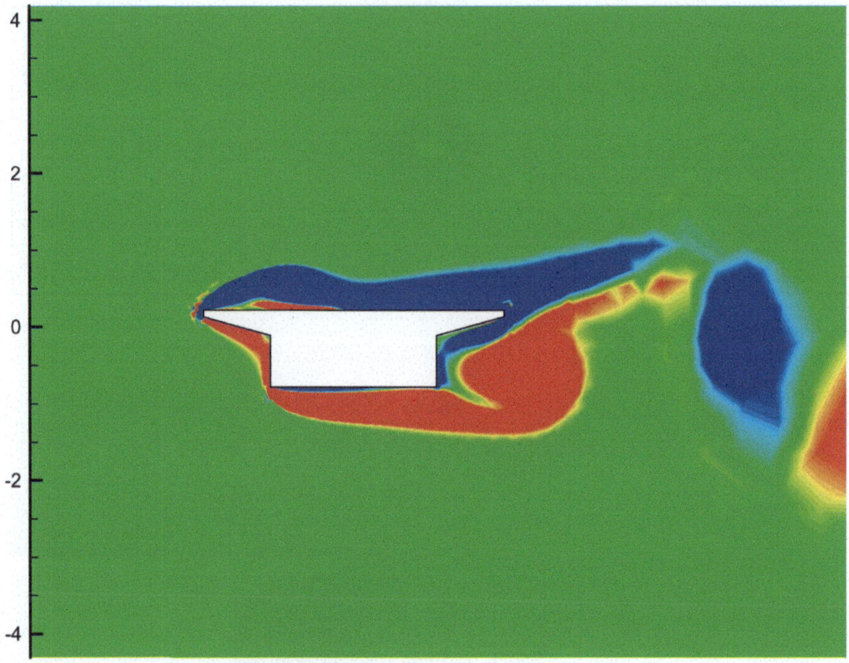

Fig. 11.7 Vorticity around bridge without wings

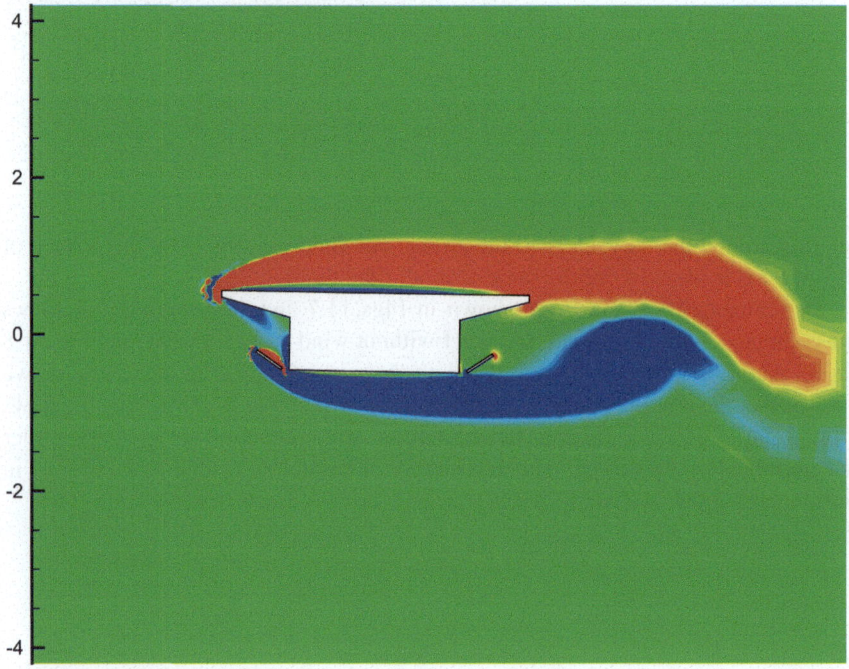

Fig. 11.8 Vorticity around bridge with wings of optimum angle

11.4 Governing Equations in the Referential Domain

Fundamentally, the governing equations must be formulated in the referential domain. The motion of a body is expressed by Eqs. (11.7) and (11.8) as

$$\boldsymbol{x} = \boldsymbol{x}(\boldsymbol{\mathfrak{x}}, t) = \boldsymbol{x}(\boldsymbol{\mathfrak{x}}(\boldsymbol{X}, t), t) \tag{11.51}$$

The velocity in the referential domain $\hat{\boldsymbol{v}}(\boldsymbol{\mathfrak{x}}, t)$ is derived as follows:

$$\hat{\boldsymbol{v}}(\boldsymbol{\mathfrak{x}}, t) = \hat{v}_\alpha \boldsymbol{\mathfrak{g}}_\alpha = v_k \frac{\partial \mathfrak{x}_\alpha}{\partial x_k} \boldsymbol{\mathfrak{g}}_\alpha = v_k \boldsymbol{g}_k = \boldsymbol{v}(\boldsymbol{x}, t) \tag{11.52}$$

where

$$\hat{v}_\alpha = v_k \frac{\partial \mathfrak{x}_\alpha}{\partial x_k} = \hat{\mathfrak{F}}_{\alpha k} v_k \tag{11.53}$$

is component of velocity in the referential domain. We will use ^ to denote the function in the referential domain. The mesh velocity $\hat{\boldsymbol{w}}(\boldsymbol{\mathfrak{x}}, t)$ in the referential domain is as follows:

$$\hat{\boldsymbol{w}}(\boldsymbol{\mathfrak{x}}, t) = \hat{w}_\alpha \boldsymbol{\mathfrak{g}}_\alpha = w_k \frac{\partial \mathfrak{x}_\alpha}{\partial x_k} \boldsymbol{\mathfrak{g}}_\alpha = w_k \boldsymbol{g}_k = \boldsymbol{w}(\boldsymbol{x}, t) \tag{11.54}$$

where

$$\hat{w}_\alpha = w_k \frac{\partial \mathfrak{x}_\alpha}{\partial x_k} = \hat{\mathfrak{F}}_{\alpha k} w_k \tag{11.55}$$

is component of mesh velocity in the referential domain. Equation (11.32) can be rewritten in the componential form as follows:

$$c_k = v_k - w_k = \mathfrak{v}_\alpha \hat{F}_{\alpha k} \tag{11.56}$$

where c_k and $\mathfrak{v}_\alpha$ are components of the convective velocity and the moving rate of mesh, respectively. Multiplying both sides of Eq. (11.56) by $\hat{\mathfrak{F}}_{\beta k}$, we get

$$\hat{\mathfrak{F}}_{\beta k} c_k = \hat{\mathfrak{F}}_{\beta k}(v_k - w_k) = \mathfrak{v}_\beta \tag{11.57}$$

using Eq. (11.18). Equation (11.57) can be rewritten as follows:

$$\mathfrak{v}_\beta = \hat{\mathfrak{F}}_{\beta k}(v_k - w_k) = \hat{v}_\beta - \hat{w}_\beta \tag{11.58}$$

Acceleration in the referential domain $\hat{\boldsymbol{a}}(\mathfrak{x}, t)$ is defined in the following form:

$$\hat{\boldsymbol{a}}(\mathfrak{x}, t) = \frac{d\hat{\boldsymbol{v}}}{dt} = \left.\frac{\partial \hat{\boldsymbol{v}}}{\partial t}\right|_{\mathfrak{x}} + \left.\frac{\partial \mathfrak{x}_\beta}{\partial t}\right|_X \frac{\partial \hat{v}_\alpha}{\partial \mathfrak{x}_\beta}\mathfrak{g}_\alpha \tag{11.59}$$

Using acceleration in the referential domain $\hat{a}_\alpha$:

$$\boldsymbol{a} = \hat{a}_\alpha \mathfrak{g}_\alpha \tag{11.60}$$

Equating Eqs. (11.59) and (11.60), and introducing Eq. (11.58), the acceleration in the referential domain is as follows:

$$\hat{a}_\alpha = \left.\frac{\partial \hat{v}_\alpha}{\partial t}\right|_{\mathfrak{x}} + (\hat{v}_\beta - \hat{w}_\beta)\frac{\partial \hat{v}_\alpha}{\partial \mathfrak{x}_\beta} \tag{11.61}$$

The conservation equation of mass is already formulated in Eq. (7.44), which is rewritten as

$$\int_V \left\{ \left.\frac{\partial \rho}{\partial t}\right|_X + (\rho v_k)_{,k} \right\} dV = 0 \tag{11.62}$$

where V is the spatial domain, whose surface is denoted by S. The first term of Eq. (11.62) is transformed into

$$\int_V \left(\left.\frac{\partial \rho}{\partial t}\right|_X \right) dV = \int_{\mathfrak{V}} \left(\left.\frac{\partial \hat{\rho}}{\partial t}\right|_{\mathfrak{x}} \right) \hat{j} d\mathfrak{V} \tag{11.63}$$

where $\mathfrak{V}$ is the referential domain, whose surface is denoted by $\mathfrak{S}$, and

$$\hat{j} = \det\left(\frac{\partial x_k}{\partial \mathfrak{x}_\alpha}\right) \tag{11.64}$$

Let unit outward normals of the surface S and of the surface $\mathfrak{S}$ be n_k and $\hat{N}_\beta$, respectively. Then the Nanson's formula is written as follows:

$$n_k dS = \hat{j}\hat{\mathfrak{F}}_{\alpha k}\hat{N}_\alpha d\mathfrak{S} \tag{11.65}$$

Using Eqs. (11.53) and (11.65), the second term of Eq. (11.62) can be transformed into

$$\begin{aligned}\int_V (\rho v_k)_{,k} dV &= \int_S (\rho v_k n_k) dS \\ &= \int_{\mathfrak{S}} (\hat{\rho}\hat{j}\hat{F}_{\alpha k}\hat{v}_\alpha \hat{\mathfrak{F}}_{\beta k}\hat{N}_\beta) d\mathfrak{S} \\ &= \int_{\mathfrak{S}} (\hat{\rho}\hat{j}\hat{v}_\beta \hat{N}_\beta) d\mathfrak{S} = \int_{\mathfrak{V}} (\hat{j}\hat{\rho}\hat{v}_\beta)_{,\beta} d\mathfrak{V}\end{aligned} \tag{11.66}$$

Based on Eqs. (11.63) and (11.66), the conservation equation of mass in the referential domain can be obtained:

$$\hat{j}\left.\frac{\partial \hat{\rho}}{\partial t}\right|_{\mathfrak{x}} + (\hat{j}\hat{\rho}\hat{v}_\beta)_{,\beta} = 0 \tag{11.67}$$

The conservation equation of momentum is formulated in Eq. (7.65), which is rewritten as follows:

$$\int_V (\rho a_k + \rho f_k + \sigma_{lk,l}) dV = 0 \tag{11.68}$$

where V is the spatial domain, the surface of which is denoted by S. The acceleration in the spatial domain is expressed as a_k, the body force is ρf_k, and total stress is σ_{kl}, which is called as the Cauchy stress. We can transform Eq. (11.68) into the equation in the referential domain. The first and second terms are transformed into

$$\int_V (\rho a_k) dV = \int_{\mathfrak{V}} (\hat{\rho}\hat{a}_k)\hat{j} d\mathfrak{V} \tag{11.69}$$

$$\int_V (\rho f_k) dV = \int_{\mathfrak{V}} (\hat{\rho}\hat{f}_k)\hat{j} d\mathfrak{V} \tag{11.70}$$

Using Eq. (11.53), the third term in Eq. (11.65) is also transformed into the following form:

$$\begin{aligned}\int_V (\sigma_{lk,l}) dV = \int_S (\sigma_{kl} n_l) dS &= \int_{\mathfrak{S}} (\hat{j}\sigma_{kl}\hat{\mathfrak{F}}_{\alpha l}\hat{N}_\alpha) d\mathfrak{S} \\ &= \int_{\mathfrak{V}} (\hat{j}\sigma_{kl}\hat{\mathfrak{F}}_{\alpha l})_{,\alpha} d\mathfrak{V}\end{aligned} \tag{11.71}$$

Based on Eqs. (11.69)–(11.71), the conservation equation of momentum in the referential domain is expressed as follows:

$$\hat{\rho}\hat{j}\hat{a}_k = \hat{\rho}\hat{j}\hat{f}_k - (\hat{j}\sigma_{kl}\hat{\mathfrak{F}}_{\alpha l})_{,\alpha} \tag{11.72}$$

If we posit

$$\hat{P}_{\alpha k} = \hat{j}\sigma_{kl}\hat{\mathfrak{F}}_{\alpha l} \tag{11.73}$$

then, Eq. (11.72) can be written as

$$\hat{\rho}\hat{j}\hat{a}_k = \hat{\rho}\hat{j}\hat{f}_k - \hat{P}_{\alpha k,\alpha} \tag{11.74}$$

In Eqs. (11.69), (11.72), and (11.73), $\hat{a}_k$ denotes component of acceleration in the referential domain. In Eq. (11.73), $\hat{P}_{\alpha k}$ corresponds to the Piola–Kirchhoff stress of the first kind in the referential domain. As shown in Eq. (7.74), the Cauchy stress σ_{kl} in the spatial domain is

$$\sigma_{kl} = -p\delta_{kl} + \lambda d_{mm}\delta_{kl} + \mu d_{kl} \tag{11.75}$$

Using the reverse relation of Eq. (11.53)

$$v_k = \hat{\mathfrak{F}}_{\alpha k}\hat{\mathfrak{v}}_\alpha \tag{11.76}$$

and the relations

$$\frac{\partial}{\partial x_l} = \frac{\partial}{\partial \mathfrak{x}_\alpha}\frac{\partial \mathfrak{x}_\alpha}{\partial x_l} = \hat{\mathfrak{F}}_{\alpha l}\frac{\partial}{\partial \mathfrak{x}_\alpha} \tag{11.77}$$

We get the deformation rate as follows :

$$\begin{aligned} d_{kl} &= \frac{1}{2}\left(\frac{\partial v_k}{\partial x_l} + \frac{\partial v_l}{\partial x_k}\right) \\ &= \frac{1}{2}\left\{\hat{\mathfrak{F}}_{\alpha l}\frac{\partial}{\partial \mathfrak{x}_\alpha}(\hat{F}_{\beta k}\hat{\mathfrak{v}}_\beta) + \hat{\mathfrak{F}}_{\beta k}\frac{\partial}{\partial \mathfrak{x}_\beta}(\hat{F}_{\alpha l}\hat{\mathfrak{v}}_\alpha)\right\} \end{aligned} \tag{11.78}$$

and

$$d_{mm} = \hat{\mathfrak{F}}_{\alpha m}\frac{\partial}{\partial \mathfrak{x}_\alpha}(\hat{F}_{\beta m}\hat{\mathfrak{v}}_\beta) \tag{11.79}$$

The governing equations in the referential domain are Eqs. (11.67) and (11.74) with appropriate boundary and initial conditions.

11.5 Summary and Conclusion

The ALE formulation is widely used to determine the free boundaries, to solve fluid–solid interaction problems, etc. We have used the spatial domain, the material domain, and the referential domain to express the movement of the mesh. Conservation laws can be expressed in the spatial domains including the effects of the movement of the referential domain. We also implemented the conservation equations in the referential domain. However, the equations are highly nonlinear and those may not be suitable for practical use. The ALE formulation is closely related to the mesh refinement. The stable computation can be performed with the remeshing procedure. In this area, a number of research topics remain to be explored, e.g., the choice of mesh velocity, solution procedures for highly nonlinear equations, mechanical conditions on the boundary of the two phase flows, which were treated in Sects. 9.4 and 8.5.8, etc.

References

Aliabadi SK, Tezduyar TE (1993) Space-time finite element computation of compressible flows involving moving boundaries and interfaces. Comput Methods Appl Mech Eng 107:209–223

Anju A, Maruoka A, Kawahara M (1997) Fluid-structure interaction problems by an arbitrary Lagrangian-Eulerian finite element method. Int J Comput Fluid Dyn 8:1–9

Arikawa T, Yamano T, Akiyama M (2010) Advanced deformation method for breaking waves by using CADMAS-SURF/3D. J Jpn Soc Civ Eng Ser B 1:6–10

Arnold DN, Brezzi F, Fortin M (1984) A stable finite element for the Stokes equation. Calcolo 23:337–344

Arbogast T, Wheeler MF (1995) A characteristic mixed finite element method for advection dominated transport problems. SIAM J Numer Anal 32(2):404–424

Baiocchi C, Brezzi F, Franca LP (1993) Virtual bubbles Galerkin-least-squares type method (GaLS). Comput Methods Appl Mech Eng 105:125–141

Beau G, Lee J, Tezduyar TE (1991) Finite element computation of compressible flows with the SUPG formulation. Adv Finite Elem Anal Fluid Dyn Asme, Fed 123:21–27

Behr M, Tezduyar TE (1994) Finite element solution strategies for large-scale flow simulations. Comput Methods Appl Mech Eng 112:3–24

Bochev P, Gunzburger M (2004) An absolute stable pressure-poisson stabilized finite element method for the Stokes equations. SIAM J Numer Anal 42(3):1189–1207

Bonet J, Wood RW (2008) Nonlinear continuum mechanics for finite element analysis. Cambridge University Press, Cambridge

Brackbill JU, Kothe DB, Zemach C (1992) Continuum method for modeling surface tension. J Comput Phys 100:335–354

Brooks AN, Hughes TJR (1982) Streamline Upwind/Petrov-Galerkin formulations for convection dominated flows with particular emphasis on the incompressible Navier-Stokes equations. Comput Methods Appl Mech Eng 32:199–259

Brezzi F, Bristeau MO, Mallet M, Roge G (1992) A relationship between stabilized finite element methods and the Galerkin method with bubble functions. Comput Methods Appl Mech Eng 96:117–129

Carey GF, Oden JT (1986) Finite elements. In: Fluid mechanics, vol 6. Prentice Hall Ltd

Cheng SW, Dey TK, Shewchuk JR (2012) Delaunay mesh generation. CRC Press, Chapman and Hall, Boca Raton

Chorin AJ (1968) Numerical solution of the Navier-Stokes equations. Math Comput 22:745–762

M. Kawahara, *Finite Element Methods in Incompressible, Adiabatic, and Compressible Flows*, Mathematics for Industry 19,
DOI 10.1007/978-4-431-55450-9

Chorin AJ (1997) A numerical method for solving incompressible viscous flow problems. J Comput Phys 135:118–125

Coleman TF, Van Loan CF (1988) Handbook for matrix computation. SIAM, Frontiers in applied mathematics

Corthwaite J (2013) Pressure poisson method for the incompressible Navier-Stokes equations using galerkin finite elements. Electronic Theses & Dissertations, Paper 831

Demokowicz L, Rachowicz W (1987) On a characteristic finite element method for compressible gas-dynamics. I. J Eng Sci 25(10):1259–1281

Donea J (1984) A Taylor-Galerkin method for convective transport problems. Int J Numer Methods Eng 20(1):101–119

Donea J, Giuliani S, Laval H, Quartapelle L (1984) Time-accurate solution of advection-diffusion problems by finite elements. Comput Methods Appl Mech Eng 45:123–145

Donea J, Huerta A (2003) Finite element method for flow problems. Wiley, New York

Donea J, Huerta A, Ponthot JP, Rodríguez-Ferran A (2004) Chapter 14, Arbitrary Lagrangian-Eulerian methods. Encyclopedia of computational mechanics. In: Stein E, de Borst R, Hughes TJR (eds) Fundamentals, vol 1. Wiley, New York

Drikakis D (2005) The artificial compressibility method. Springer, New York

Duarte F, Gormaz R, Natesan S (2004) Arbitrary Lagrangian-eulerian Method For Navier-stokes Equations With Moving Boundaries. Comput Methods Appl Mech Eng 193:4819–4836

Elkadri E, N. E, Soulaimani A, Deschênes C, (2000) A finite element formulation of comprehensible flows using various set of independent variables. Comput Methods Appl Mech Eng 181:161–189

Eringen AC (1967) Mechanics of continua. Wiley, New York

Fortin M, Fortin A (1985) Newer and newer elements for incompressible flow, finite elements in fluids, vol 6. Wiley, New York

Franca LP, Farhat C (1995) Bubble functions prompt unusual stabilized methods. Comput Methods Appl Mech Eng 123:259–299

Fumoto K, Kusuhara S, Oryu T, Suzawa M, Yoshizumi F, Hirai Y (2005) Large-scale wind tunnel test of super long suspension bridge with slotted one-box girder. In: Proceedings of 6th Asia-pacific conference wind engineering

Ghia U, Ghia KN, Shin CT (1982) High-Re solutions for incompressible flow using the Navier-Stokes equations and a multigrid method. J Comp Phys 48(1):387–411

Golub GH, Van Loan CF (2013) Matrix computations, 4th edn. The John Hopkins University Press, Baltimore

Gresho PM, Chan ST, Upson C, Lee RL (1984) A modified finite element method for solving the time-dependent incompressible Navier-Stokes equations. Int J Numer Methods Fluids, vol 4, Part 1: Theory, pp 557–597, Part 2: Applications, pp 619–640

Gresho PM, Sani RL (2000) Incompressible flow and the finite element method, vols 1 and 2. Wiley, New York

Guermond J-L, Quartapelle L (1998) On stability and convergence of projection methods based on pressure poisson equation. Int J Numer Methods Fluids 26:1039–1053

Hayashi M, Hatanaka K, Kawahara M (1991) Lagrangian finite element method for free surface Navier-Stokes flow using fractional step methods. Int J Numer Methods Fluid 13:805–840

Hirt CW, Nicols BD (1981) Volume of fluid (VOF) method for dynamics of free boundaries. J Comput Phys 39:201–225

Hopman T (2002) Introduction to indicial notation. http://jjmc.bol.ucla.edu/indicial2.pdf

Hughes TJR, Liu WK, Zimmermann TK (1981) Lagrangian-Eulerian finite element formulation for incompressible viscous flows. Comput Methods Appl Mech Eng 29:329–349

Hughes TJR, Franca LP, Balestra M (1986) A new finite element formulation for computational fluid dynamics: V. Circumventing the Babuska-Brezzi condition: a stable Petrov-Galerkin formulation of the Stokes problem accommodating equal-order interpolations. Comput Methods Appl Mech Eng 59:85–99

Hughes TJR, Mallet M (1986) A new finite element formulation for computational fluid dynamics:III. Comput Methods Appl Mech Eng 58:305–328

Hughes TJR, Scovazzi G, Tezduyar TE (2010) Stabilized methods for compressible flows. J Sci Comput 43(3):343–368

Hughes TJR, Tezduyar TE (1984) Finite element methods for first-order hyperbolic systems with particular emphasis on the compressible euler equations. Comput Methods Appl Mech Eng 45:217–284

Iannelli J (2006) Characteristic finite element methods in computational fluid dynamics. Springer, New York

Jackson CP, Cliffe KA (1981) Mixed interpolation in primitive variable finite element formulations for incompressible flows. Int J Numer Methods Eng 17:1659–1688

Jiang CB, Kawahara M (1993) A three-step finite element method for unsteady incompressible flows. Comput Mech 11:355–370

Kawahara M, Yoshimura N, Nakagawa K, Ohsaka K (1976) Steady and unsteady finite element analysis of incompressible viscous fluid flow. Int J Numer Methods Eng 10:437–456

Kawahara M, Hirano H (1983a) A finite element method for high Reynolds number viscous fluid flow using two step explicit scheme. Int J Numer Methods Fluid 3:137–163

Kawahara M, Hirano H (1983b) Two step explicit finite element method for high Reynolds number viscous fluid flow. Proc Jpn Soc Civil Eng 329:127–140

Kawahara M, Hirano H, Kodama T (1984) Two-step explicit finite element method for high Reynolds number flow passed through oscillating body. Finite element in fluids, vol 5. Wiley, pp 227–262

Kawahara M, Miwa T (1984) Finite element analysis of wave motion. Int J Numer Methods Eng 20:1193–1210

Kawahara M, Anju A (1988) Lagrangian finite element method for solitary wave propagation. Comput Mech 3:299–307

Kurahashi T, Hikichi R, Koguchi H (2012) Study on two phase flow in micro channel based on experiments and numerical examinations. In: 10th world congress on computational mechanics, São Paulo, Brazil

Leigh DC (1968) Nonlinear continuum mechanics. McGraw-Hill Inc, New York

Lewis RW, Nithiarasu P, Seetharamu KN (2004) Fundamentals of the finite element method for heat and fluid flow. Wiley, Chichester

Lo SH (2015) Finite element mesh generation. CRC Press, Boca Raton

Löhner R (2008) Applied computational fluid dynamics techniques. Wiley, Bracknell

Madsen PA, Schäffer HA (2005) A discussion of artificial compressibility. Coast Eng 53:93–98

Malan AG, Lewis RW, Nithiarasu P (2002) An improved unsteady, unstructured, artificial compressibility, finite volume scheme for viscous incompressible flows: part i. theory and implementation. Int J Numer Methods Eng 54(5):695–714

Maruoka A, Hirano H, Shimura M (2001) Three dimensional numerical flow simulation around parallel rectangular cylinders. Int J Comput Fluid Dyn 15:47–56

Maruoka A, Uchiyama I, Kawahara M (2016) Propagation of solitary wave analysis by acoustic velocity method. Int Conf Hydrosci Eng (to be published)

Matsumoto J, Kawahara M (2001) Shape identification for fluid-structure interaction problem using improved bubble element. Int J Comput Fluid Dyn 15:33–45

Matsumoto J, Khan AA, Wang SY, Kawahara M (2002) Shallow water flow analysis with moving boundary technique using least-squares bubble function. Int J Comput Fluid Dyn 16:129–134

Matsumoto J (2005) A relationship between stabilized FEM and bubble function element stabilization method with orthogonal basis for incompressible flows. J Appl Mech Jpn Soc Civil Eng 8:233–242

Matsumoto J, Kawahara M (2006) A fractional step method for incompressible viscous flow based on bubble function element stabilization method. Int J Comput Fluid Dyn 20(3–4):145–155

Mase GT, Mase GE (1999) Continuum mechanics for engineers, 2nd edn. CRC Press LIC, Boca Raton

Maury B (1996) Characteristic ALE method for the unsteady 3D Navier-Stokes equations with a free surface. Int J Comput Fluid Dyn 6:175–188

Nakajima S, Kawahara M (2000) Three dimensional analysis for incompressible viscous flow by hexahedral bubble element. In: ECCOMAS 2000. Barcelona, Spain

Nakajima S, Kawahara M (2010) Two dimensional shape optimization using partial control and finite element method for compressible flows. Comput Methods Appl Mech Eng 199:3111–3124

Nakajima S, Kawahara M (2010) New Finite Element Formulation Based On Bubble Function Interpolation For The Transient Compressible Euler Equations. Int J Numer Method Biomed Eng 26:1030–1049

Nasu S, Nojima K, Kawahara M (2013) SUPG finite element method for adiabatic flows. Comput Math Appl 66:250–268

Nojima K, Kawahara M (2011) A research of effect of the initial shape on the drag-reducing shape determination problem. In: 16th international conference finite elements in flow problems. München

Nomura T (1992) Application of predictor-corrector method to ALE finite element analysis of flow-structure interaction problems and associated computational techniques. Proc Jpn Soc Civil Eng 455:55–63

Nomura T, Hughes TJR (1992) An arbitrary Lagrangian-Eulerian finite element method for interaction of fluid and a rigid body. Comput Methods Appl Mech Eng 95:115–138

Oden JT (1970) A finite-element analogue of the Navier-Stokes equations. Proceedings of ASCE, Journal of the engineering mechanics 96(EM4):529–534

Okamoto T, Kawahara M (1992) Two-dimensional sloshing analysis by the arbitrary Lagrangian-Eulerian finite element method. In: Proceedings of the Japan Society of Civil Engineers, vol 441/I-18, pp 39–48

Okauchi K, Ito M, Miyata T (1977) Wind resistant structures, Maruzen Co. (in Japanese)

Okumura H (2014) A study of CUDA parallel computation and visualization on-boad GPU for the three-dimensional free surface flow simulater CADMAS-SURF/3D for designing seawall coastal and maritime structures. In: Proceedings of information technology centre, Toyama University

Okumura H, Kawahara M (2000) Shape optimization of body located in incompressible Navier-Stokes flow based on optimal control theory. Comput Methods Eng Sci 1(2):71–77

Okumura H, Kawahara M (2003) A new stable bubble element for in compressible fluid flow based on a mixed Petrov-Galerkin finite element formulation. Int J Comput Fluid Dyn 17(4):275–282

Okumura H, Arikawa T (2014) A study of CUDA/MPI parallel computations for CADMAS-SURF/3D. In: Proceedings of 24 international ocean and polar engineering conference. Busan, Korea

Pierre R (1988) Simple C^0 approximation for the computation of incompressible flows. Comput Methods Appl Mech Eng 68:205–227

Pierre R (1995) Optimal selection of the bubble function in the stabilization of the P_1-P_1 element for the Stokes prolblem. SIAM, J Numer Anal Eng 32:1210–1224

Pironneau O (1982) On the transport-diffusion algorithm and its applications to the Navier-Stokes equations. Numer Math 38:309–332

Pironneau O (1988) Finite element methods for fluids. Wiley, Chichester

Ramaswamy B, Kawahara M (1987) Arbitrary Lagrangian-Eulerian finite element method for unsteady, convective, incompressible viscous free surface fluid flow. Int J Numer Methods Fluids 7:65–87

Ramaswamy B, Kawahara M (1987) Arbitrary Lagrangian-Eulerian finite element method for unsteady, convective, incompressible viscous free surface fluid flow. Int J Numer Methods Fluids 7:1053–1075

Sawanobori H, Kawahara M (2013) Shape determination of wind-resistent wings attached to an oscillating bridge using adjoint equation method. Comput Methods Appl Mech Eng 254:254–270

Sedov LI (1971) A course in continuum mechanics, vol. 1: basic equations and analytical techniques, Wolters-Noordhoff Publishing, Groningen

Shewchuk JR (2012) Lecture notes on delaunay mesh generation. http://www.cs.berkeley.edu/~jrs/meshpapers/delnotes.pdf

Shi DY, Wang XL (2009) A low order anisotropic non comforming characteristic finite element method for a convection dominated transport problem. Appl Math Comput 213(2):411–418

Simo JC, Armero F, Taylor C (1995) Galerkin finite element methods with bubble for advection dominated incompressible Navier-Stokes. Int J Numer Methods Eng 38:1475–1506

Soulaimani A, Saad Y (1998) An arbitrary Lagrangian-Eulerian finite element method for solving three-dimensional free surface flows. Comput Methods Appl Mech Eng 162:79–106

Street RL, Camfield FE (1966) Observations and experiments on solitary wave deformation. In: Coastal engineering proceedings 10, ASCE

Taylor C, Hood P (1973) A numerical solution of the Navier-Stokes equations using the finite element technique. Comput Fluids 1:73–100

Temam R (1968) Une Méthode d'Approximation de la Solution des Équations de Navier-Stokes. Bull Soc Math 96:115–152

Temam R (2001) Navier-Stokes equations: theory and numerical analysis. American Mathematical Society

Terachi K, Kawahara M (2010) Shape optimization of a body located in viscous flows using acoustic velocity method. Chuo University, Internal Paper in Kawahara Laboratory

Tezduyar TE, Mittal S, Ray SE, Shih R (1992) Incompressible flow computations with stabilized bilinear and linear equal-order-interpolation velocity-pressure elements. Comput Methods Appl Mech Eng 95:221–242

Tezduyar TE, Osawa Y (2000) Finite element stabilization parameters computed from element matrices and vectors. Comput Methods Appl Mech Eng 190:411–430

Tezduyar TE, Senga M (2006) Stabilization and shock-capturing parameters in SUPG formulation of compressible flows. Comput Methods Appl Mech Eng 195:1621–1632

Tezduyar TE, Takizawa K, Brummer T, Chenand PR (2011) Space-time fluid-structure interaction modeling of patient-specific cerebral aneurysms. Int J Numer Methods Biomed Eng 27:1665–1710

Uchiyama I, Kawahara M (2015) Finite element analysis of adiabatic flows. In: 13th U.S. national congress on computational mechanics, San Diego, USA

Vellando P, Puertas J, Colominas I (2002) SUPG stabilized finite element resolution of the Navier-Stokes equations, applications to water treatment engineering. Comput Methods Appl Mech Eng 191:5899–5922

Watanabe S, Fumoto K (2008) Aerodynamic study of slotted box girder using computational fluid dynamics. Int J Wind Eng Ind Aerodyn 96:1885–1894

Watkins DS (2010) Fundamentals of matrix computations, 3rd edn. Wiley, New York

Yamada T (1995) A bubble element for inviscid flow. Finite Ele Fluids 9:1567–1576

Yamada T (1998) A bubble element for the compressible Euler equations. Int J Comput Fluid Dyn 9:273–283

Zienkiewicz OC, Taylor RL, Nithiarasu P (2014) The finite element method for fluid Dynamics, 7th edn. Elsevier

Zienkiewicz OC, Taylor RL, Zhu JZ (2014) The finite element method: its basis and fundamentals, 7th edn. Elsevier

Index

M. Kawahara, *Finite Element Methods in Incompressible, Adiabatic, and Compressible Flows*, Mathematics for Industry 19,
DOI 10.1007/978-4-431-55450-9

The manufacturer's authorised representative in the EU is Springer Nature Customer Service Centre GmbH, Europaplatz 3, 69115 Heidelberg, Germany. If you have any concerns regarding our products, please contact ProductSafety@springernature.com

Printed and bound by CPI Group (UK) Ltd, Croydon, CR0 4YY
12/08/2026
02194531-0002